Sprinkler Fitting

Level One

Trainee Guide
Fourth Edition

NCCER

AFSA
American Fire Sprinkler Association

Pearson

NCCER
President and Chief Executive Officer: Boyd Worsham
Vice President – Innovation and Advancement: Jennifer Wilkerson
Chief Learning Officer: Lisa Strite
Sprinkler Fitting Project Managers: Lauren Corley, Veronica Westfall
Senior Manager of Projects: Chris Wilson
Senior Production Manager: Erin O'Nora
Testing/Assessment Project Manager: Elizabeth Schlaupitz
Testing and Assessment Project Coordinator: Chelsi Causley
Lead Technical Writer: Veronica Westfall
Art Manager: Carrie Pelzer
Permissions Specialists: Kay Lewis, Amanda Smith
Desktop Publishing Manager: James McKay
Production Specialists: Gene Page, Eric Caraballoso, Daphney Milian
Digital Content Coordinator: Rachael Downs
Managing Editor: Graham Hack
Editors: Jordan Hutchinson, Karina Kuchta, Hannah Murray, Zi Meng

Pearson
Manager of Project Management: Vanessa Price
Employability Solutions Coordinator: Monica Perez
Senior Content Producer: Alexandrina B. Wolf
Content Producers: Alma Dabral, Aurodeep Mukherjee

Composition: NCCER
Printer/Binder: Lakeside Book Company
Cover Printer: Lakeside Book Company
Text Fonts: Palatino and Univers
Content Technologies: Gnostyx

Cover Image
Cover image of activated sprinkler courtesy of Tyco/Johnson Controls, Inc.

Credits and acknowledgments for content borrowed from other sources and reproduced, with permission, in this textbook appear at the end of each module.

AFSA
American Fire Sprinkler Association

ISBN-13: 978-0-13-747549-0

Pearson

Preface

To the Trainee

Sprinkler fitting is dedicated to the preservation of life and property, and carries with it a tremendous amount of responsibility. The work of a sprinkler fitter may save lives and thousands of dollars in property. Therefore, it is essential that sprinkler fitters demand excellence and precision from themselves and coworkers.

Sprinkler fitters may install new systems, retrofit old buildings with updated sprinkler systems, or repair existing systems. Accordingly, the demand for sprinkler fitters grows with new construction and code changes. All of this results in higher demand for trained workers which generates higher wages. Stay on top of industry developments, code changes, and emerging technologies, to broaden opportunities and earning potential.

NCCER wishes you success as you embark on your first year of training in the sprinkler fitting trade, and we hope that you'll continue your training outside of this series. By taking advantage of training opportunities as they arise, you'll demonstrate initiative and a desire to learn—qualities that are present in the industry's best professionals.

We invite you to visit the NCCER website at **https://www.nccer.org** for information on the latest product releases and training.

Your feedback is welcome. You may email your comments to **curriculum@nccer.org** or send general comments and inquiries to **info@nccer.org**.

New with *Sprinkler Fitting Level One*

NCCER and Pearson are pleased to present *Sprinkler Fitting Level One*. In partnership with the American Fire Sprinkler Association (AFSA), this book has been updated and enhanced to provide the most comprehensive apprenticeship training in the sprinkler industry. AFSA is the voice of sprinkler fitting professionals and the leading authority in the sprinkler industry for education, training, consulting, information, representation, and advocacy.

Some of the features of this edition include the following:

- Revised by subject matter experts who sit on the code-making panels of the National Fire Protection Association (NFPA) to reflect the latest edition of NFPA 13, *Standard for the Installation of Sprinkler Systems*, as well as all other applicable codes and standards
- Updated with hundreds of new images to represent the latest in tools, technology, and materials used in the sprinkler industry, with input and review by major manufacturers, including Johnson Controls, Inc.; Reliable Automatic Sprinkler Co., Inc.; Victaulic; The Viking Corporation; and more
- Instructional material revised by leading educators and contractors in the sprinkler industry to reflect the most pertinent information and practical examples for sprinkler apprentices that can be applied immediately, both in the classroom and on the job
- New shop drawings and 3D images from NICET-certified sprinkler system layout technicians to reflect actual project drawings

NCCER Standardized Curricula

NCCER is a not-for-profit 501(c)(3) education foundation established in 1996 by the world's largest and most progressive construction companies and national construction associations. It was founded to address the severe workforce shortage facing the industry and to develop a standardized training process and curricula. Today, NCCER is supported by hundreds of leading construction and maintenance companies, manufacturers, and national associations. The NCCER Standardized Curricula was developed by NCCER in partnership with Pearson, the world's largest educational publisher.

Some features of the NCCER Standardized Curricula are as follows:

- An industry-proven record of success
- Curricula developed by the industry, for the industry
- National standardization providing portability of learned job skills and educational credits
- Compliance with the Office of Apprenticeship requirements for related classroom training (*CFR 29:29*)
- Well-illustrated, up-to-date, and practical information

NCCER also maintains the NCCER Registry, which provides transcripts, certificates, and wallet cards to individuals who have successfully completed a level of training within a craft in NCCER's Curricula. *Training programs must be delivered by an NCCER Accredited Training Sponsor in order to receive these credentials.*

For information on NCCER's credentials and the NCCER Registry, contact NCCER Customer Service at 1-888-622-3720 or visit **https://www.nccer.org**.

Special Features

In an effort to provide a comprehensive and user-friendly training resource, this curriculum showcases several informative features. Whether you are a visual or hands-on learner, these features are intended to enhance your knowledge of the construction industry as you progress in your training. Some of the features you may find in the curriculum are explained below.

Introduction

This introductory page, found at the beginning of each module, lists the module Objectives, Performance Tasks, and Trade Terms. The Objectives list the knowledge you will acquire after successfully completing the module. The Performance Tasks give you an opportunity to apply your knowledge to real-world tasks. The Trade Terms are industry-specific vocabulary that you will learn as you study the module.

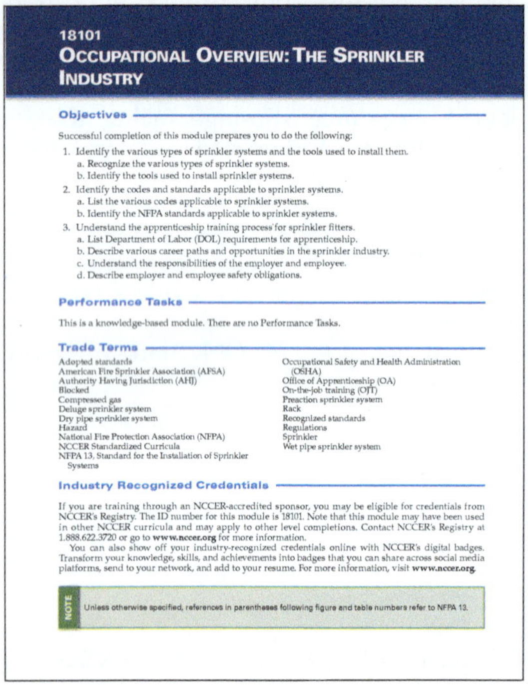

18101
OCCUPATIONAL OVERVIEW: THE SPRINKLER INDUSTRY

Objectives

Successful completion of this module prepares you to do the following:

1. Identify the various types of sprinkler systems and the tools used to install them.
   a. Recognize the various types of sprinkler systems.
   b. Identify the tools used to install sprinkler systems.
2. Identify the codes and standards applicable to sprinkler systems.
   a. List the various codes applicable to sprinkler systems.
   b. Identify the NFPA standards applicable to sprinkler systems.
3. Understand the apprenticeship training process for sprinkler fitters.
   a. List Department of Labor (DOL) requirements for apprenticeship.
   b. Describe various career paths and opportunities in the sprinkler industry.
   c. Understand the responsibilities of the employer and employee.
   d. Describe employer and employee safety obligations.

Performance Tasks

This is a knowledge-based module. There are no Performance Tasks.

Trade Terms

Adopted standards
American Fire Sprinkler Association (AFSA)
Authority Having Jurisdiction (AHJ)
Blocked
Compressed gas
Deluge sprinkler system
Dry pipe sprinkler system
Hazard
National Fire Protection Association (NFPA)
NCCER Standardized Curricula
NFPA 13, Standard for the Installation of Sprinkler Systems
Occupational Safety and Health Administration (OSHA)
Office of Apprenticeship (OA)
On-the-job training (OJT)
Preaction sprinkler system
Rack
Recognized standards
Regulations
Sprinkler
Wet pipe sprinkler system

Industry Recognized Credentials

If you are training through an NCCER-accredited sponsor, you may be eligible for credentials from NCCER's Registry. The ID number for this module is 18101. Note that this module may have been used in other NCCER curricula and may apply to other level completions. Contact NCCER's Registry at 1.888.622.3720 or go to **www.nccer.org** for more information.

You can also show off your industry-recognized credentials online with NCCER's digital badges. Transform your knowledge, skills, and achievements into badges that you can share across social media platforms, send to your network, and add to your resume. For more information, visit **www.nccer.org**.

NOTE
Unless otherwise specified, references in parentheses following figure and table numbers refer to NFPA 13.

Trade Features

Trade features present technical tips and professional practices based on real-life scenarios similar to those you might encounter on the jobsite.

Bowline Trivia

Some people use this saying to help them remember how to tie a bowline: "The rabbit comes out of his hole, around a tree, and back into the hole."

Figures and Tables

Photographs, drawings, diagrams, and tables are used throughout each module to illustrate important concepts and provide clarity for complex instructions. Text references to figures and tables are emphasized with *italic* type.

Figure 31 Marking a cutting line.

Notes, Cautions, and Warnings

Safety features are set off from the main text in highlighted boxes and categorized according to the potential danger involved. Notes simply provide additional information. Cautions flag a hazardous issue that could cause damage to materials or equipment. Warnings stress a potentially dangerous situation that could result in injury or death to workers.

NOTE
Nameplates must be posted on each material handling device. The nameplate must indicate

CAUTION
It is essential to note the revision designation on a construction drawing and to use only the latest

WARNING!
Saw teeth are very sharp. Use gloves and do not handle the saw teeth with bare hands. When cutting with a saw, ensure that your fingers remain clear of the teeth at all times.

Going Green

Going Green looks at ways to preserve the environment, save energy, and make good choices regarding the health of the planet. Through the introduction of new construction practices and products, you will see how the greening of America has already taken root.

> **Going Green**
>
> **Big Horn Wind Farm**
>
> The Big Horn wind farm uses 133 GE turbines, each rising 119 meters (389 feet) from the ground to the top of the blade tips. The wind farm uses only a

Did You Know?

The *Did You Know*? features offer hints, tips, and other helpful bits of information.

> **Did You Know?**
>
> **How Blueprints Started**
>
> The process for making blueprints was developed in 1842 by an English astronomer named Sir John F. Herschel. The method involved coating a paper with a special chemical. After the coating dried, an original hand drawing was placed on top of the paper. Both papers were then covered with a piece of glass and set in the sunlight for about an hour. The coated paper was developed much like a photograph. After a cold-water wash, the coated paper turned blue, and the lines of the drawing remained white.

Step-by-Step Instructions

Step-by-step instructions are used throughout to guide you through technical procedures and tasks from start to finish. These steps show you not only how to perform a task but also how to do it safely and efficiently.

> Perform the following steps to erect this system area scaffold:
>
> *Step 1* Gather and inspect all scaffold equipment for the scaffold arrangement.
>
> *Step 2* Place appropriate mudsills in their approximate locations.
>
> *Step 3* Attach the screw jacks to the mudsills.
>
> *Step 4* Adjust the screw jacks to near their lowest position.
>
> *Step 5* Determine the location of the highest base.

Trade Terms

Each module presents a list of Trade Terms that are discussed within the text and defined in the Glossary at the end of the module. These terms are denoted in the text with **bold, blue type** upon their first occurance. To make searches for key information easier, a comprehensive Glossary of Trade Terms from all modules is located at the back of this book.

> During a rigging operation, the **load** being lifted or moved must be connected to the apparatus, such as a crane, that will provide the power for movement. The connector—the link between the load and the apparatus—is often a sling made of synthetic, chain, or **wire rope** materials. This section focuses on three types of slings:
>
> - Synthetic slings
> - Alloy steel chain slings
> - Wire rope slings

Section Review

Section Review questions can be found at the end of each section to test your knowledge of the content.

> **1.0.0 Section Review**
>
> 1. The type of sprinkler system that might be found in an aircraft hangar is a _____.
>   a. dry pipe sprinkler system
>   b. wet pipe sprinkler system
>   c. preaction sprinkler system
>   d. deluge sprinkler system
> 2. A type of ladder that provides a platform for standing is a(n) _____.
>   a. A-frame
>   b. extension trestle ladder
>   c. podium ladder
>   d. movable leg ladder

Review Questions

Review Questions are provided to reinforce the knowledge you have gained. This makes them a useful tool for measuring what you have learned.

> **Review Questions**
>
> 1. Identification tags for slings must include the _____.
>   a. type of protective pads to use
>   b. type of damage sustained during use
>   c. color of the tattle-tail
>   d. manufacturer's name or trademark
> 2. The type of wire rope core that is susceptible to heat damage at relatively low temperatures is the _____.
>   a. fiber core
>   b. strand core
>   c. independent wire rope core
>   d. metallic link supporting core
> 3. Synthetic slings must be inspected _____.
>   a. once every month
>   b. visually at the start of each work week
>   c. before every use
>   d. once wear or damage becomes apparent
> 4. An alloy steel chain sling must be removed from service if there is evidence that _____.
>
> 7. Chain hoists are able to lift heavy loads by utilizing a _____.
>   a. rope and pulley system
>   b. rigger's strength
>   c. stationary counterweight
>   d. gear system
> 8. Before attempting to lift a load with a chain hoist, make sure that the _____.
>   a. hoist is secured to a come-along
>   b. load is properly balanced
>   c. tag lines are properly anchored
>   d. tackle is connected to its power source
> 9. A hitch configuration that allows slings to be connected to the same load without using a spreader beam is a _____.
>   a. double-wrap hitch
>   b. choker hitch
>   c. bridle hitch
>   d. basket hitch
> 10. To make the emergency stop signal that is

NCCER Standardized Curricula

NCCER's training programs comprise more than 80 construction, maintenance, pipeline, and utility areas and include skills assessments, safety training, and management education.

Boilermaking
Cabinetmaking
Carpentry
Concrete Construction
Construction Craft Laborer
Construction Technology
Core: Introduction to Basic Construction Skills
Drywall
Electrical
Electronic Systems Technician
Heating, Ventilating, and Air Conditioning
Heavy Equipment Operations
Heavy Highway Construction
Hydroblasting
Industrial Coating and Lining Application Specialist
Industrial Maintenance Electrical and Instrumentation Technician
Industrial Maintenance Mechanic
Instrumentation
Ironworking
Manufactured Construction Technology
Masonry
Mechanical Insulating
Millwright
Mobile Crane Operations
Painting
Painting, Industrial
Pipefitting
Pipelayer
Plumbing
Reinforcing Ironwork
Rigging
Roofing
Scaffolding
Sheet Metal
Signal Person
Site Layout
Sprinkler Fitting
Tower Crane Operator
Welding

Maritime

Maritime Industry Fundamentals
Maritime Electrical
Maritime Pipefitting
Maritime Structural Fitter
Maritime Welding
Maritime Aluminum Welding

Green/Sustainable Construction

Building Auditor
Fundamentals of Weatherization
Introduction to Weatherization
Sustainable Construction Supervisor
Weatherization Crew Chief
Weatherization Technician
Your Role in the Green Environment

Energy

Alternative Energy
Introduction to the Power Industry
Introduction to Solar Photovoltaics
Power Generation Maintenance Electrician
Power Generation I&C Maintenance Technician
Power Generation Maintenance Mechanic
Power Line Worker
Power Line Worker: Distribution
Power Line Worker: Substation
Power Line Worker: Transmission
Solar Photovoltaic Systems Installer
Wind Energy
Wind Turbine Maintenance Technician

Pipeline

Abnormal Operating Conditions, Control Center
Abnormal Operating Conditions, Field and Gas
Corrosion Control
Electrical and Instrumentation
Field and Control Center Operations
Introduction to the Pipeline Industry
Maintenance
Mechanical

Safety

Fall Protection Orientation
Field Safety
Safety Orientation
Safety Technology

Supplemental Titles

Applied Construction Math
Tools for Success

Management

Construction Workforce Development Professional
Fundamentals of Crew Leadership
Mentoring for Craft Professionals
Project Management
Project Supervision

Spanish Titles

Acabado de concreto: nivel uno (*Concrete Finishing Level One*)
Aislamiento: nivel uno (*Insulating Level One*)
Albañilería: nivel uno (*Masonry Level One*)
Andamios (*Scaffolding*)
Carpintería: Formas para carpintería, nivel tres (*Carpentry: Carpentry Forms, Level Three*)
Currículo básico: habilidades introductorias del oficio (*Core Curriculum: Introductory Craft Skills*)
Electricidad: nivel uno (*Electrical Level One*)
Herrería: nivel uno (*Ironworking Level One*)
Herrería de refuerzo: nivel uno (*Reinforcing Ironwork Level One*)
Instalación de rociadores: nivel uno (*Sprinkler Fitting Level One*)
Instalación de tuberías: nivel uno (*Pipefitting Level One*)
Instrumentación: nivel uno, nivel dos, nivel tres, nivel cuatro (*Instrumentation Levels One through Four*)
Orientación de seguridad (*Safety Orientation*)
Paneles de yeso: nivel uno (*Drywall Level One*)
Seguridad de campo (*Field Safety*)

Acknowledgments

This curriculum was revised as a result of the vision and leadership of the following sponsors:

Cen-Cal Fire Systems, Inc.
Fire & Life Safety America
Johnson Controls, Inc.
Lubrizol Advanced Materials, Inc.
North East Fire Protection Systems, Inc.
Piper Fire Protection
Reliable Automatic Sprinkler Co., Inc.
Security Fire Sprinkler
Strickland Fire Protection
Victaulic
The Viking Corporation
Wolfe Fire Protection, Inc.

This curriculum would not exist were it not for the dedication and unselfish energy of those volunteers who served on the Authoring Team. A sincere thanks is extended to the following:

Charles Bamford
Leslie Clounts
John Denhardt
Robert Dunn
Kevin Kelly
Mac McGaffagan
Steven Rasch
Matt Rouleau
Taylor Schumacher
Byron Weisz

NCCER Partners

NCCER partnering organizations are national associations and organizations that share a common interest in the goals and objectives of NCCER. To learn more about NCCER business partners, go to **https://www.nccer.org/about-us/partners**.

You can also scan this code using the camera on your phone or mobile device to view these partnering organizations.

Contents

Module Six
Underground Pipe

Describes the installation of underground piping systems. Covers ductile iron and PVC piping and components, including risers, backflow preventer assemblies, hydrants, and restraints. Also covers trench safety requirements, backfilling, and testing. (Module ID 18106; 12.5 Hours)

Glossary

This course map shows all of the modules in *Sprinkler Fitting Level One.* The suggested training order begins at the bottom and proceeds up. Skill levels increase as you advance on the course map. The local Training Program Sponsor may adjust the training order.

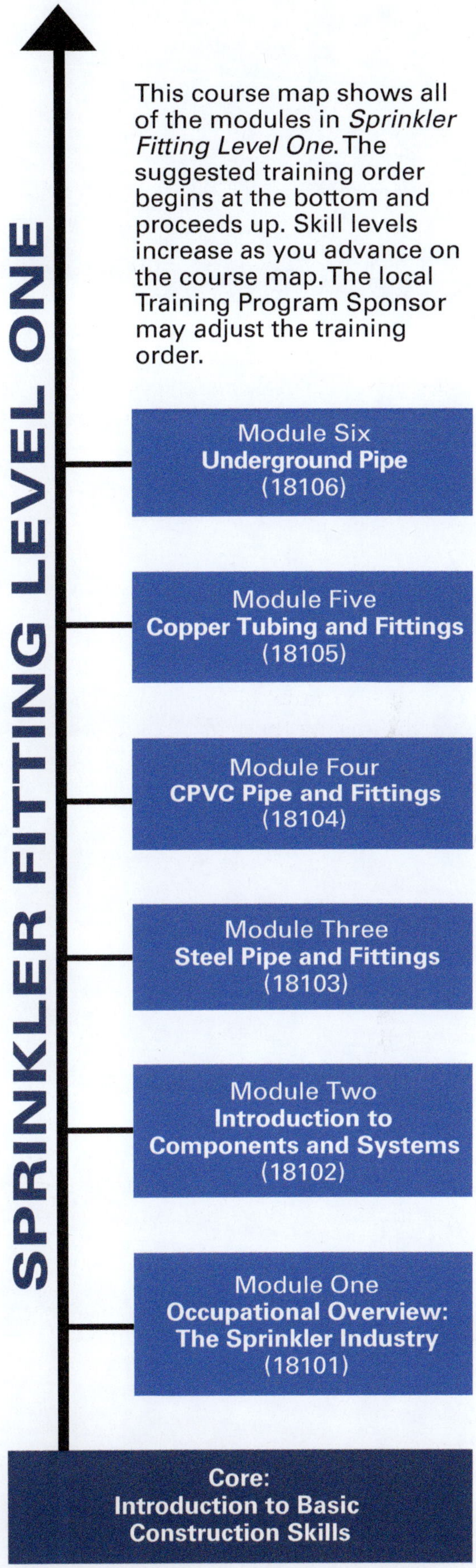

Occupational Overview: The Sprinkler Industry

Overview

This module describes the basic types of sprinkler systems, the tools used to install them, and the codes and standards that apply to these systems. It also covers the apprenticeship requirements for sprinkler fitters, employer and employee responsibilities, and career paths in the sprinkler industry.

Module 18101

Trainees with successful module completions may be eligible for credentialing through the NCCER Registry. To learn more, go to **www.nccer.org** or contact us at 1.888.622.3720. Our website, **www.nccer.org**, has information on the latest product releases and training.

Your feedback is welcome. You may email your comments to **curriculum@nccer.org**, send general comments and inquiries to **info@nccer.org**, or fill in the User Update form at the back of this module.

This information is general in nature and intended for training purposes only. Actual performance of activities described in this manual requires compliance with all applicable operating, service, maintenance, and safety procedures under the direction of qualified personnel. References in this manual to patented or proprietary devices do not constitute a recommendation of their use.

18101 V4.0

16 2025

18101

Occupational Overview: The Sprinkler Industry

Objectives

Successful completion of this module prepares you to do the following:

1. Identify the various types of sprinkler systems and the tools used to install them.
   a. Recognize the various types of sprinkler systems.
   b. Identify the tools used to install sprinkler systems.
2. Identify the codes and standards applicable to sprinkler systems.
   a. List the various codes applicable to sprinkler systems.
   b. Identify the NFPA standards applicable to sprinkler systems.
3. Understand the apprenticeship training process for sprinkler fitters.
   a. List Department of Labor (DOL) requirements for apprenticeship.
   b. Describe various career paths and opportunities in the sprinkler industry.
   c. Understand the responsibilities of the employer and employee.
   d. Describe employer and employee safety obligations.

Performance Tasks

This is a knowledge-based module. There are no Performance Tasks.

Trade Terms

Adopted standards
American Fire Sprinkler Association (AFSA)
Authority Having Jurisdiction (AHJ)
Blocked
Compressed gas
Deluge sprinkler system
Dry pipe sprinkler system
Hazard
National Fire Protection Association (NFPA)
NCCER Standardized Curricula
NFPA 13, Standard for the Installation of Sprinkler Systems
Occupational Safety and Health Administration (OSHA)
Office of Apprenticeship (OA)
On-the-job training (OJT)
Preaction sprinkler system
Rack
Recognized standards
Regulations
Sprinkler
Wet pipe sprinkler system

Industry Recognized Credentials

If you are training through an NCCER-accredited sponsor, you may be eligible for credentials from NCCER's Registry. The ID number for this module is 18101. Note that this module may have been used in other NCCER curricula and may apply to other level completions. Contact NCCER's Registry at 1.888.622.3720 or go to **www.nccer.org** for more information.

You can also show off your industry-recognized credentials online with NCCER's digital badges. Transform your knowledge, skills, and achievements into badges that you can share across social media platforms, send to your network, and add to your resume. For more information, visit **www.nccer.org**.

NOTE

Unless otherwise specified, references in parentheses following figure and table numbers refer to NFPA 13.

Contents

Figures and Tables

SECTION ONE

1.0.0 SPRINKLER SYSTEM TYPES AND INSTALLATION TOOLS

Objective

Identify the various types of sprinkler systems and the tools used to install them.

a. Recognize the various types of sprinkler systems.
b. Identify the tools used to install sprinkler systems.

Trade Terms

Deluge sprinkler system: A type of sprinkler system that uses open sprinklers rather than closed sprinklers. When heat triggers the detection system, the deluge valve operates, producing immediate water flow through all the sprinklers in a given area.

Dry pipe sprinkler system: Sprinkler systems that are filled with pressurized air or nitrogen. Water flows into the system only when a sprinkler is activated.

Hazard: A possible source of danger or potential injury.

National Fire Protection Association (NFPA): An organization that publishes codes and standards with the goal of preventing loss of life and property.

NFPA 13, Standard for the Installation of Sprinkler Systems: The National Fire Protection Agency (NFPA) standard that provides the minimum requirements for the safe installation of sprinkler systems.

Occupational Safety and Health Administration (OSHA): An agency of the US Department of Labor established by the Occupational Safety and Health Act of 1970. OSHA's mission is to ensure safe and healthful working conditions for working men and women by setting and enforcing standards as well as providing safety training, outreach, education, and assistance.

Preaction sprinkler system: A sprinkler system employing sprinklers attached to a piping system containing air or nitrogen that may or may not be under pressure. A supplemental detection system is installed in the same area as the system sprinklers.

Sprinkler: A heat-sensing device used in sprinkler systems to discharge water onto a fire.

Wet pipe sprinkler system: The simplest, most common type of sprinkler system. Wet pipe sprinkler systems are always filled with water, compared to dry pipe sprinkler systems that contain no water until the system is activated.

Sprinkler systems are one of the most important methods for protecting life and property. They provide protection 24 hours a day, seven days a week, even when building occupants are asleep or absent from the premises. They have a success rate worldwide of over 99 percent, a major factor in the insurance industry's insistence on installation of sprinklers in high-risk areas. However, sprinkler systems provide adequate protection only if designed and installed properly. Sprinkler systems are designed and installed based on specific codes and standards from organizations like the **National Fire Protection Association (NFPA)**, with worker safety considerations governed by the **Occupational Safety and Health Administration (OSHA)**.

The first sprinkler system was invented in 1806 by an Englishman, John Carey. He came up with the idea for a heat-operated device for putting out fires that delivered water through perforated pipes. The first sprinkler was invented by Major Stewart Harrison of London, England. In 1864, Harrison also invented an automatic fire extinguishing system, but could not interest British businesses in buying it and did not take out a patent. Perforated pipe systems proved unsuccessful in the long run due to excessive water damage, exhaustion of water supplies, and most of all because they were manual rather than automatic.

The first practical application of sprinklers is credited to Americans Henry Parmelee and Frederick Grinnell. The modern sprinkler industry came about due to the pioneering automatic sprinkler patented by Henry Parmelee in August 1874. A second design was introduced that was the initial step toward commercialization. Three years later, Grinnell and Parmelee agreed to manufacture the Parmelee sprinklers on a royalty basis. After a great deal of research, in October 1881 Grinnell received four patents on the pendent sprinkler, known as the first *sensitive sprinkler*, which turned out to be the first standard-orifice sprinkler. Sprinklers underwent continuous development, and by January 1896, it is estimated that about 25,000 fires had been controlled and/or extinguished by Grinnell sprinklers.

From the earliest days of perforated pipe to today's complex and specialized components, sprinkler systems have undergone continuous change and improvement. Sprinkler systems not only protect property but have been proven to save lives as well.

The principles of sprinkler operation are relatively simple. A closed piping system is generally run overhead from a water source to the areas needing protection. Sprinklers are fitted to the pipes at appropriate intervals. A thermal element in the sprinkler prevents water release by holding a seal closed until the sprinkler's activation temperature is reached, which is typically between 135°F and 286°F. When the thermal element senses its activation temperature, it deforms, releases the seal, and discharges water over its area of coverage. Only the sprinkler or sprinklers that sense heat will activate. All others remain closed, thereby minimizing the water damage to areas where there is no fire and limiting the amount of water required.

Sprinkler system design considerations include sprinkler spacing; anticipated water flow; building size, contents, and use; and the type of space in which the sprinklers are located (enclosed roof spaces, floor space, etc.).

Sprinkler systems have a control valve placed at the point where the water enters the system. This valve is normally open and used for emergency shutoff and system maintenance. When a sprinkler activates, water flows through the control valve and past the alarm device to operate an alarm bell or bell and strobe.

1.1.0 Sprinkler System Types

The four fundamental types of sprinkler systems include the **wet pipe sprinkler system**, **dry pipe sprinkler system**, **preaction sprinkler system**, and **deluge sprinkler system**. Each of these systems is designed for specific applications, situations, and conditions.

1.1.1 Wet Pipe Sprinkler Systems

Wet pipe sprinkler systems, defined in **NFPA 13, Standard for the Installation of Sprinkler Systems**, are the simplest and most reliable sprinkler systems (*Figure 1*). The operation of a single sprinkler activates a wet pipe sprinkler system. Note that only those sprinklers activated by heat will open. All others remain closed.

In wet pipe sprinkler systems, the pipes are always filled with water. When heat activates the thermal element in a sprinkler, water is discharged onto the fire. Wet pipe sprinkler systems are simple, reliable, have fewer components, and experience fewer problems than the other system types. They are also the most common and are preferred in most instances because they put water onto the fire quickly.

Wet pipe sprinkler system installation and maintenance costs are lower because less installation and maintenance time is required for these systems. Wet pipe sprinkler systems can be modified easily; no additional work is required for the control and detection circuits required by other types of systems. There is less downtime following a fire because usually all that is required is to replace the activated sprinklers and turn on the water. For these reasons, wet pipe sprinkler systems are typically considered first.

However, a wet pipe sprinkler system is not a good choice for any location likely to freeze. Freezing can damage the piping and can render the system inoperable. A dry pipe or preaction system is a better choice for protection of areas with freezing temperatures.

Did You Know?

Beginning in 1895, complaints of improperly installed sprinklers and inconsistencies in the insurance rating industry became a concern to the fire insurance industry as well as sprinkler manufacturers and installers. These concerns led to the creation of the NFPA in 1896 by five insurance companies and Frederick Grinnell.

1.1.2 Dry Pipe Sprinkler Systems

Dry pipe sprinkler systems are filled with pressurized air or nitrogen, not water. The system has a dry pipe valve that is typically held closed by air pressure. When heat activates a sprinkler, the sprinkler releases the air pressure, and the dry pipe valve operates. Water enters through the dry pipe valve and discharges via the activated sprinklers. The only significant advantage of dry pipe sprinkler systems is in areas subject to freezing temperatures.

Dry pipe sprinkler systems have several disadvantages compared to wet pipe sprinkler systems: additional complexity, more costly installation and maintenance, a design resistant to change, greater chance of corrosion, and slower reaction time. These disadvantages are due to having more control and air pressure components; limited system size; a time delay before water is released onto the fire; and the requirement that the system be completely drained after discharge.

> **Did You Know?**
>
> - Fires are usually controlled or contained with only one or two sprinklers.
> - In most systems, one sprinkler at a time activates.
> - Most sprinkler systems are designed to control or contain a fire, not necessarily to fully extinguish the fire.

1.1.3 Preaction Sprinkler Systems

A preaction sprinkler system uses sprinklers attached to a piping system containing air or nitrogen that may or may not be under pressure. A supplemental detection system installed near the sprinklers is used to open the preaction valve, which then triggers the flow of water through the system.

Preaction sprinkler systems use more complex components than wet pipe sprinkler systems. These components increase the initial cost and the amount of time required to perform system inspection, testing, and maintenance. As a result, higher costs and additional labor are required to keep the system working properly.

1.1.4 Deluge Sprinkler Systems

A deluge sprinkler system uses normally open sprinklers rather than normally closed sprinklers, as in a wet pipe sprinkler system. When heat triggers the detection system, the deluge valve is released, producing immediate water flow through all of the sprinklers in a given area. Deluge sprinkler systems are typically found in places such as aircraft hangers and chemical plants, where high volumes of water and fast reaction are required.

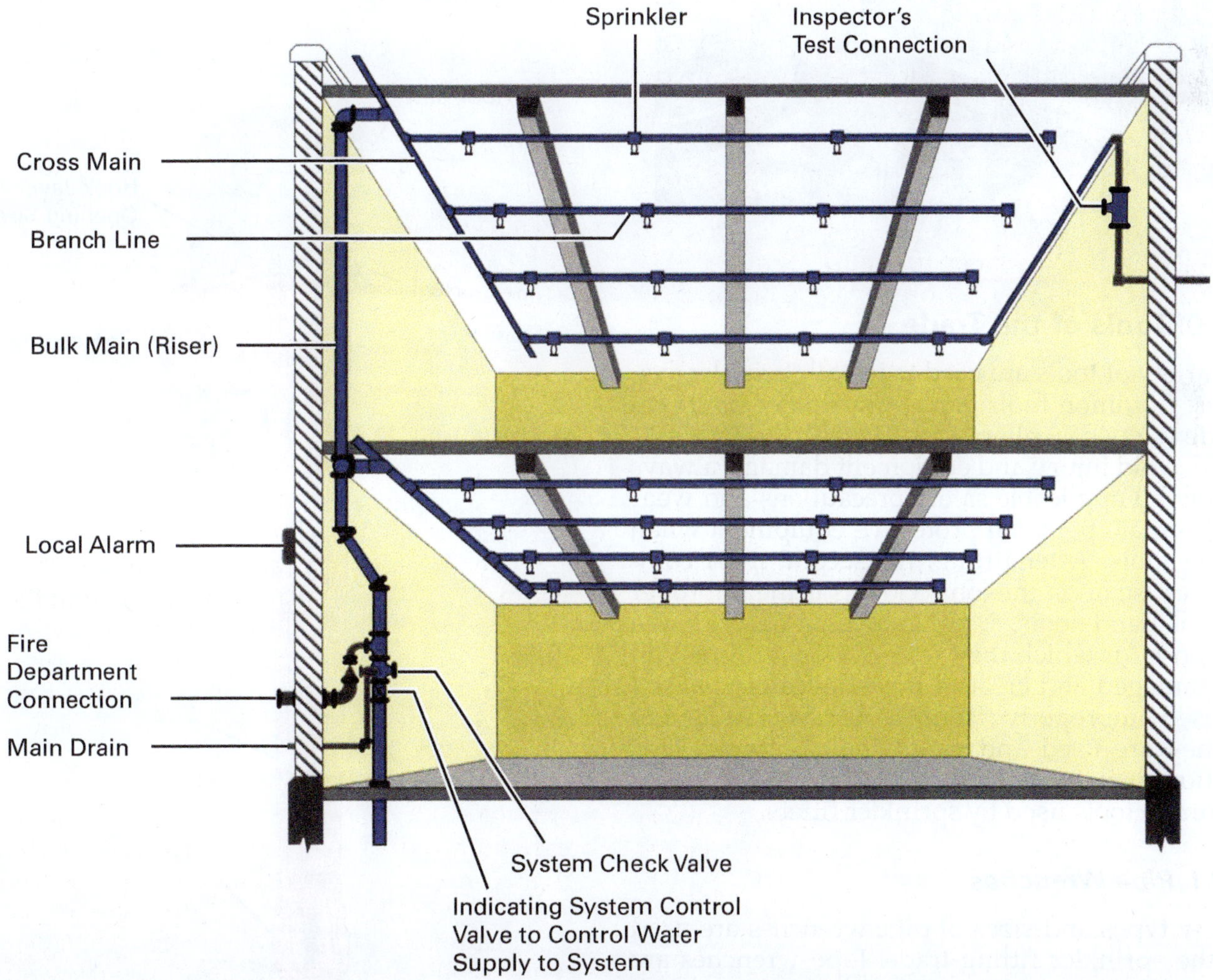

Figure 1 Typical wet pipe sprinkler system.

Shop Drawings

Shop drawings are a very important part of the installation process. A sprinkler apprentice must understand and be able to read piping plans, as well as understand where to find information on the drawings.

A sprinkler installation begins with drawings that have been prepared by a competent sprinkler layout technician or professional engineer. Following the shop drawings with little or no variation is very important. Proposed deviations must be submitted to the layout technician because any changes may impact the successful operation of the system.

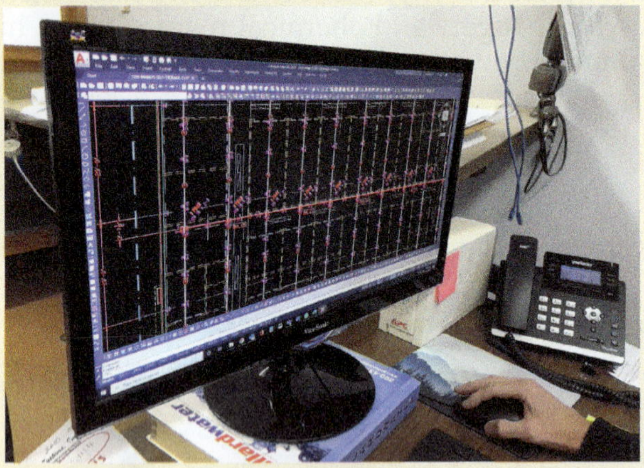

1.2.0 Tools of the Trade

A variety of tools are used to install sprinkler systems. Common tools used by sprinkler fitters include wrenches, pliers, saws, levels, and ladders.

To avoid injury and equipment damage, always follow all applicable safety precautions and wear appropriate personal protective equipment when using tools. After lifting, the second most common cause of on-the-job accidents is the improper use of hand tools. Use hand tools only for the purpose for which they were designed. Tools can be damaged and create a potential safety **hazard** if used improperly. Hand tools also last longer if they are used and maintained properly. This section explains the purpose and proper use of common tools used by sprinkler fitters.

1.2.1 Pipe Wrenches

Many types and sizes of pipe wrenches are used in the sprinkler fitting trade. Pipe wrenches are heavy-duty wrenches made of steel or aluminum. They are used to grip and turn pipe or to grip and hold pipe stationary. It is critical to select the right size pipe wrench for a specific job. A pipe wrench that is too small will not hold the pipe firmly and a handle that is too short will not provide enough leverage. A pipe wrench that is too large can strip the threads, break the pipe or fitting, or cause excessive marring or scratching of the pipe. Sprinkler fitters use several types of pipe wrenches, including the following:

- Straight pipe wrenches
- Offset pipe wrenches
- Chain wrenches
- Strap wrenches

Straight Pipe Wrenches

Straight pipe wrenches have two jaws. The stationary jaw is known as the *heel* and the movable jaw is known as the *hook jaw*. Both jaws have teeth cut into them to give the jaws a firm grip on the pipe. The teeth on the hook jaw face inward and the teeth on the heel face outward. *Figure 2* shows the components of a straight pipe wrench.

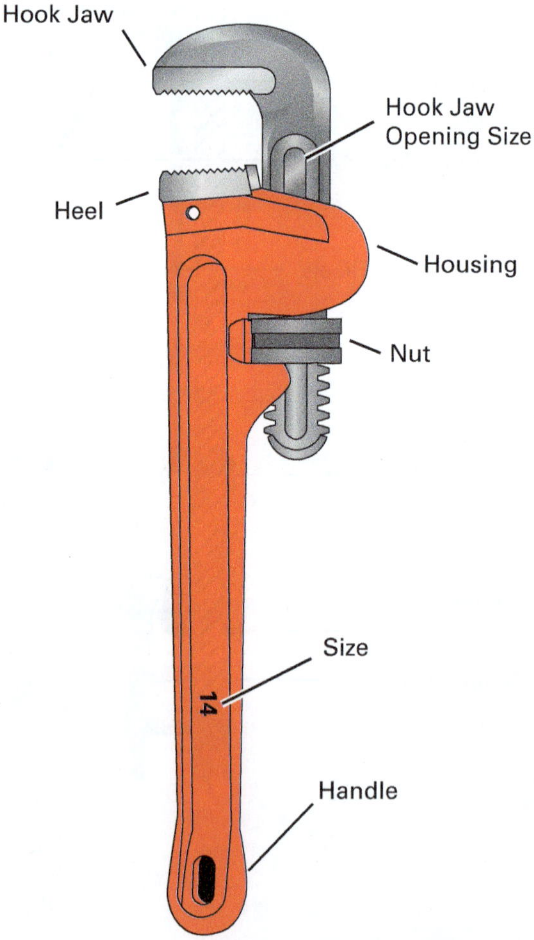

Figure 2 Components of a straight pipe wrench.

The hook jaw of a straight pipe wrench is adjusted by turning the nut located below the housing. A spring inside the housing allows the hook jaw to float in the housing. This causes the hook jaw to open slightly as you apply the wrench to the pipe. Straight pipe wrenches are sized by the length measured from the inside top of the hook jaw to the end of the handle when the jaws of the wrench are fully open. Straight pipe wrenches range in size from 6" to 48". The size does not refer directly to the size pipe it fits, but each size wrench has recommended pipe capacities. *Table 1* shows pipe wrench sizes and capacities.

To use a pipe wrench, grip the pipe with the wrench so the jaw opening faces you and you can tighten the pipe or fitting by pulling the handle towards your body. If there is not enough space to position the wrench this way, face the jaw opening away from you and push on the handle to tighten or remove the pipe. Be sure to brace yourself when pushing on the wrench so that you do not fall if the wrench slips off the pipe.

Follow these steps to tighten a fitting onto the threaded end of a pipe:

Step 1 Select a pipe wrench suitable for the size of pipe being used.

Step 2 Inspect the wrench for excessive wear on the jaw teeth and corners, and for any noticeable damage to the handle. Repair or replace the wrench, if necessary.

Step 3 Secure the pipe in a chain vise, allowing the threaded end to extend from the vise by about 8".

Step 4 Apply pipe thread compound or Teflon® tape and hand-tighten the fitting onto the pipe.

Step 5 Position your body so that the fitting is on the left side.

Step 6 Turn the nut on the pipe wrench until the jaws are opened slightly wider than the diameter of the fitting.

NOTE

Many straight pipe wrenches have pipe sizes marked on the hook jaw just above the housing. If your wrench has these marks, line up the correct mark with the top of the housing.

Step 7 Hold the pipe wrench by the handle so that the open jaws are facing toward you.

Step 8 Slide the jaws onto the shoulder of the fitting.

NOTE

Be sure the jaws fit snugly onto the shoulder of the fitting. The shoulder should be centered inside the jaws, and there should be a gap between the fitting and the back of the hook jaw.

Step 9 Pull the handle toward your body to tighten the fitting.

Step 10 Push the handle in the opposite direction to release the jaws from the fitting.

Step 11 Regrip the fitting with the wrench and pull the wrench toward your body again.

Step 12 Repeat *Steps 10* and *11* until the fitting is tight on the pipe.

CAUTION

Do not overtighten the fitting on the pipe as the threads may strip.

Step 13 Remove the wrench from the pipe fitting.

Step 14 Clean the wrench with a wire brush, lubricate it, and inspect it for any damage.

Step 15 Close the jaws and store the wrench.

Table 1 Pipe Wrench Sizes and Capacities

Wrench Size (Handle Length)	Capacity	Recommended for Pipe Sizes
6	1/8–1/2	1/8
8	1/8–3/4	1/4–3/8
10	1/8–1	1/2–3/4
14	1/4–1 1/2	1
18	1/4–2	1 1/4–1 1/2
24	1/2–2 1/2	2–2 1/2
36	1/2–3 1/2	3
48	1–5	4–5

Note: All dimensions in inches.

You may have to install or remove a fitting from a piece of pipe where a vise is not available, or the fitting is part of a piping system. In this case, use two pipe wrenches. Follow these steps to tighten a fitting without using a vise:

Step 1 Position yourself with the fitting on the right side.

Step 2 Place one pipe wrench on the pipe with the jaw opening facing you.

Step 3 Place the other pipe wrench on the fitting with the jaw opening facing away from you. *Figure 3* shows the use of two pipe wrenches.

Step 4 Push on the handle of the wrench that is on the pipe while at the same time pulling on the handle of the wrench that is on the fitting.

Step 5 Regrip the pipe and the fitting when you have turned them as far as you can.

Step 6 Repeat Steps 2 through 5 until the fitting is tight.

Step 7 Remove the wrenches from the pipe and fitting.

Step 8 Clean the wrenches with a wire brush, lubricate them, and inspect for any damage.

Step 9 Close the jaws and store the wrenches.

Offset Pipe Wrenches

An offset pipe wrench operates in the same way as a straight pipe wrench. It is made the same way and has similar components. The only difference is that the hook jaw and heel are offset from the handle at either 45 degrees or 90 degrees. This wrench is useful when working in tight spaces, such as in areas where pipes are located close to a ceiling or wall, or in other areas where you cannot grip and turn a pipe using a straight pipe wrench. *Figure 4* shows offset pipe wrenches. Follow the same steps for using straight pipe wrenches when using offset pipe wrenches.

Chain Wrenches

Chain wrenches are used in the same manner as pipe wrenches. A chain wrench is a pipe wrench with a fixed, serrated jaw. A length of steel chain, attached to the pipe wrench, wraps around a pipe and connects to the other side of the jaw. This wrench holds the pipe firmly and distributes the bite of the jaw evenly. It also allows you to tighten or loosen the pipe without removing the chain. *Figure 5* shows a chain wrench.

A chain wrench is especially useful in close quarters. Follow these steps to use a chain wrench:

Step 1 Place the jaw of the wrench against the pipe.

Step 2 Wrap the chain around the pipe and lock the chain into the wrench frame.

CAUTION

Be sure the chain is tight around the pipe to ensure a good grip.

Step 3 Push the handle to tighten the pipe; pull the handle to loosen the pipe.

Step 4 Slide the chain wrench around the pipe to regrip the pipe before turning again.

Step 5 Repeat *Steps 3* and *4* until the pipe is tightened or loosened enough to remove by hand.

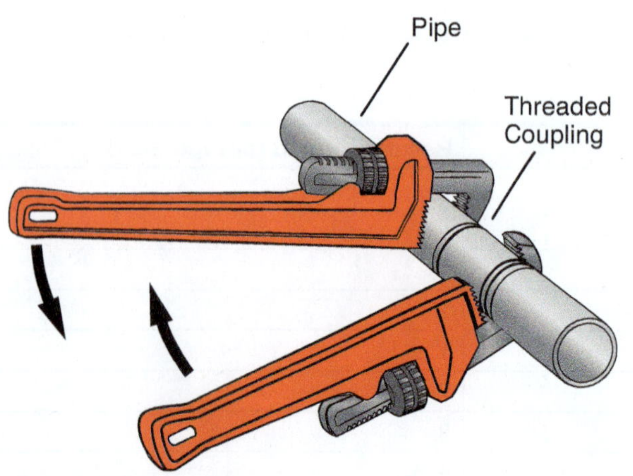

Figure 3 Tightening pipe.

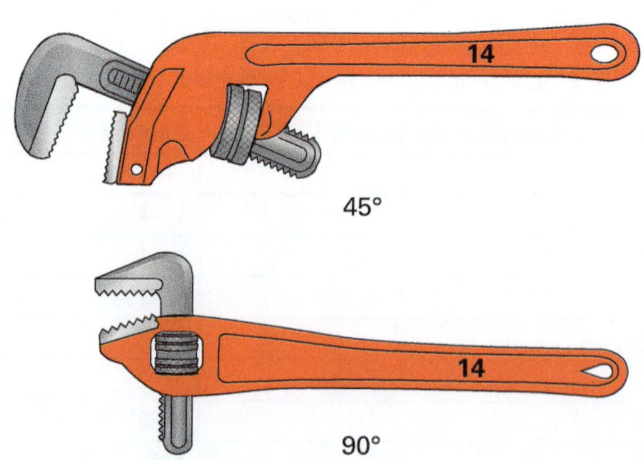

Figure 4 Offset pipe wrenches.

Figure 5 Chain wrench.

Step 6 Remove the wrench from the pipe.

Step 7 Clean the wrench with a wire brush, lubricate it, and inspect it for any damage.

Step 8 Store the wrench.

Strap Wrenches

A strap wrench is similar to a chain wrench except that a heavy-duty cloth or leather strap replaces the chain. A strap wrench is used to remove or install polished or plated valves and fittings that could be cosmetically damaged by other types of wrenches. The strap serves as the jaw of the wrench and is adjustable to provide a tight, secure grip. *Figure 6* shows a strap wrench. Follow the same steps for using chain wrenches when using strap wrenches.

CAUTION

Standard wrenches and vises cannot be used on finished surfaces, such as brass or chrome.

Use and Care of Pipe Wrenches

Follow these guidelines to use and care for pipe wrenches:

- Use pipe wrenches only to turn pipe and fittings. Do not use a pipe wrench to bend, raise, or lift a pipe.
- Do not use the pipe wrench as a hammer.
- Do not drop or throw a pipe wrench. You might cause injury to yourself or others, and you might break or crack the wrench.
- Check the teeth of the wrench often and keep them clean and sharp. This will keep the wrench from slipping on the pipe.
- Apply penetrating oil to the wrench threads to prevent them from sticking.
- Never use a cheater or extension bar on a pipe wrench handle to increase leverage. This may cause the wrench components or the pipe to bend or break.
- Use the proper size wrench for the pipe.
- Ensure that the jaws are fully seated on the pipe before turning.
- Close and open the jaws by turning the threaded wheel.
- Pull on the wrench instead of pushing it. If pushing is necessary, use the base of your palm. This avoids skinning the knuckles if the pipe moves suddenly.
- Apply the torque at right angles to the pipe. Side force can spring the jaws and can break the grip unexpectedly.
- Double-check to ensure that the applied force will close the jaws instead of opening them.
- Use a steady application of force and do not jerk on the handle or use a hammer on it.
- Clean all wrenches after each use and lubricate the moving parts.
- Repair or replace the pipe wrench if the jaws are sprung.
- Never use a pipe wrench on nuts or bolts. The jaws of a pipe wrench will strip the edges off a nut or bolt and make it difficult to move the nut or bolt with a standard wrench.

1.2.2 Sprinkler Wrenches

Each type of sprinkler must be installed using a special wrench designed by the sprinkler manufacturer to match the specific sprinkler make and model. These wrenches are designed to fit snugly on the sprinkler and not slip off and damage the sprinkler. Many sprinkler failures are the result of using improper wrenches. Often, the damage is so small that it is not clearly visible. Only after catastrophic water damage and laboratory analysis is the damage revealed. *Figure 7* shows some typical sprinkler wrenches.

Figure 6 Strap wrench.

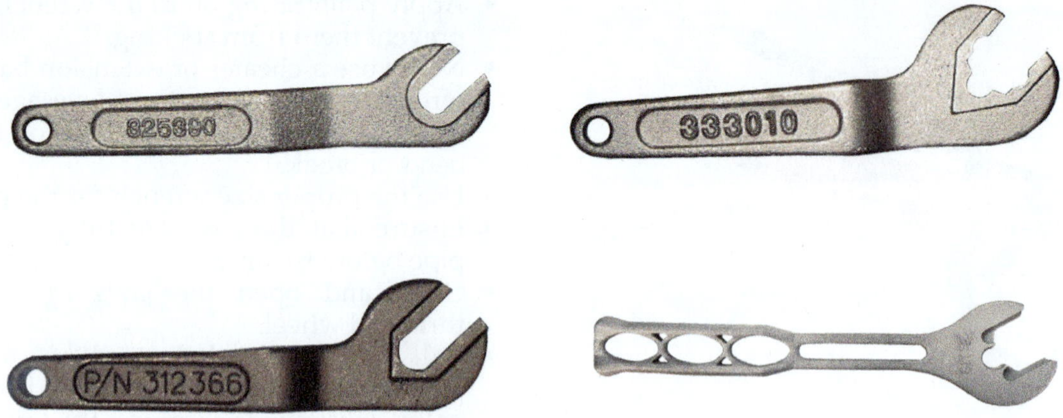

Typical Open-Ended Wrenches

Typical Socket Wrenches

Figure 7 Sprinkler wrenches.

1.2.3 Torque Wrenches

Torque wrenches (*Figure 8*) measure resistance to turning. They are used when installing fasteners that must be tightened in sequence without distorting the workpiece. Use a torque wrench only when a torque setting is specified for a particular fastener.

Torque specifications are usually stated in inch-pounds for small fasteners or foot-pounds for large fasteners.

How to Use a Torque Wrench

Follow these steps to use a torque wrench properly:

Step 1 Refer to the manufacturer's installation instructions for the object to be fastened to determine how many inch-pounds or foot-pounds of torque are required. Set the controls on the wrench to the desired torque level (wrench models vary).

Step 2 Refer to the manufacturer's instructions to determine the proper torque sequence (which fastener comes first, second, and so on). Using the wrong sequence may damage the object being fastened.

Step 3 Place the torque wrench on the fastener, such as a bolt. Hold the head of the wrench with one hand to support the bolt and make sure it is properly aligned.

Step 4 Watch the torque indicator and/or listen for the click (depending on the model of the wrench) as you tighten the bolt.

CAUTION

A torque wrench can cause injury and property damage if used incorrectly. Always follow the manufacturer's instructions for safety, maintenance, and calibration. Never use the wrench as a ratchet or as anything other than its intended purpose. Always store the wrench in its case to prevent damage.

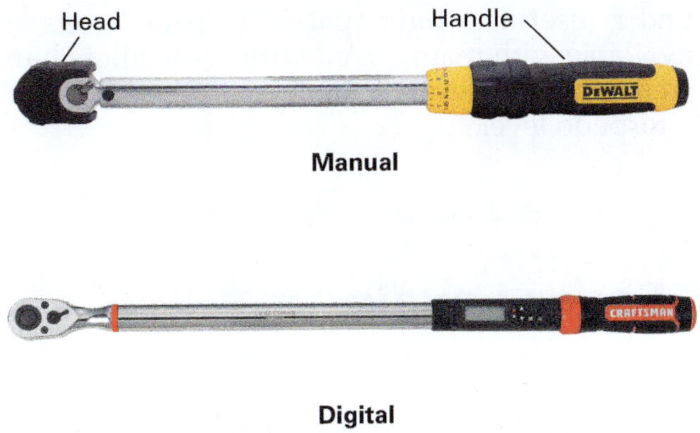

Figure 8 Torque wrenches.

How to Calculate Torque When Using an Adapter

When you use an adapter or extension, the torque wrench becomes longer. The extra length and the applied torque will be greater than the torque indicator. Use this formula to determine the correct torque:

$$\text{Preset torque} = \frac{\text{length of torque wrench} \times \text{desired torque}}{\text{length of torque wrench} + \text{length of extension}}$$

To determine the length of a torque wrench, measure the distance from the center of the square drive of the wrench to the center of the handle. To determine the length of the extension, measure the distance from the center of the square drive of the extension to the center of the bolt or nut. Be sure to measure only the length that is parallel to the handle.

Torque Wrench Calibration

The calibration of torque wrenches must be checked periodically to guarantee accuracy. The following terms, sometimes used in manufacturers' service literature, must be understood when using a torque wrench:

- *Breakaway torque* — The torque required to loosen a fastener. This is generally lower than the torque to which it has been tightened. For a given size fastener, there is a direct relationship between tightening torque and breakaway torque. This relationship is determined by an actual test. Once known, the tightening torque can be checked by loosening and checking breakaway torque.
- *Set or seizure* — In the last stages of rotation in reaching a final torque, seizing or setting of the fastener may occur. When this happens, there is usually a noticeable popping sound and vibration. To break the set, back off and then again apply the tightening torque. Accurate torque settings cannot be made if the fastener is seized.
- *Run-down resistance* — The torque required to overcome the resistance of associated hardware, such as locknuts and lockwashers, when tightening a fastener. To obtain the proper torque value where tight threads on locknuts produce a run-down resistance, add the resistance to the required torque value. Run-down resistance must be measured on the last rotation or as close to the makeup point as possible.

> **CAUTION**
>
> Tighten fasteners only small amounts at a time, following the proper tightening sequence for the bolt pattern and fastener type before reaching the final torque. Failure to follow the correct tightening sequence or overtightening can result in damage to the fasteners, any sealing gaskets, or the object being fastened.

1.2.4 Pliers

Many types of pliers are used by sprinkler fitters. The most common pliers are tongue-and-groove pliers and slip-joint pliers.

Tongue-and-groove pliers, sometimes called *slip-groove pliers* or *Channellocks®*, are versatile utility pliers that work well on objects of all shapes. They range in size from $4\frac{1}{2}$" to 16", with a jaw capacity of $\frac{7}{8}$" to $4\frac{1}{4}$". Pliers with long handles provide more leverage. Tongue-and-groove pliers are available with straight jaws or curved jaws. The jaws are joined with a heavy-duty nut and bolt and are self-locking. They are especially useful for adjusting hanger rods. *Figure 9* shows straight jaw tongue-and-groove pliers.

Figure 9 Straight jaw tongue-and-groove pliers.

Because of their shape, curved jaws are best for use with pipe and are preferable for the sprinkler fitter. *Figure 10* shows curved jaw tongue-and-groove pliers.

Slip-joint pliers are all-purpose tools that are useful for gripping and turning round objects, such as hanger rods. Slip-joint pliers are available in thin nose and bent nose shapes. The bent nose type has a nose that is bent at a 30-degree angle, which keeps the hand clear of a flat surface while using the tool. Slip-joint pliers range in length from 5" to 10", with jaw-opening capacities from ¾" to 1½". *Figure 11* shows slip-joint pliers.

1.2.5 Levels

A level is a device used to determine the trueness of a line. Some levels have liquid-filled vials containing a bubble and two lines. Level occurs when the bubble is centered between the two lines. The longer the level, the more accurate the measurement.

The most important level for the sprinkler fitter is a torpedo level. A torpedo level has three vials. A fitter can check pipes for vertical trueness, horizontal level, and true 45-degree angles with the torpedo level. The torpedo level is usually 9" long and is useful in tight spaces. Torpedo levels are available with magnetized bottoms to allow hand freedom while leveling steel pipe. *Figure 12* shows a torpedo level.

Figure 10 Curved jaw tongue-and-groove pliers.

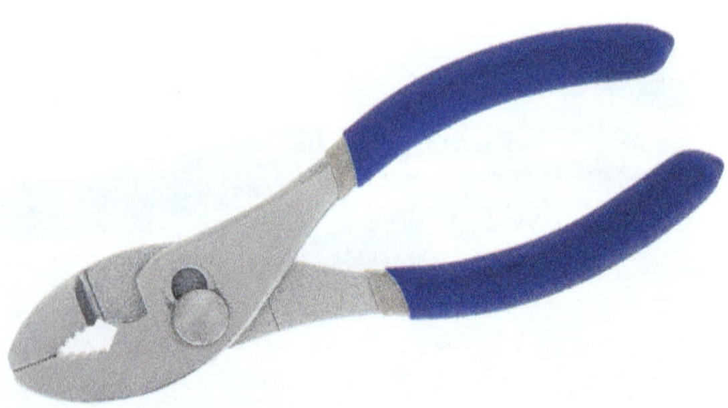

Figure 11 Slip-joint pliers.

Laser Levels

Laser levels are commonly used to determine level and plumb. These levels do not use liquid-filled vials, but instead use low-power lasers that project either single or multiple continuous beams. Lasers are generally battery-operated and self-leveling. They are often used in large, open areas where measurement to a common reference line is necessary.

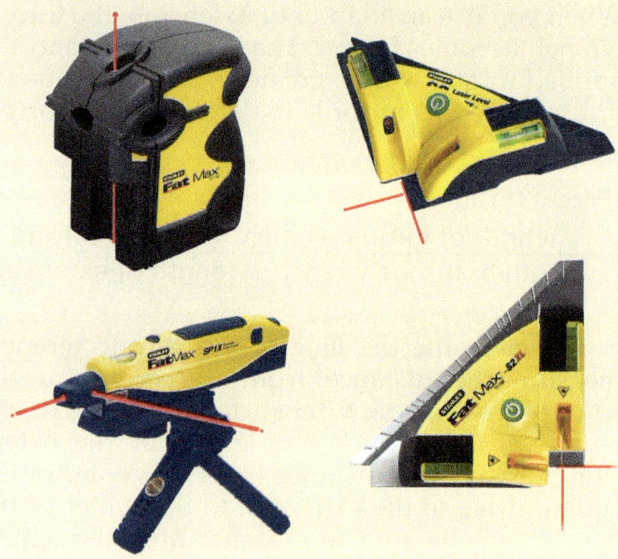

Figure Credit: Image property of Stanley Black & Decker. Used with permission.

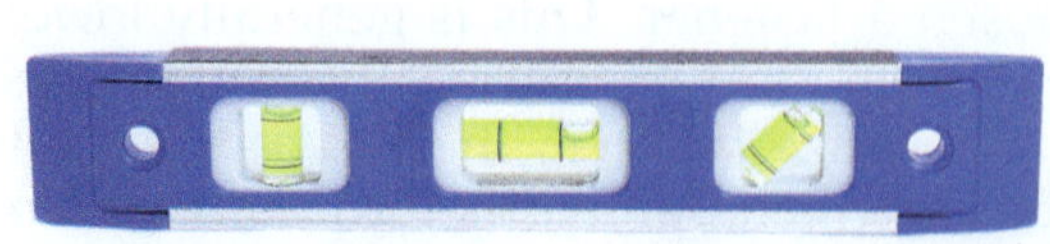

Figure 12 Torpedo level.

Portable Band Saws

Portable band saws are common where large amounts of metal cutting is necessary. These tools are electrically driven and have a metal cutting blade that is circular and moves in one direction around two wheels.

Figure Credit: Image property of Stanley Black & Decker. Used with permission.

1.2.6 Hacksaws

Hacksaws are utility saws that are used to cut metal that is too heavy for bolt cutters or tin snips. The saw blades can be changed out to accommodate any sawing need. Replacement blades should be carried in case a blade breaks. The teeth of the blade should be curved forward, away from the handle, and the blade should be tight. *Figure 13* shows hacksaws.

Hacksaws cut on the push stroke. To use a hacksaw, apply pressure on the push stroke and lift on the return stroke. It is best to saw with long, slow, steady strokes.

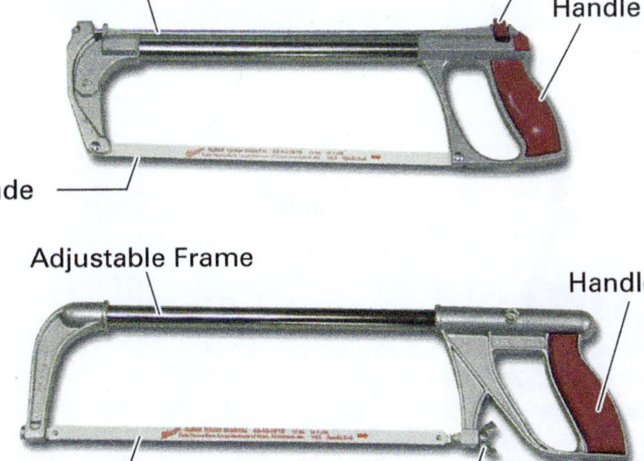

Figure 13 Hacksaws.

1.2.7 Toolboxes

A good quality toolbox is needed to ensure that small hand tools are properly stored and protected. The toolbox should have compartments so tools can be separated by size and type. When tools are well organized, time will not be wasted looking for one tool. A lock is recommended to prevent tools from being stolen.

1.2.8 Ladders

Sprinklers often work overhead and use a variety of ladders. Common types include A-frame, extension trestle, and podium ladders. Always follow the manufacturer's instructions when setting up and using any type of ladder.

A-Frame Ladders

The most common type of ladder is the simple A-frame ladder (*Figure 14*). A-frame ladders are available in many different sizes. They are typically constructed of wood, aluminum, or fiberglass.

Extension Trestle Ladders

Extension trestle ladders, commonly referred to as *extension* or *strut ladders*, are three-piece ladders that allow for work at heights higher than the usual A-frame ladder. An extension trestle ladder (*Figure 15*) has a primary adjustable vertical ladder in the center, with legs or trestles extending on either side. They are usually constructed of wood, aluminum, or fiberglass.

Figure 14 A-frame ladder.

Podium Ladders

A podium or *platform* ladder (*Figure 16*) provides a stable working platform with a safety guard rail at the top that allows for work at a fixed height in any direction. Some podium ladders are equipped with casters that must be locked before climbing the ladder.

Ladder Safety

Observe the following basic safety rules when using ladders:

- Always face the ladder when ascending or descending.
- Maintain three-point contact when climbing a ladder.
- Use only one rung or step at a time.
- When using a ladder, always step in the middle of the rung so your weight is evenly distributed.
- Never reach beyond arm's length when working from a ladder.
- Never stand on the top two steps of a stepladder or the third rung of an extension ladder.
- Never use a ladder horizontally as a platform.
- Never use a painted ladder or paint one. Paint can hide defects that may lead to an accident.

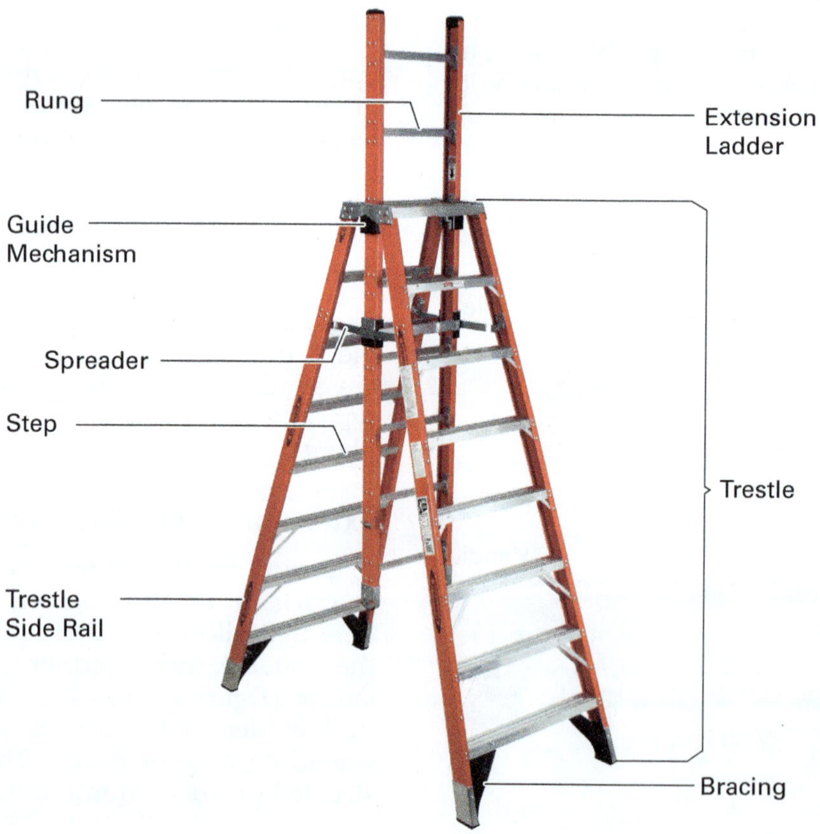

Figure 15 Extension trestle ladder.

Movable Leg Ladders

Movable leg, or *multi-position* ladders, are common in the industry. These ladders are collapsible and more easily carried and stored when compared to A-frame or extension ladders. Movable leg ladders can be adjusted for use on uneven surfaces such as stairs or extended to reach high areas only accessible with an extension trestle ladder. They are many different ladders in one.

Figure Credit: Courtesy of Werner

Figure 16 Podium ladder.

- OSHA requires the use of fall protection whenever a worker is exposed to a fall of 6' or more. Use scaffolding or personnel lifts to reduce hazards when installing overhead piping systems and components.
- Ensure that the ladder is resting against a solid surface and that each leg is firmly positioned on a solid base.
- Ensure that the spreaders on A-frame and similar ladders are always fully extended, with the braces in the locked position.
- Extension ladders must be secured and extended 3' above the landing.
- Never exceed the duty rating of a ladder. This includes your own weight plus the weight of any tools, supplies, or equipment that you may be carrying and/or storing on the ladder.

1.0.0 Section Review

1. The type of sprinkler system that might be found in an aircraft hangar is a _____.
   a. dry pipe sprinkler system
   b. wet pipe sprinkler system
   c. preaction sprinkler system
   d. deluge sprinkler system

2. A type of ladder that provides a platform for standing is a(n) _____.
   a. A-frame
   b. extension trestle ladder
   c. podium ladder
   d. movable leg ladder

SECTION TWO

2.0.0 CODES AND STANDARDS

Objective

Identify the codes and standards applicable to sprinkler systems.

a. List the various codes applicable to sprinkler systems.
b. Identify the NFPA standards applicable to sprinkler systems.

Trade Terms

Adopted standards: Standards that have been adopted through statutory or administrative laws in a specific geographic area. Once a document is adopted by reference in a state statute or local ordinance, it has the same force and effect as if published therein in its entirety.

Authority Having Jurisdiction (AHJ): An organization, office, or individual responsible for enforcing the requirements of a code or standard, or for approving equipment, materials, an installation, or a procedure.

Recognized standards: Standards that are generally accepted as relevant in civil law, even in areas where no standards have been formally adopted.

Sprinkler systems are designed and installed based on specific codes and standards from organizations such as NFPA, with worker safety considerations governed by OSHA. Both codes and standards serve to protect life, health, and property and to ensure the quality of construction. A code specifies what must be done, while a standard describes how it should be done. Codes and standards have no legal status unless they are adopted by state law or local ordinance.

Codes and standards are usually generated by non-profit, third-party organizations on a consensus basis. Consensus basically means that all concerned parties agree. Consensus is a democratic process in which all interested parties have an opportunity to contribute their opinion. These opinions are harmonized into a consistent document and then accepted by a vote of the members of the sponsoring organization.

The codes and standards addressed in this module are primarily those that affect construction and the installation and maintenance of fire protection systems. A municipal code package typically consists of the following:

- Building code
- Electrical code
- Excavating and grading code
- Fire prevention code
- Gas code
- Housing code
- Mechanical code
- One- and two-family dwelling code
- Plumbing code
- Swimming pool code
- Unsafe building abatement code

In the sprinkler industry, the primary concern is with the effects of the building codes and the fire codes. However, there are provisions in the electrical codes, mechanical codes, and plumbing codes that may affect the work of a fitter indirectly. There are many types of codes and standards that can apply to a system or job. The decision as to which codes and standards apply is made by the **Authority Having Jurisdiction (AHJ)**. Each job has at least one AHJ. The AHJ is the person or agency that decides the acceptability of the systems and devices being installed in accordance with code requirements.

The AHJ can include one or all of the following: a building inspector; a fire code inspector; another federal, state, local, or regional authority; a designated insurance company representative; or the property owner. The AHJ determines which rules apply and ultimately resolves any conflicts between any conflicting codes or rules. The AHJ is the organization, office, or individual responsible for approving equipment, an installation, or a procedure.

Most communities adopt one of the nationally recognized building codes and related codes and standards. Most also have local ordinances that modify or add to these documents, which can have a significant effect on sprinkler work. Some jurisdictions do opt to write their own building code. Some communities use zoning ordinances to require sprinkler installations when they are not required by other documents.

The general rules are indicated in this module, but local rules take precedence. These can change from one side of the street to another, and if a building is built across municipal or state lines, part of the building may be covered by one set of rules and part by a different set of rules. It will be up to the sprinkler contractor to understand and operate under the local rules.

Knowing why things must be done a certain way, or where the standard or code can be found, is part of your training as a sprinkler fitter. The base reference for the sprinkler fitting industry is NFPA 13, *Standard for the Installation of Sprinkler Systems.*

2.1.0 Codes

Building codes and other codes normally provide minimum safety requirements. They typically describe circumstances under which the use of a given type of system, equipment, or component is required, depending on the use and occupancy of a building or structure. Generally, codes are written so that they can easily be adopted into law.

Useful building codes are complex legal documents that evolve through experience from application. This makes the nationally recognized model building codes an attractive source for a useful code. Building codes that are national in scope provide minimum standards to guard the life and safety of the public by regulating and controlling the design, construction, and quality of materials used in modern construction. They have also come to govern the use and occupancy, location of a type of building, and the ongoing maintenance of all buildings and facilities. Once adopted by a local jurisdiction, these national building codes then become law. To meet more stringent requirements and/or local needs, it is common for localities to change or add new requirements to any adopted national code requirements. The provisions of the national building codes apply to the construction, alteration, movement, demolition, repair, structural maintenance, and use of any building or structure within the local jurisdiction.

The national building codes are the legal instruments that enforce public safety in the construction of human habitation and assembly structures. They are used not only in the construction industry, but also by the insurance industry for compensation appraisals and claims adjustments, and by the legal industry for court litigation.

The principal objective of a building code is to regulate the quality of building construction in a community. Community may mean village, city, township, parish, county, or state. Canada has a national building code. Provisions of building codes are administered by the local building officials, and in some instances, by state building officials. Building codes regulate new construction by establishing limits on the height and the floor area of buildings. The degree of combustibility of the construction materials used in a building is another major consideration. A building code also deals with separations between buildings. The minimum requirements for the fire rating of walls and partitions, finish details, etc. are established by usage categories.

Generally, building codes administer new construction and alterations. Approximately 80 percent of the content of building codes deals with fire protection and life safety. Building codes list installation requirements for sprinkler and standpipe systems. Certain occupancies must be sprinklered. Other occupancies can have their heights and/or areas increased by the installation of sprinkler systems that are not otherwise required. These mandatory sprinklering requirements and the available sprinkler incentives are important.

2.1.1 Model Building Codes

There are two active model building codes: The *International Building Code*® (IBC®) developed by the International Code Council (ICC) and the *Building Construction and Safety Code*® (NFPA 5000), developed by the National Fire Protection Association. See *Figure 17.* State and local governments may also have additional requirements that must be met. As a sprinkler fitter, you must be familiar with the local and national codes in force at each installation location.

Codes are revised and updated on a regular basis, usually every two or three years. Always check to make sure that you are working with the most current codes that have been adopted in your area.

Did You Know?

The ICC also has additional codes, including the *International Plumbing Code (IPC), International Mechanical Code (IMC), International Fire Code (IFC),* and several others.

Figure 17 Model building codes.

- Address minimum safety requirements
- Describe circumstance(s) under which the use of a given type of system, equipment, or component is required
- Written so they can be easily adopted into law

2.1.2 NFPA 101®, Life Safety Code®

The *Life Safety Code*® (NFPA 101®) is a mixture of building code requirements and fire code requirements (*Figure 18*). It addresses the construction problem strictly from a life safety standpoint, and there are specific requirements within it for sprinkler systems. Most states and the federal government have adopted NFPA 101® for use. There are conflicts between the *Life Safety Code* and the building codes that may require resolution where both codes are imposed.

Did You Know?

Anyone can submit a recommended change to an NFPA code. Visit **www.nfpa.org** for more information.

Figure 18 Life Safety Code®.

2.1.3 Fire Codes

Each model building code has an associated fire code or fire prevention code. There are two primary fire codes, as shown in *Figure 19*. The *International Fire Code* (IFC) is often used with the IBC, while the *Fire Code* (NFPA 1) is often used in conjunction with NFPA 5000. Usually, a community will adopt the fire code that is compatible with its building code. However, there are communities that have adopted a building code but have not adopted a fire code. Some communities write their own fire code separate and apart from the model codes.

Generally, a fire prevention code covers existing buildings, not new construction. It is primarily a maintenance code. The fire codes deal with operation, fire prevention, and maintenance of fire protection systems. Fire codes are usually administered and enforced by the AHJ.

Conformance to the fire code is usually a requirement before a municipality will issue a Certificate of Occupancy (CO). By law, in many communities, the permanent utilities will not be connected without a CO.

Figure 19 Fire codes.

2.2.0 Standards

Standards detail how the protection specified by the code is to be achieved. Performance standards specify functions and capabilities of hardware and conditions under which the equipment must operate.

2.2.1 Types of Standards

There are many types of standards. The more common types are defined as follows:

- *International standards* — Mainly used to enhance trade, these standards use a common language and establish equivalent measurements.
- *National standards* — These standards usually define a method of manufacture or design. Two organizations that establish national standards are the American National Standards Institute (ANSI) and the National Electrical Manufacturers Association (NEMA).
- *Regional and local standards* — Both regional and local standards establish requirements that conform to state and local needs.
- *Safety standards* — These standards establish requirements related to safety issues.
- *Installation standards* — These standards establish requirements related to how the installation of systems and equipment is to be achieved. NFPA 13 is an example of an installation standard.
- *Industry standards* — Standards developed over time that form the foundation of how most companies in an industry deal with complexities relating to systems and devices. For example, Underwriters Laboratories® (UL) has developed standards on most aspects of the telecommunications and electronic security industries, including equipment, installation, and maintenance procedures. To be Listed by UL, a company must demonstrate that its products meet the standards for each category.
- *Company standards* — Most companies have developed specific procedures for handling many aspects of your job. These procedures are usually based on applicable codes and standards, as well as other good practice methods.
- *Manufacturer's instructions* — Instructions that come with the equipment. It is important to remember that these instructions are part of the approval or Listing. If the instructions are not followed, the approval or Listing is void.

2.2.2 Adopted and Recognized Standards

Adopted standards are standards that have been adopted through statutory or administrative laws in a specific geographic area. Once a document is adopted by reference in a state statute or local ordinance, it has the same force and effect as if published therein in its entirety.

Recognized standards are standards that are generally accepted as relevant in civil law, even in areas where no standards have been formally adopted. Where no applicable code or standard has been officially implemented, it is common for the courts to compare actual circumstances in a case with nationally recognized standards.

2.2.3 NFPA Codes and Standards

The standards published by NFPA are almost universally accepted as guides for the design and installation of sprinkler systems. These are produced by the consensus process in the volunteer standards system used in the United States. However, these standards have international representation with significant participation from Canada.

While there are many standards that apply to sprinkler fitting, two of the primary standards are NFPA 13, *Standard for the Installation of Sprinkler Systems* and NFPA 25, *Standard for the Inspection, Testing, and Maintenance of Water-Based Fire Protection Systems* (*Figure 20*).

Most communities codify these standards through adoption by local ordinance. This process makes these consensus standards law in that community. A knowledge of NFPA standards therefore becomes critical to the design and installation of fire protection systems. The

Figure 20 Common sprinkler system standards.

Did You Know?

Codes and standards are living documents. Most of these are on a three- to five-year revision cycle, although there are annual meetings and annual changes that may be adopted by ordinance. The revision years are staggered among the standards so that they are not all revised in the same calendar year. This means a building that is 20 years old may be under the standards of 20 years ago or it may be under current standards. The year of issue of the applicable NFPA standard must be determined to know the requirements for a specific project.

standards identify the spacing, location, and position of sprinklers; acceptable materials; and the requirements for installing and maintaining these systems.

Not all localities adopt new standards when they are released. The model building codes cite specific editions of NFPA standards. When these building codes are adopted, specific editions of the NFPA standards are also adopted. If the local ordinances or fire code do not adopt different years of issue, by default, the issue that is law is determined by the building code. It can become a problem as to which edition of an NFPA standard should be used. Frequently, the standard of latest issue is used by the fire marshal when the actual, legally adopted issue, is somewhat older.

Small changes in a structure are usually permitted without upgrading the whole building to the current standard. However, if the changes become too expensive, some building codes will require that the entire building be brought up to current standards. There are also retroactive laws that require all buildings within a jurisdiction to be brought up to a current standard.

The NFPA standard system covers a wide range of subjects. These standards have a uniform format. Chapter 1 of each standard has a scope definition that defines the applicability of that standard. Use of the standard beyond its defined scope can result in liability if a problem develops. Those kinds of deviations should be covered by proper documentation (i.e., written permission of the AHJ). The NFPA standards most significant to the sprinkler industry are listed below.

Standard for the Installation of Sprinkler Systems (NFPA 13)

NFPA 13 was initiated in 1896 as the first publication of NFPA. It has been periodically updated since that time. NFPA 13 is the key standard for the design and installation of sprinkler systems. Its stated purpose is to provide a "reasonable degree of protection for life and property." To provide this, it specifies minimum requirements.

When a sprinkler system is to be installed, this standard provides the engineering inputs so that technicians can design effective sprinkler systems and, by following this same guide, mechanics can install properly functioning sprinkler systems. All of the other sprinkler system standards in the United States depend upon this standard for the basics of pipe, fittings, valves, hangers, devices, and sprinkler locations.

Design is based on occupancies that include Light Hazard, Ordinary Hazard, Extra Hazard, and storage. The old design approach of pipe schedules is usable under very limited conditions. The more recent design criteria uses density/area curves or specific design criteria. The design criteria includes sprinkler spacing, location of sprinklers relative to the ceiling or roof deck, and position of the sprinklers relative to obstructions. These factors vary according to occupancy and type of construction.

This standard goes into substantial detail to define various types of sprinkler systems and where and how they are to be used. It also defines acceptable pipes, fittings, valves, hangers, and methods for testing and installation. This assures—as best as possible—functional, reliable, and economic sprinkler systems. Many of the hardware-related requirements of NFPA 13 apply to the other NFPA standards and are referenced by these standards. NFPA 13 is thus the most important and basic sprinkler system standard in the NFPA series.

NFPA 13 covers the proper design and installation of typical sprinkler systems for most types of fire hazards. Coverage includes the following items:

- Design considerations
- Water supplies
- Equipment requirements
- System flow rates
- Sprinkler location and position

- General storage and rack storage of materials
- Specifically identified criteria for special occupancy hazards
- Minimum sizes for sprinklers used in storage applications
- Separation requirements between early suppression fast response and other types of sprinkler systems
- Rules for protecting sprinkler systems against seismic events

Standard for the Installation of Sprinkler Systems in One- and Two-Family Dwellings and Manufactured Homes (NFPA 13D)

NFPA 13D covers the design and installation of sprinkler systems for protection against fire hazards in one- and two-family dwellings and manufactured homes. NFPA 13D was originally issued in 1975. NFPA 13D is on the same revision schedule as NFPA 13 and is processed by the same committee.

Standard for the Installation of Sprinkler Systems in Low-Rise Residential Occupancies (NFPA 13R)

NFPA 13R covers the design and installation of sprinkler systems for protection against fire hazards in residential occupancies up to and including four stories in height. This standard was first issued in 1989. It, too, is processed by the NFPA 13 committee and follows the same revision cycle as NFPA 13.

Standard for the Installation of Standpipe and Hose Systems (NFPA 14)

NFPA 14 covers the installation of standpipes and hose systems to ensure an adequate and reliable water supply during a fire emergency. However, it does not cover requirements for periodic inspection, testing, and maintenance of these systems. This standard was first issued in 1968.

Standard for the Installation of Stationary Pumps for Fire Protection (NFPA 20)

NFPA 20 is concerned with the requirements for selecting and installing pumps supplying liquid for private fire protection. Included in this standard are liquid supplies; suction, discharge, and auxiliary equipment; power supplies; electric drive and control; diesel engine drive and control; steam turbine drive and control; and acceptance tests and operations.

Fire pumps are used when the normal automatic water supply cannot furnish the pressure needed to counter a fire. The first issue of NFPA 13, published in 1896, included information on steam and rotary fire pumps. Subsequently, NFPA 20 was developed to deal with the selection and installation of stationary pumps used for firefighting purposes.

Not covered are system liquid capacity and pressure requirements; periodic inspection, testing, and maintenance of fire pump systems; or the requirements for installing wiring of fire pump units.

Standard for Water Tanks for Private Fire Protection (NFPA 22)

NFPA 22 specifies the minimum requirements for the design, construction, installation, and maintenance of tanks and other equipment providing water for private fire protection. This includes the following:

- Gravity, suction, pressure, and embankment-supported coated-fabric suction tanks
- Towers
- Foundations
- Pipe connections and fittings
- Valve enclosures
- Tank filling
- Protection against freezing

Standard for the Installation of Private Fire Service Mains and Their Appurtenances (NFPA 24)

NFPA 24 focuses on the minimum requirements for installing private fire service mains and their equipment. The standard covers the following seven areas:

- Automatic sprinkler systems
- Open sprinkler systems
- Water spray fixed systems
- Foam systems
- Private hydrants
- Monitor nozzles or standpipe systems relative to water supplies
- Hose houses

NFPA 24 applies to combined service mains that carry water for fire service and other uses. This standard does not apply to mains under the control of the water utility.

Standard for the Inspection, Testing, and Maintenance of Water-Based Fire Protection Systems (NFPA 25)

NFPA 25 sets the minimum requirements for inspecting, testing, and maintaining water-based fire protection systems, both land-based and marine. The standard covers sprinkler, standpipe and hose, fixed water spray, water mist, and foam water systems. Also covered are the water supplies that are part of the systems, including private fire service mains and equipment, fire pump and water storage tanks, and valves that control system flow. Other areas covered include impairment handling, reporting, and obstruction investigation.

As a standard, NFPA 25 covers all of the water-based fire extinguishing systems. It has been coordinated with the applicable design requirements of the related NFPA standards to provide a complete system inspection and maintenance standard. This standard will be accessory to fire codes as an inspection and maintenance requirement, if adopted. It is required to be given to the owner, usually at the end of the project.

NFPA 25 does not apply to systems installed improperly, and corrective actions are outside the scope of this standard and the responsibility of the property owner. This standard does not typically apply to sprinkler systems designed and installed to the requirements of NFPA 13D.

> **NOTE**
> The requirements for water mist systems can also be found in NFPA 750, *Standard on Water Mist Fire Protection Systems*.

2.0.0 Section Review

1. The NFPA fire code is also known as _____.
   a. NFPA 1
   b. NFPA 13
   c. NFPA 24
   d. NFPA 25

2. NFPA 25 provides information on _____.
   a. installing private fire service mains and their equipment
   b. inspecting, testing, and maintaining water-based fire protection systems
   c. installing tanks and other equipment providing water for private fire protection
   d. selecting and installing pumps supplying liquid for private fire protection

Section Three

3.0.0 Apprenticeship Training and Career Paths

Objective

Understand the apprenticeship training process for sprinkler fitters.

a. List Department of Labor (DOL) requirements for apprenticeship.
b. Describe various career paths and opportunities in the sprinkler industry.
c. Understand the responsibilities of the employer and employee.
d. Describe employer and employee safety obligations.

Trade Terms

American Fire Sprinkler Association (AFSA): A non-profit, international association representing merit shop sprinkler contractors, dedicated to the educational advancement of its members and promotion of the use of sprinkler systems.

Blocked: Secured for storage in tiers.

Compressed gas: Gas stored under pressure in cylinders.

NCCER Standardized Curricula: Standardized construction education materials produced by NCCER.

Office of Apprenticeship (OA): The US Department of Labor office that sets the minimum standards for training programs across the country.

On-the-job training (OJT): Job-related training acquired while working.

Rack: Device used to support pipe in tiers.

Regulations: Rules and specifications for the sprinkler industry provided by various organizations.

Apprentice training goes back thousands of years, and its basic principles have not changed in that time. First, it is a means for individuals entering the craft to learn from those who have mastered the craft. Second, it focuses on learning by doing—real skills versus theory. Although some theory is presented in this coursework, it is always presented in a way that helps the trainee understand the purpose behind the skill that is to be learned.

To address the training needs of the professional communities, the **American Fire Sprinkler Association (AFSA)** and NCCER developed a four-year sprinkler fitter training program. This program uses the minimum Department of Labor (DOL) standards as a foundation for comprehensive curricula that provide trainees with in-depth classroom and **on-the-job training (OJT)** experience.

3.1.0 DOL Apprenticeship Requirements

The DOL **Office of Apprenticeship (OA)** sets the minimum standards for training programs across the country. These programs rely on mandatory classroom or correspondence courses of instruction and on-the-job training (OJT). They require 144 hours of classroom or related instruction per year and 2,000 hours of OJT per year.

In a typical OA sprinkler fitter apprentice program, trainees spend 576 hours in classroom or related instruction and 8,000 hours in OJT before receiving journey certificates.

This NCCER Standardized Curriculum provides trainees with industry-driven training and education. It adopts a purely competency-based teaching philosophy. This means that trainees must demonstrate to the instructor that they possess the understanding and the skills necessary to perform the hands-on tasks that are covered in each module before they can advance to the next stage of the curriculum.

When the instructor is satisfied that a trainee has successfully demonstrated the required knowledge and skills for a module, that information is sent to NCCER and kept in the Registry system. The NCCER Registry can then confirm training and skills for workers as they move from state to state, company to company, or even within a company (see *Appendix A*).

AFSA offers a correspondence course using the same curriculum. The AFSA course is highly flexible, combining correspondence course study, support for supplemental classroom training, and on-the-job training guidance. Test materials are provided online to allow student testing on a periodic basis determined by the employer. The AFSA Sprinkler Fitter correspondence course has been approved for use in state or federal apprenticeship programs in all 50 states. AFSA sample apprenticeship credentials are shown in *Appendix B*.

Whether you enroll in an NCCER program or another OA-approved program, ensure that you work for an employer or sponsor who supports a nationally standardized training program that includes credentials to confirm your skill development.

All apprenticeship standards prescribe certain work-related or on-the-job training. This on-the-job training is broken down into specific tasks in which the apprentice receives hands-on training during the period of the apprenticeship. In addition, a specified number of hours is required in each task. The total number of on-the-job training hours for the sprinkler fitter apprenticeship program is traditionally 8,000, which amounts to about four years of training.

The apprentice must log all work time and turn it in to the apprenticeship committee so that accurate time control can be maintained. After each 1,000 hours of related work, the apprentice may receive a pay increase as prescribed by the apprenticeship standards.

The course materials, related instruction, and work-related training will not always run concurrently due to such reasons as layoffs, type of work needed to be done in the field, and so on. Furthermore, apprentices with special job experience or coursework may obtain credit toward their classroom requirements. This reduces the total time required in the classroom while maintaining the total 8,000-hour on-the-job training requirement. These special cases will depend on the type of program and the regulations and standards under which it operates.

Informal on-the-job training provided by employers is usually less thorough than that provided through a formal apprenticeship program. The degree of training and supervision in this type of program often depends on the size of the employing firm. A small contractor may provide training in only one area, while a large company may be able to provide training in several areas.

For those entering an apprenticeship program, a high school or technical school education is desirable, as are courses in shop, mechanical drawing, and general mathematics. Manual dexterity, good physical condition, and quick reflexes are important. The ability to solve problems quickly and accurately and to work closely with others is essential. You must have a high concern for safety.

The prospective apprentice must submit certain information to the apprenticeship committee. This may include the following:

- Aptitude test (General Aptitude Test Battery or GATB Form Test) results (usually administered by the local Employment Security Commission)
- Proof of educational background (candidate should have school transcripts sent to the committee)
- Letters of reference from past employers and friends
- Results of a physical examination
- Proof of age
- If the candidate is a veteran, a copy of Form DD214
- A record of technical training received that relates to the construction industry and/or a record of any pre-apprenticeship training

Child Labor Laws

Federal law establishes the minimum standards for workers under the age of 18. Some municipal jurisdictions may enforce stricter regulations. Employers are required to abide by the laws that apply to them.

The child labor provisions of the Fair Labor Standards Act forbid employers from using illegal child labor, and also forbid companies from doing business with any other business that does. The DOL investigates alleged abuses of the law. In such cases, employers must provide proof of age for their employees.

In addition to the child labor provisions, employers in the construction trades are required to follow the DOL's Child Labor Bulletin No. 101, *Child Labor Requirements in Nonagricultural Occupations Under the Fair Labor Standards Act*. Bulletin No. 101 does the following:

- Explains the coverage of the child labor provisions
- Identifies minimum age standards
- Lists the exemptions from the child labor provisions
- Sets out employment standards for 14- and 15-year-old workers
- Defines the work that can be performed in hazardous occupations
- Provides penalties for violations of the child labor provisions
- Recommends the use of age certificates for employees

- High school diploma or General Equivalency Diploma (GED)

The apprentice must do the following:

- Wear proper safety equipment on the job
- Purchase and maintain tools of the trade as needed and required by the contractor
- Submit a monthly on-the-job training report to the committee
- Report to the committee if a change in employment status occurs
- Attend classroom-related instruction and adhere to all classroom regulations, such as attendance requirements

3.2.0 Sprinkler Fitter Careers

Once you complete your initial training and gain job experience, there are many career opportunities available to you in the sprinkler industry. *Figure 21* shows general career paths for sprinkler fitters.

If you have leadership qualities, you can become the supervisor of a sprinkler fitter installation crew. Some experienced sprinkler fitters become instructors at contractor or vocational training schools. Others can become quality control inspectors, salespeople, or estimators.

With further education and training, a sprinkler fitter can become a sprinkler system layout technician, also known as a *designer.* Layout technicians must know specific sprinkler design requirements and how those requirements fit into the design of the overall structure.

Depending on your level of skill, experience, education, and interest, you may want to be a sprinkler contractor. Sprinkler contractors work with architects, general contractors, code officials, and sometimes building owners to provide fire protection systems for a structure.

Sprinkler fitters are employed all over the United States and the world. As a sprinkler fitter, you will probably work for a sprinkler contractor doing various types of commercial sprinkler installations. Sprinkler systems are now required in most commercial structures, resulting in a continuing demand for sprinkler fitters. A journey-level sprinkler fitter will earn as much as, and sometimes more than, a college-educated person.

There is a future in sprinkler fitting. During construction booms there is a strong demand for sprinkler fitters. But even when the economy weakens, there is always a demand for fitters to service, maintain, and repair existing systems, and to retrofit existing buildings that lack sprinkler systems. If you continue your education and can master the building and fire codes, you may become an inspector, superintendent, or manager.

3.2.1 Sprinkler Fitter Work

Sprinkler fitters perform a wide variety of tasks relating to the installation, repair, and maintenance of all types of sprinkler systems. Fitters must also have the skills and knowledge to install and maintain related systems such as standpipes, carbon dioxide systems, foam systems, and dry chemical systems.

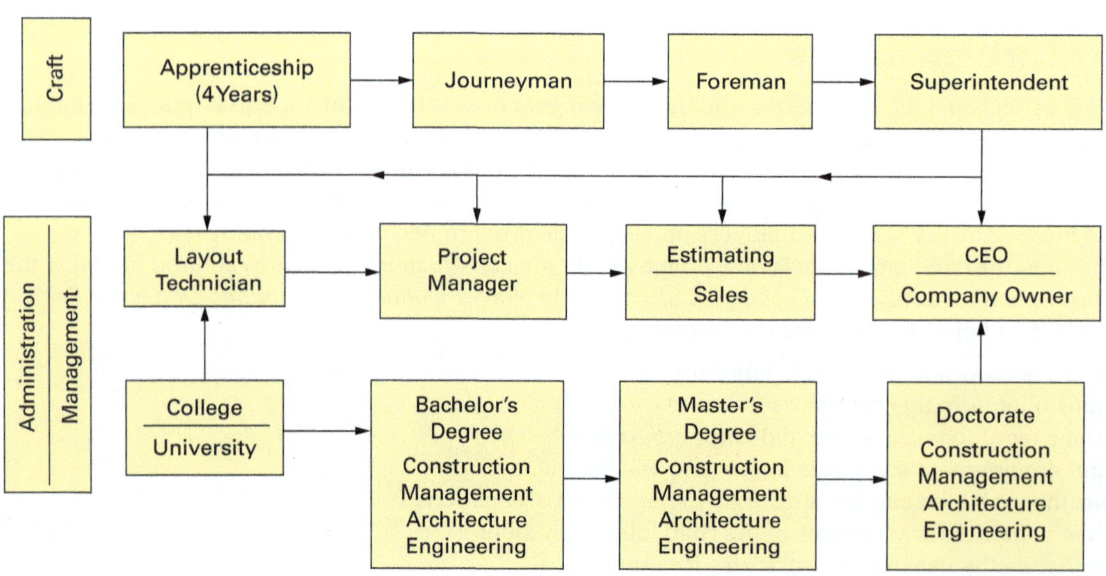

Figure 21 General career paths for sprinkler fitters.

Did You Know?

In addition to the installation of sprinkler systems in new construction, sprinkler fitters also retrofit existing installations. This turn-of-the-century bank building renovation involved the ingenious placement of catwalks that allowed the crew to install sprinklers in the vaulted glass ceiling without disturbing the historic building architecture.

As an apprentice, you most likely will be fabricating, installing, testing, maintaining, or repairing sprinkler systems in industrial facilities, department stores, office buildings, hotels, schools, hospitals, and residences. You will install wet pipe, dry pipe, preaction, and deluge sprinkler systems, along with a variety of related systems.

As a journeyman sprinkler fitter, you must also be able to read and interpret shop drawings and specifications, prepare cost estimates for clients, select the type and size of pipe required, and arrange piping to provide fire protection. You will install hangers to support the piping system, as well as solder, thread, and solvent weld pipe joints. You will also test systems for leaks and inspect, maintain, and repair piping, fixtures, and controls including hydrants, pumps, and sprinkler connections.

Many sprinkler fitters go through a four-year apprenticeship, leading to certification or licensing as a sprinkler fitter. As part of the apprenticeship program, you receive instruction and training via textbooks, classes, and lab work. The work is done both indoors and outdoors. There is a substantial amount of climbing, kneeling, standing, and walking on scaffolding. Because of this activity, a sprinkler fitter must be able to move about with ease and be in good physical condition.

If you want to be a sprinkler fitter, you should not have a fear of heights and should have a good sense of balance. You must be safety conscious for yourself and for your coworkers. You need to be good with your hands and be willing to do hard physical labor when required.

You should be mechanically inclined, able to read construction and shop drawings, and be able to use power tools to cut and thread pipe both safely and efficiently.

3.2.2 NCCER

NCCER was created to address the critical workforce shortage facing the construction industry and to develop industry-driven standardized craft training programs with portable credentials. NCCER is a not-for-profit education foundation established by the nation's leading construction companies. NCCER was created to provide the industry with standardized construction education materials, the **NCCER Standardized Curricula** (the sprinkler fitting modules are part of this series), and a system for tracking and recognizing students' training accomplishments—NCCER's Registry. See *Appendix A* for samples of NCCER's apprentice training credentials, certificate, and transcript.

NCCER also offers accreditation, instructor certification, and skills assessments. NCCER is committed to developing and maintaining a training process that is internationally recognized, standardized, portable, and competency-based.

Working in partnership with industry and academia, NCCER has developed a system for program accreditation that is similar to those found in institutions of higher learning. NCCER's accreditation process ensures that students receive quality training based on uniform standards and criteria. These standards are outlined in NCCER's Accreditation Guidelines and must be adhered to by NCCER Accredited Training Sponsors.

There are over 6,000 NCCER-accredited training and assessment locations across the United States offering training in more than 70 craft areas. Millions of craft professionals and construction managers have received quality construction education through NCCER.

A craft instructor is a journeyman craft professional or career and technical educator trained and certified to teach NCCER's Standardized Curricula. This network of certified instructors ensures that NCCER training programs will meet the standards of instruction set by the industry. More information is available at **www.nccer.org**.

3.2.3 The American Fire Sprinkler Association

The American Fire Sprinkler Association (AFSA) is committed to be a relentless force for the merit shop sprinkler contractor in educating our members and advocating for a world that is safe from loss of life and property damage. AFSA was organized as a non-profit association in 1981 to provide the merit shop sprinkler contractor with education, training, consulting, information, representation, and many more services that have expanded over the years. See *Appendix B* for an example of AFSA's apprentice training credentials.

Membership is open to contractors, manufacturers, suppliers, designers, authorities having jurisdiction, and students. Currently, AFSA represents nearly 1,200 companies and organizations in the United States and throughout the world. AFSA believes that the installation of sprinklers could save thousands of lives and billions of dollars that are lost to fire each year. Increasingly, public officials are realizing this.

AFSA acts as a liaison with other national associations involved in fire safety. Working together with government agencies, such as the Federal Emergency Management Agency (FEMA), the NFPA, and the Home Fire Sprinkler Coalition, AFSA helps to increase awareness concerning the dangers of fire.

To assist sprinkler contractors in maintaining a high-quality workforce, AFSA co-sponsored with NCCER the development of this standardized training program specifically for apprentices in the sprinkler trade. In addition, AFSA offers the following specialized training:

- A two-week Beginning Sprinkler System Planning School offering a comprehensive practical approach to preparing a sprinkler system drawing
- Inspection, Testing, and Maintenance (ITM) Inspector Development Program, a combination of online, classroom, and lab instruction using NFPA 25 standard as the basis, with the goal of preparing beginner trainees for a National Institute for Certification in Engineering Technologies (NICET) Level II certification
- Computer-based courses offering topics including sprinkler system layout, hydraulic calculations, seismic bracing, and more
- Frequent live webinars on various subjects such as NFPA codes and standards updates, system layout, installations, etc.
- Recorded webinars, an expanding library of previously recorded live webinars
- Training tools for project management, estimating and bidding, foremanship, and on-the-job safety

3.3.0 Employer/Employee Responsibilities

In order to be successful, the professional must be able to use current trade materials, tools, and equipment to finish the task quickly and efficiently. A sprinkler fitter must be adept at adjusting methods to meet each situation. The successful sprinkler fitter must continuously train to remain knowledgeable about technical advancements and to gain the skills to use them. A professional never takes chances with regard to personal safety or the safety of others.

3.3.1 Traits of a Professional

The word *professionalism* is a broad term that describes the desired overall behavior and attitude expected in the workplace. Professionalism is too often absent from the construction site and the various trades. Most people would argue that professionalism must start at the top to be successful. It is true that management support of professionalism is important to its success in the workplace, but it is more important that individuals recognize their own responsibility for professionalism.

Professionalism includes honesty, productivity, safety, civility, cooperation, teamwork, clear and concise communication, being on time and prepared for work, and regard for one's impact on one's coworkers. It can be demonstrated in a variety of ways every minute you are in the workplace. Most important is that you do not tolerate the unprofessional behavior of coworkers. This is not to say that you shun the unprofessional worker; instead, you work to demonstrate the benefits of professional behavior.

Professionalism is a benefit to both the employer and the employee. It is a personal responsibility. Our industry is what each individual chooses to make of it; choose professionalism and the industry image will follow. Professionals pride themselves in performing a job well, and in being punctual and dependable. Each job is completed in a professional way, never by cutting corners or reducing materials. A valued professional maintains work attitudes and ethics that protect property such as tools and materials belonging to employers, customers, and other trades from damage or theft at the shop or jobsite.

Social Media and Electronic Devices

It is important to maintain a professional attitude regarding social media and the use of electronic devices. Put your phone away and focus on the job at hand during working hours. When sending texts or emails, use complete sentences rather than abbreviations to ensure that your message is easily understood. Do not use company-provided electronic devices for personal use. Know your company's policy for posting jobsite photos on social media, as many companies prohibit it. Consider that your employer and customers may run across or even search for your personal social media accounts. A good rule of thumb is to avoid posting anything you would not want a future employer to see.

Honesty

Honesty and personal integrity are important traits of the successful professional. Honesty and success go hand-in-hand. It is not simply a choice between right and wrong, but a choice between success and failure. Dishonesty will always catch up with you. Whether you are stealing materials, tools, or equipment from the jobsite or simply lying about your work, it will not take long for your employer to find out. Of course, you can always go and find another employer, but this option will ultimately run out on you.

If you plan to be successful and enjoy continuous employment, consistency of earnings, and being sought after as opposed to seeking employment, then start out with the basic understanding of honesty in the workplace and you will reap the benefits. Honesty means more, however, than just not taking things that do not belong to you. It means giving a fair day's work for a fair day's pay. It means carrying out your side of a bargain. It means that your words convey true meanings and actual happenings. Our thoughts as well as our actions should be honest. Employers place a high value on an employee who is strictly honest.

Loyalty

Employees expect employers to look out for their interests, to provide them with steady employment, and to promote them to better jobs as openings occur. Employers feel that they, too, have a right to expect their employees to be loyal to them—to keep their interests in mind, to speak well of them to others, to keep any minor troubles strictly within the company, and to keep absolutely confidential all matters that pertain to the business. Both employers and employees should keep in mind that loyalty is not something to be demanded; rather, it is something to be earned.

Willingness to Learn

Every company has its own way of doing things. Employers expect their workers to be willing to learn these ways. Adapting to change and being willing to learn new methods and procedures as quickly as possible are key. Sometimes, a change in safety regulations or the purchase of new equipment makes it necessary for even experienced employees to learn new methods and operations. Employees often resent having to accept improvements because of the retraining that is involved. However, employers will undoubtedly think they have a right to expect employees to put forth the necessary effort. Methods must be kept up to date in order to meet competition and show a profit. It is this profit that enables the owner to continue in business and provide jobs and advancement for the employees.

Willingness to Take Responsibility

Most employers expect their employees to see what needs to be done, then go ahead and do it. It is very tiresome to have to ask again and again that a certain job be done. It is obvious that having been asked once, an employee should assume the responsibility from then on. Once the responsibility has been delegated, the employee should continue to perform the duties without further direction. Every employee has the responsibility for working safely.

Willingness to Cooperate

To cooperate means to work together. In our modern business world, cooperation is the key to getting things done. Learn to work as a member of a team with your employer, supervisor, and fellow workers in a common effort to get the work done efficiently, safely, and on time.

Rules and Regulations

People can work well together only if there is some understanding about what work is to be done, when and how it will be done, and who will do it. Rules and regulations are a necessity in any work situation and should be so considered by all employees.

Tardiness and Absenteeism

Tardiness means being late for work, and absenteeism means being off the job for one reason or another. Consistent tardiness and frequent absences are an indication of poor work habits, unprofessional conduct, and a lack of commitment.

We are all creatures of habit. What we do once, we tend to do again, unless the results are too unpleasant. The habit of always being late may have begun back in our school days when we found it hard to get up in the morning. This habit can get us into trouble at school, and it can continue to get us into trouble when we are through with school and go to work.

Our work life is governed by the clock. We are required to be at work at a definite time, and so is everyone else. Failure to get to work on time results in confusion, lost time, and resentment on the part of those who do come on time. In addition, it may lead to penalties up to and including dismissal. Although it may be true that a few minutes out of a day are not very important, you must remember that a principle is involved. It is your obligation to be at work at the time indicated. We agree to the terms of work when we accept the job. Perhaps you can see this more clearly from the point of view of the boss. Supervisors cannot keep track of people if they come in at any time they please. It is not fair to others to ignore tardiness. Failure to be on time may hold up the work of fellow workers. Better planning of your morning routine will often keep you from being delayed and so prevent a late arrival. In fact, arriving a little early indicates your interest and enthusiasm for your work, which is appreciated by employers. The habit of being late can stand in the way of promotion.

It is sometimes necessary to take time off from work. No one should be expected to work when sick or when there is serious trouble at home. However, it is possible to get into the habit of letting unimportant and unnecessary matters keep us from the job. This results in lost production and hardship on those who try to carry on the work. Again, there is a principle involved. The person who hires us has a right to expect us to be on the job unless there is some very good reason for staying away. Certainly, we should not let some trivial reason keep us home. We should not stay up nights until we are too tired to go to work the next day. If we are ill, we should use the time at home to do all we can to recover quickly. This, after all, is no more than most of us would expect of a person we had hired to work for us, and on whom we depend to do a certain job.

If it is necessary to stay home, then at least phone the office early in the morning so that the boss can find another worker for the day. Time and again, employees have remained home without sending any word to the employer. This is the worst possible way to handle the matter. It leaves those at work uncertain about what to expect. They have no way of knowing whether you have merely been held up and will be in later, or whether immediate steps should be taken to assign your work to someone else. Courtesy alone demands that you let the boss know if you cannot come to work.

The most frequent causes of absenteeism are illness, death in the family, accidents, personal business, and dissatisfaction with the job. Some of the causes are legitimate and unavoidable, while others can be controlled. One can usually plan to carry on most personal business affairs after working hours. Frequent absences will reflect unfavorably on a worker when promotions are being considered.

Employers sometimes resort to docking pay, demotion, and even dismissal in an effort to control tardiness and absenteeism. No employer likes to impose restrictions of this kind. However, in fairness to those workers who do come on time and who do not stay away from the job, an employer is sometimes forced to discipline those who will not follow the rules.

Setting Goals

Goal setting is very important to achieving success. Having short-term and long-term goals is critical to improving your value on the jobsite. An employee's wages are dependent on how the employee excels at work. Setting goals and working toward them are the keys to improving your performance at the jobsite and gaining improved wages and job satisfaction.

The goals that people set for themselves must include pride, professionalism, reputation, and recognition. If employees do not take pride in their work, it will affect their reputation and their company's reputation.

Employee Reviews and Evaluations

Some employers have annual or semi-annual employee reviews to evaluate job performance. The employer evaluates the quality of performance relative to wages. When actual job performance is greater than or equal to expected job performance, a raise and/or promotion is a reward for a job well done. Attitude on the job will also be considered. Developing and maintaining a positive attitude now will pay off in the future. *Figure 22* shows a typical employee review form.

3.3.2 Human Relations

Most people underestimate the importance of working well with others. There is a tendency to pass off human relations as nothing more than common sense. What exactly is involved in human relations? One response would be to say that part of human relations is being friendly, pleasant, courteous, cooperative, adaptable, and sociable.

Making Human Relations Work

As important as the previously noted characteristics are for personal success, they are not enough. Human relations is much more than just getting people to like you. It is also knowing how to handle difficult situations as they arise.

Human relations is knowing how to work with supervisors who are often demanding and sometimes unfair. It involves understanding your own personality traits and those of the people you work with. Building sound working relationships in various situations is important. If working relationships have deteriorated for one reason or another, restoring them is essential. Human relations is learning how to handle frustrations without hurting others.

Human Relations and Productivity

Effective human relations is directly related to productivity. Productivity is the key to business success. Every employee is expected to produce at a certain level. Employers quickly lose interest in an employee who has a great attitude but produces very little. There are work schedules to be met and jobs that must be completed.

All employees, both new and experienced, are measured by the amount of quality work they can safely turn out. The employer expects every employee to do his or her share of the workload.

Ethical Principles for Members of the Construction Trades

- *Honesty* — Be honest and truthful in all dealings. Conduct business according to the highest professional standards. Faithfully fulfill all contracts and commitments. Do not deliberately mislead or deceive others.
- *Integrity* — Demonstrate personal integrity and the courage of your convictions by doing what is right even where there is pressure to do otherwise. Do not sacrifice your principles because it seems easier.
- *Loyalty* — Be worthy of trust. Demonstrate fidelity and loyalty to companies, employers and sponsors, coworkers, and trade institutions and organizations.
- *Fairness* — Be fair and just in all dealings. Do not take undue advantage of another's mistakes or difficulties. Fair people are open-minded and committed to justice, equal treatment of individuals, and tolerance for and acceptance of diversity.
- *Respect for others* — Be courteous and treat all people with equal respect and dignity.
- *Obedience* — Abide by laws, rules, and regulations relating to all personal and business activities.
- *Commitment to excellence* — Pursue excellence in performing your duties, be well informed and prepared, and constantly try to increase your proficiency by gaining new skills and knowledge.
- *Leadership* — By your own conduct, seek to be a positive role model for others.

Employee ____________________ **Fitter** ____________________ **Apprentice** ____________________

Employer ____________________ **Evaluation Date** ____________________

Factors	UNSATISFACTORY 1 2 3 4 5	MARGINAL 6 7 8 9 10	GOOD 11 12 13 14 15	VERY GOOD 16 17 18 19 20	EXCELLENT 21 22 23 24 25
1. QUALITY OF WORK	Is very slip-shod with frequent, avoidable errors.	Makes frequent errors, work often careless, leaves loose ends.	Is generally satisfactory, sometimes uncorrected errors slip through.	Is almost always accurate and neat. Corrects own errors.	Work is always of superior quality.
2. COOPERATIVENESS	Seems unable to cooperate with others. Is argumentative over every innovation.	Often fails to cooperate/frequently disagreeable, difficult to get along with.	Gets along well with others most of the time, is sometimes obstructive.	Often cooperates well/willing to try new methods.	Works effectively with co-workers and supervisors, falls in readily with new ideas.
3. JOB KNOWLEDGE	Does not know enough about the job.	More knowledge of job would be desirable.	Knows job well enough to get along.	Thorough knowledge of work.	Has excellent mastery of work.
4. INITIATIVE	Routine worker, usually waits to be told what to do, needs constant supervision.	Often at a loss in other than routine situations, requires frequent checking.	Does regular work without waiting for direction. Requires some supervision.	Resourceful; needs minimal supervision, alert to ways to improve work.	Always gets on with job on his own, highly ingenious.
5. INDUSTRIOUSNESS	Socializes on the job whenever possible and/or excessive phone use.	Frequently neglects work, needs to improve work attitude.	Usually sticks to the job but with occasional wandering.	Conscientious most of the time.	Can always be relied on to get things done.
6. QUANTITY OF WORK	Very slow.	Below average in output.	Turns out required volume, seldom more.	Above average producer.	Unusual output, exceptionally fast, does more than required.
7. ABILITY TO LEARN	Unable to grasp job without constant re-instruction.	Learns slowly.	Learns moderately fast and checks occasionally with supervisor.	Learns fast and remembers well.	Unusually quick and complete, full grasp.
8. PAPERWORK	Fails to turn in completed paperwork.	Paperwork is turned in, but is often incomplete.	Usually turns in paperwork on time. Sometimes needs prompting.	Paperwork is turned in, but sometimes lacks signatures.	Paperwork is always filled out neatly and accurately, and signed.
9. ATTENDANCE	Excessively absent or tardy.	Frequently absent or tardy.	Occasionally absent or tardy.	Rarely absent or tardy, then only for good cause.	Never absent or tardy.

ATTENDANCE RECORD

For Rating Period: Times Tardy: __________ Excused Absence: __________ Unexcused Absence: __________

Length of Time Employed: Years: __________ Months: __________ Days: __________

Remarks: ____

____

Signature of Employee ____________________

Date ____________________

Signature of Superintendent or Foreman ____________________

Title ____________________ Date ____________________

Figure 22 Typical employee review form.

However, doing one's share is not enough. If you are to be productive, you must do your share (or more than your share) without antagonizing your fellow workers. You must perform your duties in a manner that encourages others to follow your example. It makes little difference how ambitious you are or how capably you perform. You cannot become the kind of employee you want to be, or the type of worker management wants you to be without learning how to work with your peers.

Employees must do everything they can to build strong, professional working relationships with fellow employees, supervisors, and clients.

Attitude

A positive attitude is essential to a successful career. First, being positive means being energetic, highly motivated, attentive, and alert. A positive attitude is essential to safety on the job. Second, a positive employee contributes to the productivity of others. Both negative and positive attitudes are transmitted to others on the job. A persistent negative attitude can spoil the positive attitudes of others. It is very difficult to maintain a high level of productivity while working next to a person with a negative attitude. Third, people favor a person who is positive. Being positive makes a person's job more interesting and exciting. Finally, the kind of attitude transmitted to management has a great deal to do with an employee's future success in the company. Supervisors can determine a subordinate's attitude by their approach to the job, reactions to directives, and the way they handle problems.

A positive attitude is far more than a smile, which is only one example of an inner positive attitude. As a matter of fact, some people transmit a positive attitude even though they seldom smile. They do this by the way they treat others, the way they look at their responsibilities, and the approach they take when faced with problems. Here are a few suggestions that will help you to maintain a positive attitude:

- Remember that your attitude follows you wherever you go. If someone makes a greater effort to be a more positive person in their social and personal lives, it will automatically help them on the job. The reverse is also true. One effort will complement the other.
- Negative comments are seldom welcomed by fellow workers on the job. Neither are they welcome on the social scene. The solution: talk about positive things and be complimentary. Constant complainers do not build healthy and fulfilling relationships.
- Look for the good things in people on the job, especially your supervisor. Nobody is perfect, and almost everyone has a few worthwhile qualities. If you dwell on people's good features, it will be easier to work with them.
- Look for the good things where you work. What are the factors that make it a good place to work? Is it the hours, the physical environment, the people, the actual work being done? Or is it the atmosphere? Keep in mind that you cannot expect to like everything. No work assignment is perfect, but if you concentrate on the good things, the negative factors will seem less important and bothersome.
- Look for the good things in the company. Just as there are no perfect assignments, there are no perfect companies. Nevertheless, almost all organizations have good features. Is the company progressive? What about promotional opportunities? Are there chances for self-improvement? What about the wage and benefit package? Is there a good training program? You cannot expect to have everything you would like, but there should be enough to keep you positive. In fact, if you decide to stick with a company for a long period of time, it is wise to look at the good features and think about them. If you think positively, you will act the same way.

Tips for a Positive Attitude

Here is a short checklist of things that you should keep in mind to help you develop and maintain a positive attitude:

- Remember that your attitude follows you wherever you go.
- Helpful suggestions and compliments are much more effective than negative ones.
- Look for the positive characteristics of your teammates and supervisors.
- You may not be able to change the negative attitude of another employee, but you can protect your own attitude from becoming negative.

3.4.0 Employer and Employee Safety Obligations

An obligation is like a promise or a contract. In exchange for the benefits of your employment and your own well-being, you agree to work safely. In other words, you are obligated to work safely. You are also obligated to make sure anyone you happen to supervise or work with is working safely. Your employer is obligated to maintain a safe workplace for all employees. Safety is everyone's responsibility.

Some employers have safety committees. If you work for such an employer, you are then obligated to that committee to maintain a safe working environment. This means two things:

- Follow the safety committee's rules for proper working procedures and practices.
- Report any unsafe equipment and conditions directly to the committee or your supervisor.

On the job, if you see something that is not safe, report it! Do not ignore it. It will not correct itself. You have an obligation to report it.

In the long run, even if you do not think an unsafe condition affects you, it does. Do not mess around; report what is not safe. Do not think your employer will be angry because your productivity suffers while the condition is corrected. On the contrary, your employer will be more likely to criticize you for not reporting a problem.

Your employer knows that the short time lost in making conditions safe again is nothing compared with shutting down the whole job because of a major disaster. If that happens, you are out of work anyway. So do not ignore an unsafe condition. In fact, OSHA regulations require you to report hazardous conditions.

This applies to every part of the construction industry. Whether you work for a large contractor or a small subcontractor, you are obligated to report unsafe conditions. The easiest way to do this is to tell your supervisor. If that person ignores the unsafe condition, report it to the next highest supervisor. If it is the owner who is being unsafe, let that person know your concerns. If nothing is done about it, report it to OSHA. If you are worried about your job being on the line, think about it in terms of your life, or someone else's, being at risk.

3.4.1 OSHA

The US Congress passed the Occupational Safety and Health Act in 1970. This act also created OSHA. It is part of the US Department of Labor. The job of OSHA is to set occupational safety and health standards for all places of employment, enforce these standards, ensure that employers provide and maintain a safe workplace for all employees, and provide research and educational programs to support safe working practices.

OSHA was adopted with the stated purpose "to assure as far as possible every working man and woman in the nation safe and healthful working conditions and to preserve our human resources." OSHA requires each employer to provide a safe and hazard-free working environment. OSHA also requires that employees comply with OSHA rules and regulations that relate to their conduct on the job. To gain compliance, OSHA can perform spot inspections of jobsites, impose fines for violations, and even stop any more work from proceeding until the jobsite is safe.

According to OSHA standards, you are entitled to on-the-job safety training. As a new employee, you are entitled to the following:

- Being shown how to do your job safely
- Being provided with the required personal protective equipment
- Being warned about specific hazards
- Being supervised for safety while performing the work

The enforcement of this Act of Congress is provided by the federal and state safety inspectors, who have the legal authority to make employers pay fines for safety violations. The law allows states to have their own safety regulations and agencies to enforce them, but they must first be approved by the US Secretary of Labor. For states that do not develop such regulations and agencies, federal OSHA standards must be obeyed.

These standards are found in OSHA Safety and Health Standards for the Construction Industry (*29 CFR, Part 1926*), sometimes called *OSHA Standards 1926*. Other safety standards that apply to construction are published in OSHA Safety and Health Standards for General Industry (*29 CFR, Parts 1900 to 1910*).

The most important general requirements that OSHA places on employers in the construction industry are as follows:

- The employer must perform frequent and regular jobsite inspections of equipment.
- The employer must instruct all employees to recognize and avoid unsafe conditions, and to know the regulations that pertain to the job so they may control or eliminate any hazards.
- No one may use any tools, equipment, machines, or materials that do not comply with *OSHA Standards 1926*.

- The employer must ensure that only qualified individuals operate tools, equipment, and machines.

Most trainees will undergo an OSHA 10-hour safety course as soon as they begin their apprenticeship. The OSHA 10-hour course covers general safety and health hazards for entry-level workers. Later in your training, you may also take the OSHA 30-hour safety course. The 30-hour course covers additional safety subjects and provides industry-specific training. It is intended for supervisors and workers in charge of the safety and health of other personnel.

3.4.2 Working Safely

A major part of a sprinkler fitter's job involves lifting, carrying, and lowering materials to be used on the job. OSHA recommends that you follow these guidelines when lifting and carrying materials:

- Lift the load with your legs instead of your back.
- Do not carry material so large or bulky that it obstructs vision.
- Make sure all walkways are free of obstacles.
- Once a secure grip is obtained, lift the load properly keeping the load close to your body.
- If more than one person is carrying the load, all team members should carry it on the same side shoulder, walk in step, and move slowly.
- When lowering the load, use your legs and not your back.
- Make sure that your fingers and toes are not in the way before setting the load down.
- Since many back and strain injuries occur early in the workday, spend a few minutes stretching your muscles before any heavy lifting.

In addition, OSHA provides strict regulations concerning the way materials are to be stored at the jobsite. Aisles and pathways must be kept clear for the safe movement of materials. Materials not needed for immediate operations should not be stored on scaffolds or runways. This creates a safety hazard and slows the work process. All storage areas used for flammable, explosive, or toxic material must be ventilated properly.

OSHA regulations state that all materials stored in tiers must be stacked, racked, **blocked**, interlocked, or otherwise secured to prevent sliding, falling, or collapse. The safest way to store pipe and other cylindrical materials is on a **rack**. When cylindrical materials are stored in stacks, they must be blocked to prevent spreading and tilting. The same size, type, and length of material should be stacked together. Neat stacks allow more efficient use of the materials.

When material is stored inside a building that is under construction, the material must not be placed within 6' of any hoistway or inside floor opening. Material must not be stored within 10' of an exterior wall that does not extend above the top of the material. These regulations are designed to prevent injuries caused by falling material.

Special safety precautions apply to **compressed gas** cylinders stored and used on the jobsite. Follow these requirements when storing and using compressed gas cylinders:

- Compressed gas cylinders must be stored in an upright position.
- The contents label must be located on the cylinder where it can easily be seen.
- Valves must be closed except when they are in use.
- Valve protection caps must be in place when the cylinders are not in use.
- When using acetylene, a key must be left on the bottle to quickly turn the gas off if necessary.
- Compressed gas cylinders should never be stored near a welding operation because they can be ignited by sparks, hot slag, or flames. If it is not possible to store the cylinders a safe distance away, a fire-resistant shield must be used.
- Do not store empty cylinders with full or partially full cylinders. Place the empty cylinders in a separate area and mark the cylinder MT. Only the gas supplier is permitted to mix gases in a cylinder.
- The owner of the cylinder is the only person permitted to refill a cylinder.
- Propane tanks must be stored in an orderly manner and must be at least 25' from floor openings.
- Signs must be posted in storage areas to warn of flammable and explosive materials.

3.0.0 Section Review

1. Entry into an apprenticeship program is likely to require a(n) _____.
   a. GATB Form Test
   b. license
   c. Scholastic Aptitude Test
   d. OSHA 40-hour course

2. An ITM course that prepares trainees for NICET certification is available through ____.
   a. OSHA
   b. NFPA
   c. ASTM
   d. AFSA

3. Which of the following is *true* regarding tardiness and absenteeism?
   a. It is never acceptable to be absent, even when you are contagious.
   b. It is okay to be a little late for work if you make up the time.
   c. If you must be absent, call in early so that your employer can find a replacement.
   d. If you are not being paid, your absence does not cost the company anything.

4. If you see a safety violation at your jobsite, you should _____.
   a. ignore it unless it affects you directly
   b. make a mental note to avoid the area in future
   c. report it to your supervisor
   d. assume it is okay if no one has been injured

Review Questions

1. The *most* common type of sprinkler system is the _____.
   a. wet pipe sprinkler system
   b. dry pipe sprinkler system
   c. preaction sprinkler system
   d. deluge sprinkler system

2. The sprinkler systems you would *most* likely consider for a below-freezing environment are _____.
   a. wet pipe and dry pipe sprinkler systems
   b. dry pipe and preaction sprinkler systems
   c. wet pipe and preaction sprinkler systems
   d. dry pipe and deluge sprinkler systems

3. The key advantage of a deluge sprinkler system is _____.
   a. it can be used anywhere
   b. it is designed for below-freezing applications
   c. high volume and fast reaction
   d. it is much cheaper to install and maintain

4. The wrench *most* useful in removing or installing polished or plated devices such as valves is _____.
   a. a pipe wrench
   b. a chain wrench
   c. a strap wrench
   d. curved jaw pliers

5. Which of the following is *true* regarding codes and standards?
   a. A code specifies what must be done, while a standard describes how it should be done.
   b. Codes and standards gain legal status in all states as soon as they are published.
   c. The decision as to which codes and standards apply is made by the sprinkler fitter.
   d. The only code that applies to sprinkler fitting is the fire code.

6. The standard that covers the selection and installation of stationary pumps supplying liquid for private fire protection is _____.
   a. NFPA 13D
   b. NFPA 13R
   c. NFPA 20
   d. NFPA 24

7. The standard that covers the design and installation of sprinkler systems in one- and two-family dwellings and manufactured homes is _____.
   a. NFPA 13D
   b. NFPA 13R
   c. NFPA 20
   d. NFPA 24

8. The standard that sets the minimum requirements for installing private fire service mains and their equipment is _____.
   a. NFPA 13D
   b. NFPA 13R
   c. NFPA 20
   d. NFPA 24

9. The standard that covers the design and installation of sprinkler systems in low-rise residential occupancies is _____.
   a. NFPA 13D
   b. NFPA 13R
   c. NFPA 20
   d. NFPA 24

10. OSHA standards for the construction industry are covered in _____.
    a. *29 CFR, Part 1910*
    b. *29 CFR, Part 1926*
    c. *29 CFR, Part 1965*
    d. *29 CFR, Part 1970*

Trade Terms Quiz

Fill in the blank with the correct term learned from your study of this module.

1. The mission of the ______________ is to ensure safe and healthful working conditions for working men and women by setting and enforcing standards as well as providing safety training, outreach, education, and assistance.
2. The sprinkler system that has a supplemental detection system installed in the same area as the system sprinklers is a(n) ______________.
3. The type of sprinkler system that uses open sprinklers rather than closed sprinklers is a(n) ______________.
4. ______________ pipe has been secured for storage in tiers.
5. The ______________ sets the minimum standards for training programs across the United States.
6. The simplest, most common type of sprinkler system is a(n) ______________.
7. A(n) ______________ is a heat-sensing device used in sprinkler systems to discharge water onto a fire.
8. ______________ refers to training acquired while working.
9. A possible source of danger or potential injury is called a(n) ______________.
10. The standardized construction education materials produced by NCCER are called ______________.
11. The person or agency that decides the acceptability of the systems and devices being installed is referred to as the ______________.
12. The ______________ is a non-profit, international association representing merit shop sprinkler contractors.
13. The rules and specifications provided to the sprinkler industry are called ______________.
14. ______________ is stored under pressure in cylinders.
15. A(n)______________ is a device used to support pipe in tiers.
16. The industry reference for installing sprinkler systems is ______________.
17. A sprinkler system having pipes filled with air or nitrogen that operates without a detection system is a(n) ______________.
18. The ______________ publishes codes and standards with the goal of preventing loss of life and property.
19. Standards that have been adopted through statutory or administrative laws in a specific geographic area are known as ______________.
20. Standards that are generally accepted as relevant in civil law, even in areas where no standards have been formally adopted, are known as ______________.

Trade Terms

Adopted standards
American Fire Sprinkler Association (AFSA)
Authority Having Jurisdiction (AHJ)
Blocked
Compressed gas
Deluge sprinkler system
Dry pipe sprinkler system
Hazard
National Fire Protection Association (NFPA)
NCCER Standardized Curricula
NFPA 13, Standard for the Installation of Sprinkler Systems
Occupational Safety and Health Administration (OSHA)
Office of Apprenticeship (OA)
On-the-job training (OJT)
Preaction sprinkler system
Rack
Recognized standards
Regulations
Sprinkler
Wet pipe sprinkler system

Cornerstone of Craftsmanship

Taylor Schumacher

Project Manager
Security Fire Sprinkler

How did you choose a career in the industry?

Working in a sprinkler warehouse was the first job I ever had. I bounced in and out of the industry over the years with various positions. My father worked for a sprinkler contracting company and was looking for some help in the warehouse for the summer. I left that job after the summer. I came back another time to the field for two years through college. Upon graduation, I was working in a bank and wanted a job that was more exciting and rewarding. I found my way back into the industry in design and project management with a sprinkler contractor.

Who inspired you to enter the industry?

My father.

What types of training have you been through?

I have a bachelor's degree and have also been through the AFSA books, AutoCAD training, Revit training, and various manufacturer training courses.

How important is education and training in construction?

Education and training is the difference between good and great.

How important are NCCER credentials to your career?

Training and the credentials that go with it is the framework needed to gain competence and to advance throughout your career.

How has training/construction impacted your life?

Construction has always surrounded my life, from projects in my personal life to paychecks in my professional life. All the training and experience builds upon each other to provide a depth of knowledge.

Tell us about your present job.

I work primarily on multifamily residential projects. I get to wear many hats and enjoy them all, including design, project management, purchasing, employee management, and training.

What do you enjoy most about your job?

I get to save lives every day and help build careers.

What factors have contributed most to your success?

Strong mentorship and a curiosity for the craft have most contributed to my success.

Would you suggest construction as a career to others? Why?

Absolutely, few other industries allow you the ability to see the work that you put out into the world realized in different buildings all over. I get to see some of the projects I have had a hand in nearly everywhere I go.

What advice would you give to those new to the field?

Work as hard as you can and learn as much as you can.

How do you define craftsmanship?

A distinct skill that produces unbeatable quality.

Trade Terms Introduced in This Module

Adopted standards: Standards that have been adopted through statutory or administrative laws in a specific geographic area. Once a document is adopted by reference in a state statute or local ordinance, it has the same force and effect as if published therein in its entirety.

American Fire Sprinkler Association (AFSA): A non-profit, international association representing merit shop sprinkler contractors, dedicated to the educational advancement of its members and promotion of the use of sprinkler systems.

Authority Having Jurisdiction (AHJ): An organization, office, or individual responsible for enforcing the requirements of a code or standard, or for approving equipment, materials, an installation, or a procedure.

Blocked: Secured for storage in tiers.

Compressed gas: Gas stored under pressure in cylinders.

Deluge sprinkler system: A type of sprinkler system that uses open sprinklers rather than closed sprinklers. When heat triggers the detection system, the deluge valve operates, producing immediate water flow through all the sprinklers in a given area.

Dry pipe sprinkler system: Dry pipe sprinkler systems are filled with pressurized air or nitrogen. Water flows into the system only when a sprinkler is activated.

Hazard: A possible source of danger or potential injury.

National Fire Protection Association (NFPA): An organization that publishes codes and standards with the goal of preventing loss of life and property.

NCCER Standardized Curricula: Standardized construction education materials produced by NCCER.

NFPA 13, Standard for the Installation of Sprinkler Systems: The organization (NFPA) that provides the basic reference to the requirements (NFPA 13) for installing sprinkler systems.

Occupational Safety and Health Administration (OSHA): The Occupational Safety and Health Administration (OSHA) is an agency of the US Department of Labor established by the Occupational Safety and Health Act of 1970. OSHA's mission is to ensure safe and healthful working conditions for working men and women by setting and enforcing standards as well as providing safety training, outreach, education, and assistance.

Office of Apprenticeship (OA): The US Department of Labor office that sets the minimum standards for training programs across the country.

On-the-job training (OJT): Job-related training acquired while working.

Preaction sprinkler system: A sprinkler system employing sprinklers attached to a piping system containing air or nitrogen that may or may not be under pressure. A supplemental detection system is installed in the same area as the system sprinklers.

Rack: Device used to support pipe in tiers.

Recognized standards: Standards that are generally accepted as relevant in civil law, even in areas where no standards have been formally adopted.

Regulations: Rules and specifications for the sprinkler industry provided by various organizations.

Sprinkler: A heat-sensing device used in sprinkler systems to discharge water onto a fire.

Wet pipe sprinkler system: The simplest, most common type of sprinkler system. Wet pipe sprinkler systems are always filled with water, compared to dry pipe sprinkler systems that contain no water until the system is activated.

Appendix A

SAMPLES OF NCCER TRAINING CREDENTIALS

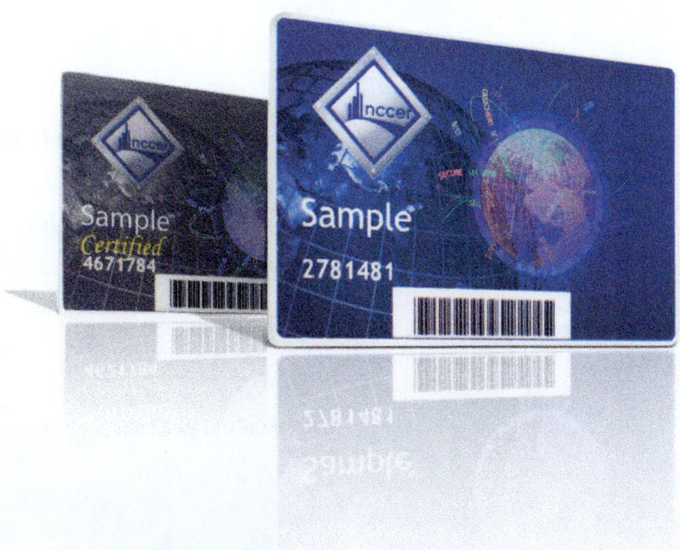

Appendix B

Samples of AFSA Apprentice Training Credentials

AMERICAN FIRE SPRINKLER ASSOCIATION

John Doe

HAS BEEN AWARDED A CERTIFICATE OF ACHIEVEMENT
IN RECOGNITION OF SUCCESSFUL COMPLETION OF ALL FOUR LEVELS OF

American Fire Sprinkler Association's
Fire Sprinkler Fitting Training Series

Four-Level Training Series in Automatic Fire Sprinkler Protection for Apprentice Fire Sprinkler Fitters

Providing a Minimum of 576 Hours of Training

May 16, 2020
Date

Marlene Garrett, CMP
Vice President Operations and Education Services

AFSA
firesprinkler.org

AFSA
American Fire Sprinkler Association

Apprenticeship Identification Card

John Doe

The above named individual is enrolled in a course of study in an Apprentice Training Series for Automatic Fire Sprinkler Systems using the curriculum registered as a part of standards approved by the Bureau of Apprenticeship and Training, U.S. Department of Labor.

N-92021
Registration No.

Marlene Garrett, CMP
Chief Operating Officer

Apprentice Credits

Level	Date Completed	CEUs Credited
Level I	5/20/2020	15.25

Student ID #	Date of Enrollment
12345	8/5/2019

Leslie Clounts
AFSA Director of Education

12750 Merit Drive, Suite 350, Dallas, TX 75251 • (214) 349-5965 • firesprinkler.org

Additional Resources

This module presents thorough resources for task training. The following reference material is suggested for further study.

NFPA 13, *Standard for the Installation of Sprinkler Systems*, Latest Edition. Quincy, MA: National Fire Protection Association.

NFPA 13D, *Standard for the Installation of Sprinkler Systems in One- and Two-Family Dwellings and Manufactured Homes*, Latest Edition. Quincy, MA: National Fire Protection Association.

NFPA 13R, *Standard for the Installation of Sprinkler Systems in Low-Rise Residential Occupancies*, Latest Edition. Quincy, MA: National Fire Protection Association.

Figure Credits

Used with permission of Johnson Controls, Figure 1

Courtesy of RIDGID Tool Company, Figure 5

Courtesy of Reed Manufacturing, Figure 6

Reliable Automatic Sprinkler Co., Inc., Figure 7 (Typical Open-Ended Wrenches, Typical Socket Wrenches, far left image)

Courtesy of Victaulic Company, Figure 7 (Typical Socket Wrenches, two images on right)

Image property of Stanley Black & Decker. Used with permission., Figure 8

iStock@vkyryl, Figure 9

iStock@Cylonphoto, Figure 10

iStock@BWFolsom, Figure 11

iStock@missphotoss, Figure 12

Courtesy of Werner, Figures 14–16

Courtesy of International Code Council, Figure 17 (International Building Codes), Figure 19 (International Fire Code)

© 2020 NFPA, reproduced with permission., Figures 17 (NFPA 5000), 18, Figure 19 (NFPA 1)

© 2018 NFPA, reproduced with permission., Figure 20 (NFPA 13)

© 2019 NFPA, reproduced with permission., Figure 20 (NFPA 25)

Courtesy of Western Fire Protection, Figures 21–22

American Fire Sprinkler Association, A02

Section Review Answer Key

Section 1.0.0

Answer	Section Reference	Objective
1. d	1.1.4	1a
2. c	1.2.8	1b

Section 2.0.0

Answer	Section Reference	Objective
1. a	2.1.3	2a
2. b	2.2.3	2b

Section 3.0.0

Answer	Section Reference	Objective
1. a	3.1.0	3a
2. d	3.2.3	3b
3. c	3.3.1	3c
4. c	3.4.0	3d

NCCER CURRICULA — USER UPDATE

NCCER makes every effort to keep its textbooks up-to-date and free of technical errors. We appreciate your help in this process. If you find an error, a typographical mistake, or an inaccuracy in NCCER's curricula, please fill out this form (or a photocopy), or complete the online form at **www.nccer.org/olf**. Be sure to include the exact module ID number, page number, a detailed description, and your recommended correction. Your input will be brought to the attention of the Authoring Team. Thank you for your assistance.

Instructors – If you have an idea for improving this textbook, or have found that additional materials were necessary to teach this module effectively, please let us know so that we may present your suggestions to the Authoring Team.

NCCER Product Development and Revision

13614 Progress Blvd., Alachua, FL 32615

Email: curriculum@nccer.org
Online: www.nccer.org/olf

❑ Trainee Guide ❑ Lesson Plans ❑ Exam ❑ PowerPoints Other ______________________

Craft / Level: Copyright Date:

Module ID Number / Title:

Section Number(s):

Description:

Recommended Correction:

Your Name:

Address:

Email: Phone:

Introduction to Components and Systems

Overview

This module provides an overview of the four main types of sprinkler systems. It also describes testing laboratories and Listing requirements. In addition, it covers sprinkler characteristics, including thermal response, spray pattern and coverage area, orientation, K-factor ratings, and temperature ratings. It also describes various types of piping, valves, and pumps.

Module 18102

Trainees with successful module completions may be eligible for credentialing through the NCCER Registry. To learn more, go to **www.nccer.org** or contact us at 1.888.622.3720. Our website, **www.nccer.org**, has information on the latest product releases and training.

Your feedback is welcome. You may email your comments to **curriculum@nccer.org**, send general comments and inquiries to **info@nccer.org**, or fill in the User Update form at the back of this module.

This information is general in nature and intended for training purposes only. Actual performance of activities described in this manual requires compliance with all applicable operating, service, maintenance, and safety procedures under the direction of qualified personnel. References in this manual to patented or proprietary devices do not constitute a recommendation of their use.

18102 V4.0

18102

Introduction to Components and Systems

Objectives

Successful completion of this module prepares you to do the following:

1. Identify the operation of the four main types of sprinkler systems.
   a. Trace the sequence of operation through a wet pipe sprinkler system.
   b. Trace the sequence of operation through a dry pipe sprinkler system.
   c. Trace the sequence of operation through a preaction sprinkler system.
   d. Trace the sequence of operation through a deluge sprinkler system.
2. Identify the nationally recognized testing laboratories that are used to evaluate sprinkler system components.
   a. Define the terms related to sprinkler codes and standards.
   b. Identify the Listing agencies that test and certify sprinkler system components.
3. Identify common sprinklers and their operating characteristics.
   a. Identify sprinkler orientations and spray patterns.
   b. Identify various sprinkler characteristics.
4. Identify different types of piping, valves, and fire pumps.
   a. Describe the installation requirements for various piping systems.
   b. Identify sprinkler system valves and waterflow switches.
   c. Describe the purpose of fire pumps.

Performance Tasks

This is a knowledge-based module. There are no Performance Tasks.

Trade Terms

C-factor
Check valve
Control valve
Drop/suspended ceiling
Electrolysis
FM Approvals®
Galvanized
Inside diameter (ID)
K-factor
Labeled
Listed
Outside diameter (OD)
Sway bracing
Test and drain valve
Underwriters Laboratories® (UL)
Water hammer

Industry Recognized Credentials

If you are training through an NCCER-accredited sponsor, you may be eligible for credentials from NCCER's Registry. The ID number for this module is 18102. Note that this module may have been used in other NCCER curricula and may apply to other level completions. Contact NCCER's Registry at 1.888.622.3720 or go to **www.nccer.org** for more information.

You can also show off your industry-recognized credentials online with NCCER's digital badges. Transform your knowledge, skills, and achievements into badges that you can share across social media platforms, send to your network, and add to your resume. For more information, visit **www.nccer.org**.

NOTE

Unless otherwise specified, references in parentheses following figure and table numbers refer to NFPA 13.

Contents

Figures and Tables

SECTION ONE

1.0.0 SYSTEM TYPES AND GENERAL OPERATION

Objective

Identify the operation of the four main types of sprinkler systems.

a. Trace the sequence of operation through a wet pipe sprinkler system.
b. Trace the sequence of operation through a dry pipe sprinkler system.
c. Trace the sequence of operation through a preaction sprinkler system.
d. Trace the sequence of operation through a deluge sprinkler system.

There are four fundamental types of sprinkler systems: wet pipe, dry pipe, preaction, and deluge. Each of these systems is designed for specific applications, situations, and conditions.

1.1.0 Wet Pipe Sprinkler System

In wet pipe systems, the pipes are always filled with water. When heat activates the thermal element in a sprinkler, water is discharged onto the fire. Wet pipe sprinkler systems are by far the most common system in use today. They are used in apartment buildings, hotels, office buildings, schools, and many other applications.

A typical sequence of operation for a wet pipe sprinkler system is shown in *Figure 1*. When a sprinkler is activated (1), the discharging water lifts the alarm valve clapper (2) and flows through the alarm port (3) to the retard chamber (4). When the retard chamber is filled, water flows to the water motor alarm and/or the optional pressure switch (5), which signals an electric alarm bell (6).

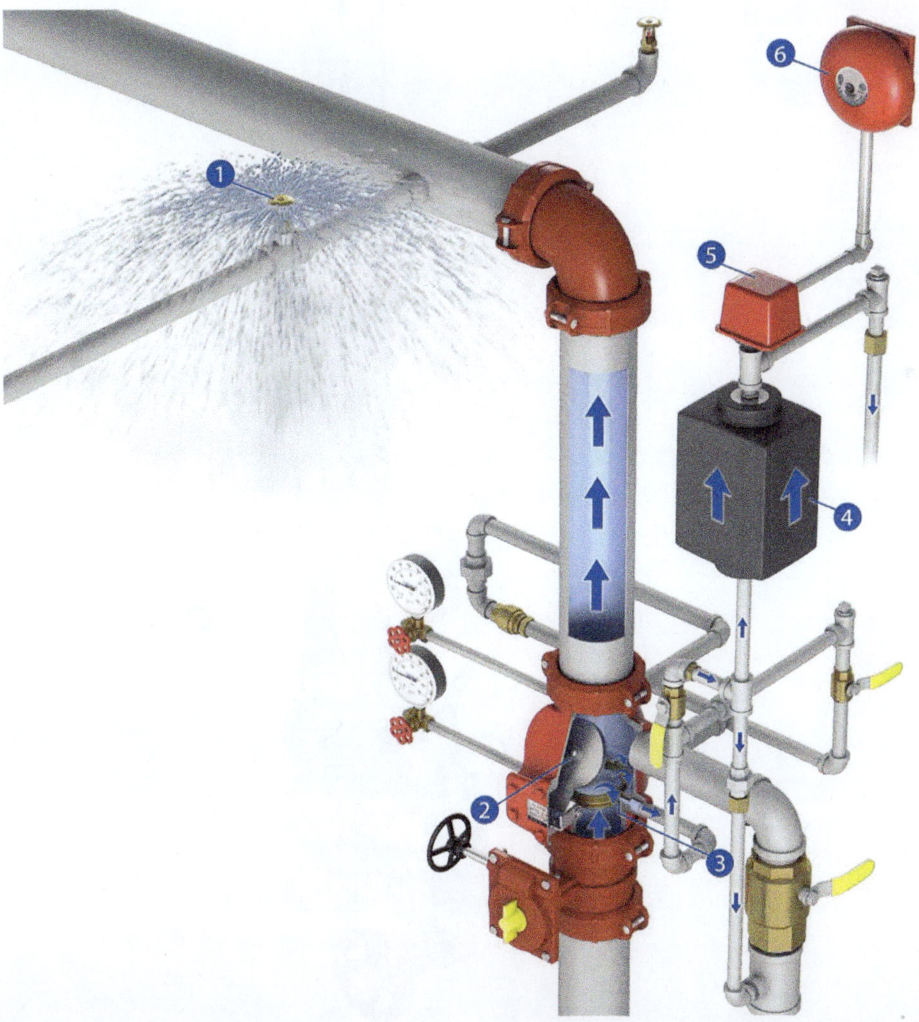

Figure 1 Wet pipe sprinkler system.

A wet system riser operates in a simpler way (*Figure 2*). When a sprinkler is activated (1) downstream of the assembly, incoming water flows through the riser check valve manifold (2), which in turn moves the paddle of the waterflow switch (3). This sequence of events initiates an electric alarm bell or horn/strobe.

1.2.0 Dry Pipe Sprinkler System

Dry pipe sprinkler systems are filled with pressurized air or nitrogen, not water. Dry pipe systems are used in applications subject to freezing temperatures, such as unheated warehouses, attic spaces, loading docks, or parking garages.

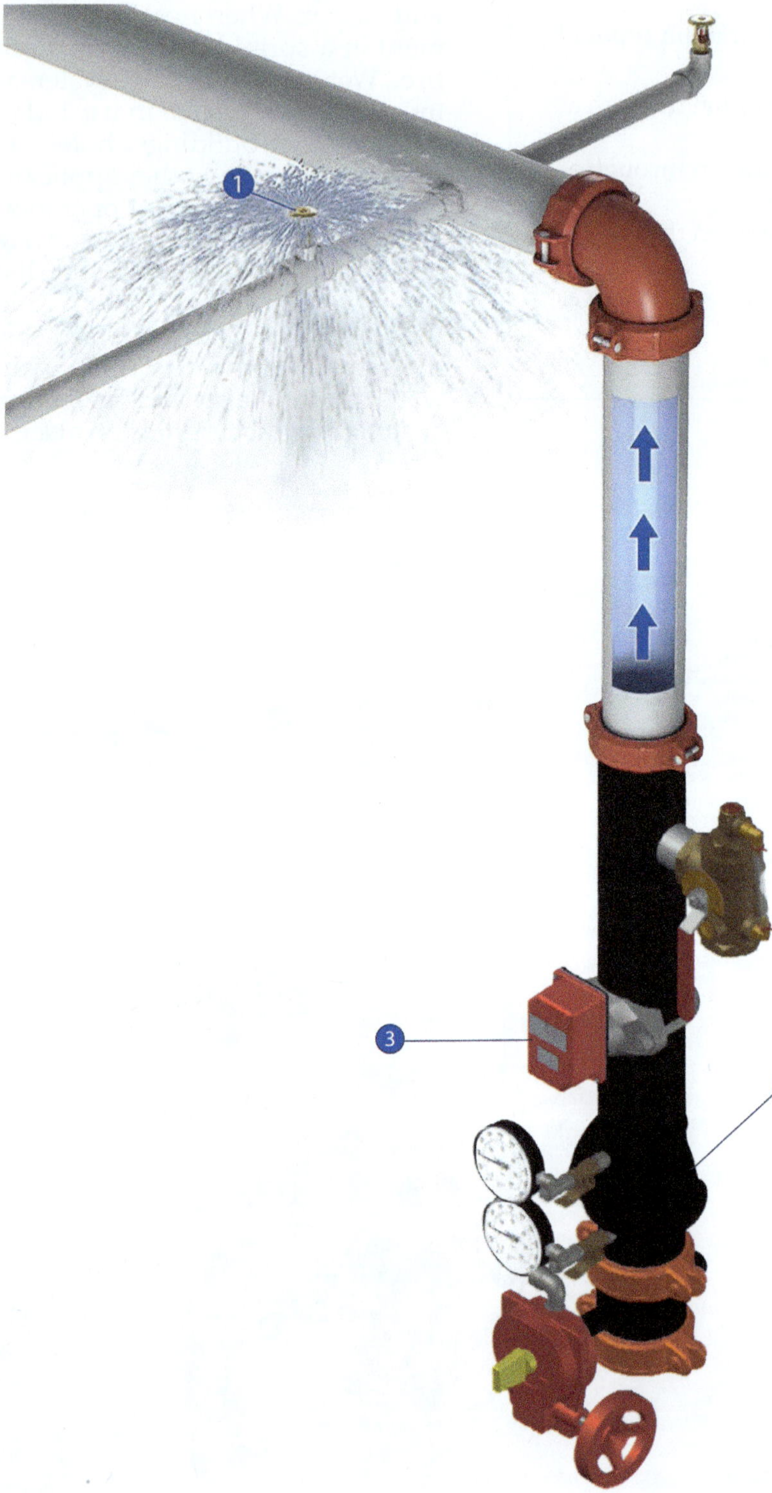

Figure 2 Wet system riser.

A typical sequence of operation for a dry pipe sprinkler system is shown in *Figure 3*. When a sprinkler is activated (1), loss of air pressure in the system allows the dry valve clapper (2) to open, filling the system with water. To speed the opening of the dry valve in large systems, an accelerator (3) can be added. An accelerator is a type of quick-opening device. An integral anti-flood device protects against accelerator flooding. Water flow from the intermediate chamber of the dry valve can cause a pressure switch (4) to activate an electric alarm and/or operate a mechanical water motor alarm (not shown).

1.3.0 Preaction Sprinkler System

A preaction system uses automatic sprinklers attached to a piping system. The piping system contains air and may or may not be under pressure. There is a supplemental detection system installed near the sprinklers.

The preaction system operates somewhat differently than a wet or dry pipe system. The alarm is given *before* the sprinkler activates to discharge water.

Preaction systems are designed to protect properties in which the danger of serious water damage from premature operation of sprinklers

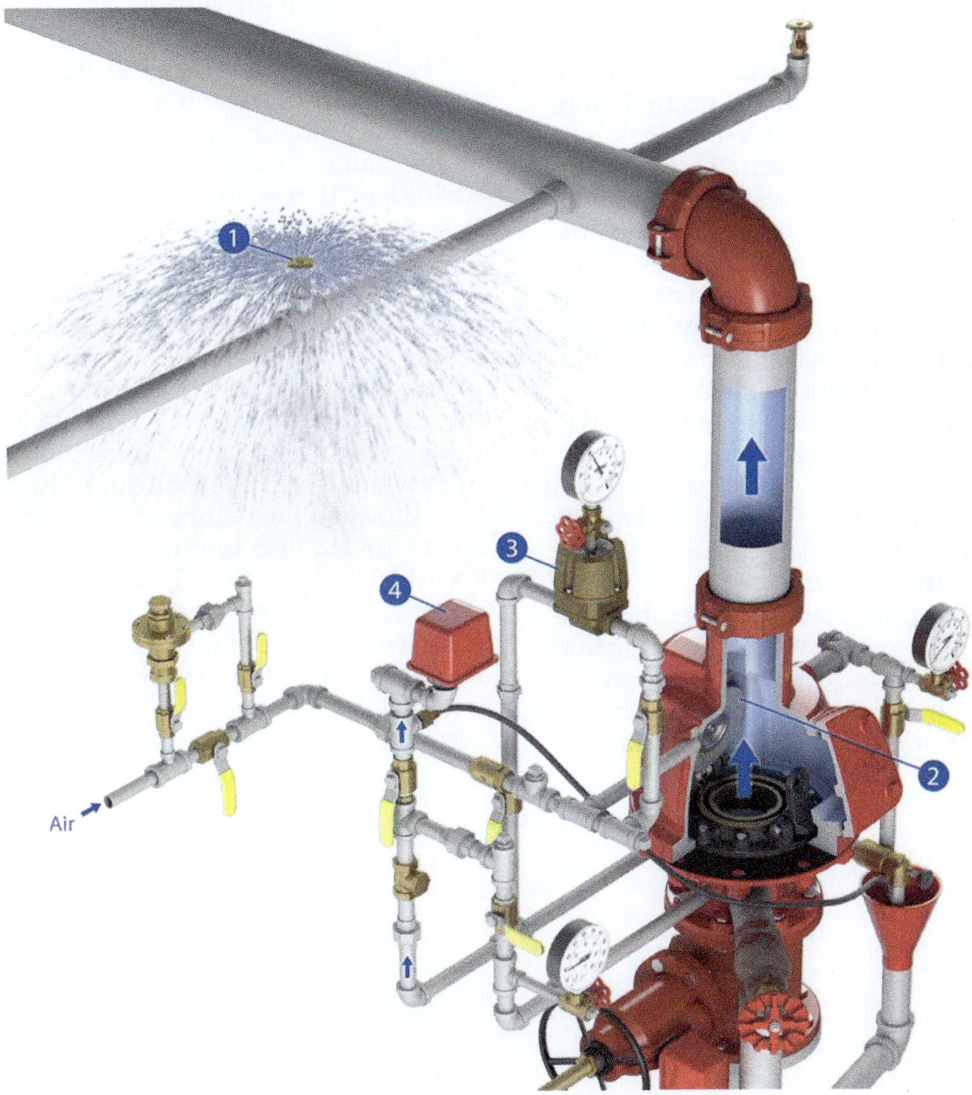

Figure 3 Dry pipe sprinkler system.

or broken pipes is unusually severe, such as museums, libraries, telecommunications rooms, and freezer warehouses. They are also used where the climate requires the use of a dry system. In a preaction system, the action of a heat- or smoke-responsive device releases a preaction valve, which causes water to fill the piping system and an advance alarm to sound before the activation of sprinklers. The alarm often gives an opportunity to extinguish the fire by manual means before a sprinkler activates.

Preaction systems may be single-interlock, non-interlock, or double-interlock types. A single-interlock system is activated by the detection system. The non-interlock system is activated by either the detection system or an activated sprinkler. Both the single-interlock and non-interlock systems require only one event to occur before water enters the system. A double-interlock system requires both events to occur before water enters the system: activation of a sprinkler as well as the detection system. When one of these events occurs, the system activation valve moves into a preset position. When the second event occurs, the valve opens, and water enters the system.

A typical sequence of operation for a single-interlock preaction sprinkler system is shown in *Figure 4*. When the detector is activated (1), a signal is sent to the release control panel (2). The panel sends appropriate alarm signals and triggers the release of the solenoid valve (3). When the solenoid opens, the priming water is relieved from the internal prime chamber assembly (4). The prime chamber assembly collapses, and water passes through the preaction deluge valve and internal check diaphragm (5). The water enters the system piping, but until a sprinkler activates (6), no water is discharged. When the preaction deluge valve operates, pressure opens the pressure-operated relief valve (PORV) (7), continuously venting the water to the priming chamber and ensuring the deluge valve remains in the open position.

1.4.0 Deluge Sprinkler System

A deluge system uses open sprinklers. When a fire or flame triggers the detection system, the deluge valve is released, which produces immediate water flow through all sprinklers in a given area. These systems are used in applications such as aircraft hangers, power plants, and chemical plants where high volume and fast reaction are required.

The purpose of a deluge system is to wet down an entire area in which a fire may originate. This is done by flowing water through all sprinklers at the same time, rather than using automatic sprinklers that would activate independently as the fire spreads.

A typical sequence of operation for a deluge sprinkler system is shown in *Figure 5*. When the independent detection system is activated (1), a signal is sent to the release control panel (2). The panel sends appropriate alarm signals and opens the normally closed (NC) solenoid (3). The deluge valve priming chamber (4) is then vented faster than water is supplied through the restricted orifice, allowing the deluge valve to open. The water enters the system piping. Water flows from all sprinklers or nozzles (5). When the deluge valve operates, pressure opens the pressure-operated relief valve (PORV) (6), continuously venting the water supply to the priming chamber and ensuring that the deluge valve remains in the open position.

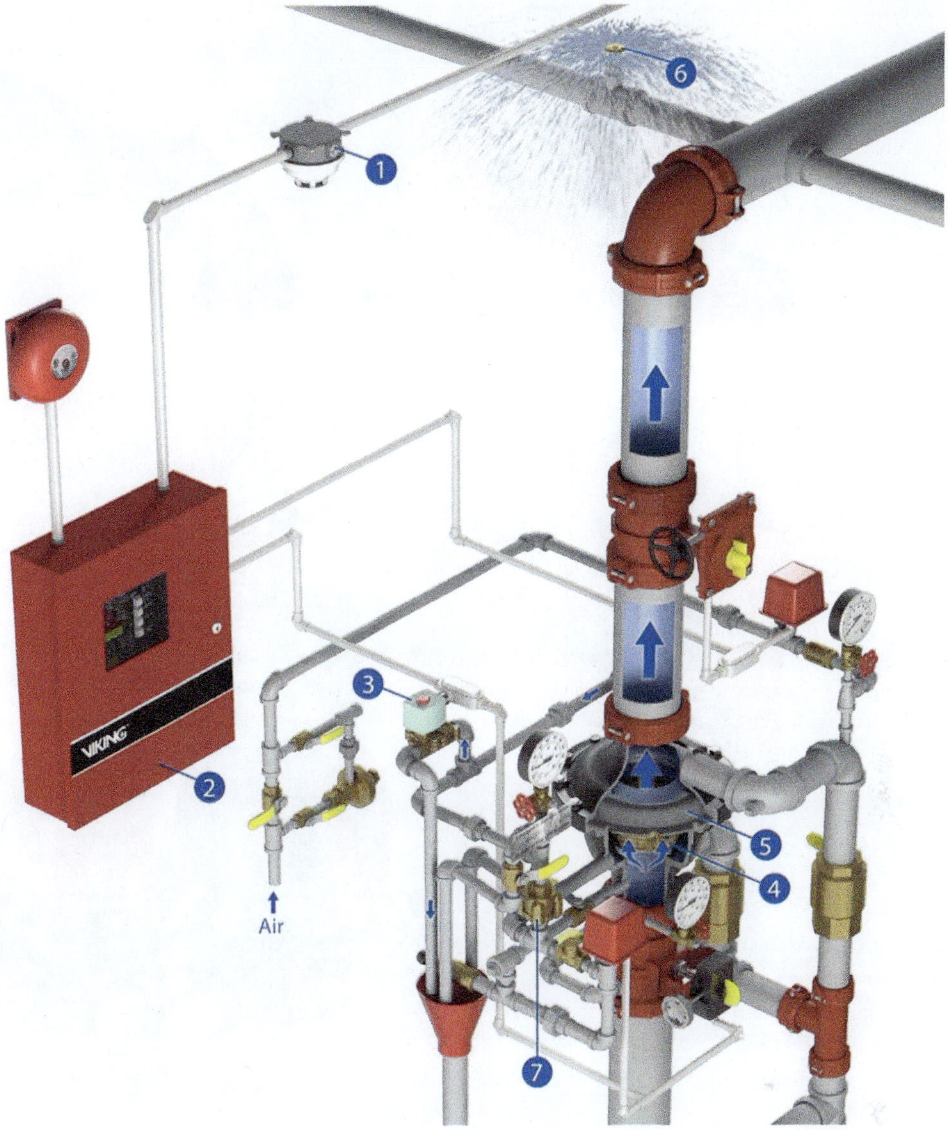

Figure 4 Preaction sprinkler system.

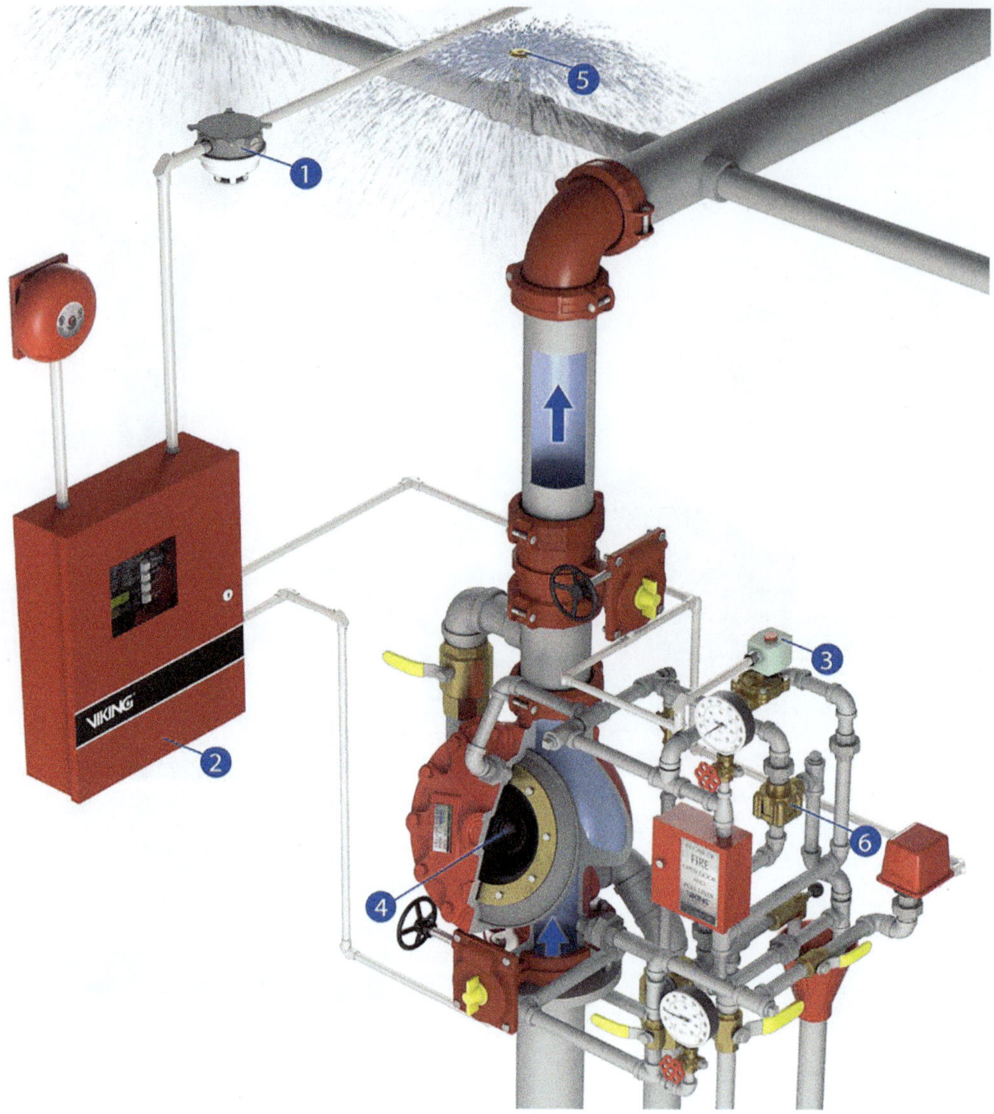

Figure 5 Deluge sprinkler system.

1.0.0 Section Review

1. The type of sprinkler system in which the filling of a retard chamber triggers an alarm is a _____.
   a. wet pipe sprinkler system
   b. dry pipe sprinkler system
   c. preaction sprinkler system
   d. deluge sprinkler system

2. A parking garage subject to freezing temperatures would be *most* likely to use a _____.
   a. wet pipe sprinkler system
   b. dry pipe sprinkler system
   c. preaction sprinkler system
   d. deluge sprinkler system

3. A museum housing priceless archives would be *most* likely to use a _____.
   a. wet pipe sprinkler system
   b. dry pipe sprinkler system
   c. preaction sprinkler system
   d. deluge sprinkler system

4. The type of sprinkler system *most* likely to be used in a chemical plant is a _____.
   a. wet pipe sprinkler system
   b. dry pipe sprinkler system
   c. preaction sprinkler system
   d. deluge sprinkler system

Section Two

2.0.0 Nationally Recognized Testing Laboratories

Objective

Identify the nationally recognized testing laboratories that are used to evaluate sprinkler system components.

a. Define the terms related to sprinkler codes and standards.
b. Identify the Listing agencies that test and certify sprinkler system components.

Trade Terms

FM Approvals®: A testing laboratory and Listing agency that certifies new products for service in particular locations and for specific applications.

Labeled: Equipment or materials to which has been attached a label, symbol, or other identifying mark of an organization that is acceptable to the Authority Having Jurisdiction (AHJ).

Listed: Included in a list of products or services published by a recognized Listing agency and acceptable to the AHJ for use in sprinkler systems for a specified purpose.

Underwriters Laboratories® (UL): A testing laboratory and Listing agency that certifies new products for service in particular locations and for specific applications.

Before you can understand how a sprinkler system works, you must learn the language of the industry and understand how it relates to the selection of specific components in the system.

2.1.0 Understanding Code Language

The basic written reference for the sprinkler industry is NFPA 13, *Standard for the Installation of Sprinkler Systems*. NFPA, working with experts in various industries, develops and publishes codes related to fire safety, electrical work, medical care facilities, and other areas. It is important that you understand the meaning of specific terms as they relate to codes and standards. These terms include the following:

- *Approved* — Acceptable to the Authority Having Jurisdiction. (Also referred to as *accepted*.)
- *Authority Having Jurisdiction (AHJ)* — An organization, office, or individual responsible for enforcing the requirements of a code or standard, or for approving equipment, materials, an installation, or a procedure.
- *Equipment Listed for the application* — When used, this term requires that equipment used in a system be properly **Listed** for the specific application. Many people assume that when an item is Listed, it is Listed for all applications, but that is not true. In the Listing process, a device becomes Listed under one or more standards, and should therefore meet the performance requirements of the standard(s) if properly installed, tested, and maintained.
- *Equivalent* — The "or equivalent" clause in most standards allows the use of systems, methods, or devices of equal or better quality, as long as sufficient documentation is provided to the AHJ to demonstrate the equivalency.
- *Labeled* — **Labeled** equipment or materials have an attached label, symbol, or other identifying mark of an organization that is acceptable to the AHJ. Labeling requires periodic inspection of the production of labeled equipment or materials, and the manufacturer indicates compliance with appropriate standards or performance in a specified manner.
- *Listed* — Equipment, materials, or services included in a list published by an organization that is acceptable to the AHJ and concerned with evaluation of products or services, that maintains periodic inspection of production of Listed equipment or materials or periodic evaluation of services, and whose Listing states that either the equipment, material, or service meets appropriate designated standards or has been tested and found suitable for a specified purpose.
- *Shall* — Indicates a mandatory requirement.
- *Should* — Indicates a recommendation or that which is advised but not required.
- *Standard* — A document, the main text of which contains only mandatory provisions using the word "shall" to indicate requirements and which is in a form generally suitable for mandatory reference by another standard or code or for adoption into law. Nonmandatory provisions shall be located in an appendix or annex, footnote, or informational note and are not considered a part of the requirements of a standard.

Did You Know?

To make sure that sprinkler systems perform adequately and are reliable, NFPA 13 requires the Listing of components crucial to system performance during a fire. Components that might malfunction during a fire, but would not adversely affect system performance, do not have to be Listed. The requirement for Listed products and materials is aimed at making sure the fire protection system will perform as it should during a fire.

2.2.0 Listing Agencies

For sprinkler systems, two of the most common Listing agencies are **Underwriters Laboratories® (UL)** and **FM Approvals®**. When these agencies test a product and find that it meets code and standard requirements and is suitable for use in a specific application, that product is said to be *Listed*. Sprinkler components must be Listed for fire protection use. Listed items may only be installed for use in fire suppression applications and must be installed in accordance with the manufacturer's instructions.

Testing laboratories are an integral part of the development of codes and standards. The NFPA and other organizations rely on testing laboratories to conduct research into fire protection equipment and its safety. These laboratories perform extensive testing of new products to make sure that they are built to meet the appropriate standards for fire safety. They receive statistics and reports from agencies all over the United States concerning fires and their causes. Upon seeing developing trends concerning the association of certain equipment and dangerous situations or circumstances, the equipment is specifically targeted for research. Reports from these laboratories are considered in the revisions of codes.

2.2.1 Underwriters Laboratories®

Underwriters Laboratories® is a non-profit organization whose principal business is to evaluate products to determine their compliance with defined standards for safety. UL evaluates electrical and mechanical products, building materials, construction systems, fire protection equipment, and marine products. UL's standards for safety establish the basis for testing products and levels of acceptability. Once a product has successfully passed testing, it may be Listed and have the UL label affixed to it. Through international agreements, UL is authorized to test and certify specific types of products for use in both US and Canadian markets.

UL's fire protection division provides insurance companies, AHJs, retailers, product manufacturers, and building owners with certification and customized testing services in many areas including:

- Fire modeling
- Ignition, smoke, and fire effluent (byproduct) testing
- Large-scale fire research

Products tested at UL are Listed in a UL product directory. The product directory is a resource used to determine the correct application of a product or system. A product directory contains the names of companies qualified to use a UL Mark on products that comply with UL requirements. Architects, AHJs, purchasing agents, and others use UL's product directories to find information about products bearing the UL mark.

As a fundamental element of UL's safety business, the fire protection division offers enhanced services including customer solutions and bundles services. The fire protection division can also develop custom certification and testing programs with technical expertise.

Underwriters Laboratories®

The Chicago World's Fair was opened in 1893 and, thanks to Edison's introduction of the electric light bulb, included a light display called the *Palace of Electricity*. But all was not perfect—wires soon sputtered and crackled, and, ironically, the Palace of Electricity caught fire. The fair's insurance company called in a troubleshooting engineer, who, after careful inspection, found faulty and brittle insulation, worn out and deteriorated wiring, bare wires, and overloaded circuits.

This engineer, William Henry Merrill, called for standards in the electrical industry and set up a testing laboratory above a Chicago firehouse to do just that. Underwriters Electrical Bureau, an independent testing organization, was born. Now known as *Underwriters Laboratories®*, it is an internationally recognized authority on product safety testing as well as safety certification and standards development.

The fire protection division's suppression team supports manufacturers, AHJs, building owners, and insurance companies by offering certification and customized testing services in the following areas:

- Automatic sprinklers
- Extinguishing system units
- Fire extinguishers
- Fire pumps and engines
- Sprinkler piping systems
- Valves for fire suppression systems
- Water mist nozzles

UL Testing

For more than 100 years, UL has been testing and certifying automatic sprinklers. The tests and certifications are based on the *Standard for Automatic Sprinklers for Fire-Protection Services* (UL 199) and *Early Suppression Fast-Response Sprinklers* (UL 1767). UL is also capable of testing sprinklers sampled from field installations as specified in NFPA 25, *Standard for the Inspection, Testing, and Maintenance of Water-Based Fire Protection Systems.* This type of testing is a service to building owners, AHJs, and insurance companies. UL provides both large-scale testing (fires) and small-scale testing (water drop size, velocity, and heat release rate).

2.2.2 FM Approvals®

FM Approvals® is a non-profit scientific research and testing organization managed by FM Global, the world's largest commercial and industrial property insurance organization. FM Approvals® provides certification for various types of products, including the following:

- Fire detection, signaling, and other electrical equipment
- Fire protection equipment
- Functional safety assessment and certification

FM Approvals® distinguishes between approval and acceptance. An *approval* is a confirmation and a subsequent Listing by FM Approvals® that a product has been examined according to applicable requirements and found suitable for use in all instances, subject to any limitations stated in the approval. Products that earn approval are Listed in the approval guide. An *acceptance* confirms that materials installed at a specific location are suitable for their intended uses. In other words, approval applies to a product, regardless of location, while acceptance is considered on a case-by-case basis. Acceptances are typically limited to properties FM Global insures.

In addition to product certification, FM Approvals® also offers specification testing. Narrower in scope than the approvals program, specification testing allows manufacturers to test a single performance characteristic of a system or assembly.

2.0.0 Section Review

1. When a code or standard uses the word "shall," it means that the instruction is _____.
   a. suggested
   b. recommended
   c. optional
   d. mandatory

2. A manufacturer who wishes to test a single system performance characteristic would submit it for _____.
   a. specification testing
   b. approval testing
   c. acceptance testing
   d. operational testing

SECTION THREE

3.0.0 SPRINKLERS

Objective

Identify common sprinklers and their operating characteristics.

a. Identify sprinkler orientations and spray patterns.
b. Identify various sprinkler characteristics.

Trade Terms

Drop/suspended ceiling: A ceiling that is not part of the structural framework of a building. This ceiling is installed by suspending it from the floor above, or from the roof, and is commonly used to provide space for services such as cables, recessed lighting, and piping.

K-factor: A mathematical constant that represents a sprinkler's discharge coefficient, which is the amount of water distributed through the sprinkler orifice at a given pressure. The larger the orifice, the larger the K-factor.

A sprinkler is a fire suppression or control device that operates automatically when its thermal element is heated to its thermal rating or above, allowing water to discharge over a specified area.

Sprinklers are generally manufactured from brass with a seat held in place by either a metal link that melts or a glass bulb containing a heat-sensitive liquid (sometimes referred to as a *frangible bulb*) that breaks at a preset temperature. The water pressure in the system forces the seat out, allowing water to discharge.

CAUTION

Sprinklers must be installed in accordance with the shop or field drawings, manufacturer's instructions, and NFPA requirements. In no case shall the area of coverage exceed 400 square feet.

3.1.0 Sprinkler Orientation and Spray Patterns

The orientation and spray pattern of a sprinkler are very important to the effective operation of the system and the application of the sprinkler. The spray pattern and orientation of the sprinkler are determined by the type of sprinkler used. Each of the following types of sprinklers is oriented in a particular manner to produce the desired spray pattern:

- Upright sprinklers
- Pendent sprinklers
- Sidewall sprinklers
- Recessed and concealed sprinklers
- Conventional sprinklers

3.1.1 Upright Sprinklers

Upright sprinklers project up toward the ceiling and have deflectors that spray the water downward in a circular pattern. They are used in mechanical rooms and similar areas to provide better coverage between ceiling obstructions, such as ductwork or beams.

Per NFPA 13, an upright sprinkler (*Figure 6*) must be installed with the frame arms parallel to (in line with) the branch line, unless specifically Listed for other orientations. This reduces the amount of shadowing (coverage areas blocked by the sprinkler frame or branch piping). The weakest part of the pattern is in line with the frame arms, so lining them up with the pipe provides the strongest possible pattern.

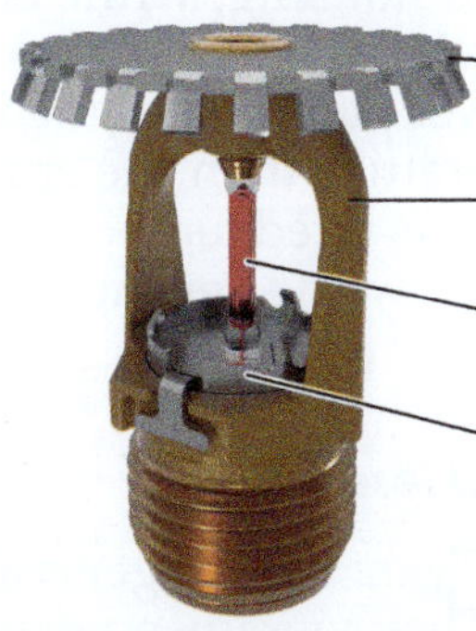

Figure 6 Upright sprinkler.

3.1.2 Pendent Sprinklers

Pendent sprinklers project down toward the floor and have deflectors that spray the water downward in a circular pattern (*Figure 7*). Pendent sprinklers are always located below the branch line or, in the case of a **drop/suspended ceiling**, at the ceiling below. The frame arms do not have to be parallel with the branch line because the pipe never interferes with the spray pattern. Like an upright sprinkler, a pendent sprinkler throws all of its water down and no water up toward the roof structure.

3.1.3 Sidewall Sprinklers

A sidewall sprinkler (*Figure 8*) is most commonly located along a wall or soffit. There are horizontal sidewall and vertical sidewall sprinklers. Horizontal sidewall sprinklers are pointed straight out from a wall and designed to throw water across a room and along the wall from which they extend. Vertical sidewall sprinklers are located in front of the wall and can be either upright or pendent types. Vertical sidewall sprinklers throw water across the room and along the wall in the same way as horizontal sidewall sprinklers. Sidewall sprinklers are used where sprinklers cannot be installed in the ceiling.

3.1.4 Recessed and Concealed Sprinklers

Recessed and concealed sprinklers are available in custom colors and are used to improve the appearance of the system in residential and commercial applications (*Figure 9*). Recessed sprinklers are partially enclosed within a housing, while concealed sprinklers completely hide the sprinkler using a cover plate. The cover plate is designed to fall away at a lower temperature than the sprinkler rating. When it reaches its rated temperature, the sprinkler will drop below the ceiling. Both types spray water in a circular pattern.

3.1.5 Conventional Sprinklers

Conventional sprinklers (*Figure 10*) are generally called *old-style sprinklers*. They were used extensively until the advent of the spray sprinkler used today. Conventional sprinklers are very simple and can be used as either uprights or pendents. They are designed to discharge approximately 40 percent of their water in the upward direction against the roof or ceiling. The other 60 percent of the spray is directed downward. They are still used in some European countries and in specific structures such as underneath piers, wharfs, and in fur storage vaults, per NFPA standards.

Figure 7 Pendent sprinkler.

Conventional sprinklers can also be used as replacement sprinklers in systems that were designed and installed with the old-style sprinklers, but they cannot be used in a new system or to replace standard spray sprinklers.

3.2.0 Sprinkler Characteristics

NFPA 13 defines the following characteristics of sprinklers that describe their ability to control or extinguish a fire:

- Thermal sensitivity
- K-factors
- Temperature ratings
- Water distribution characteristics
- Installation orientation
- Special service conditions

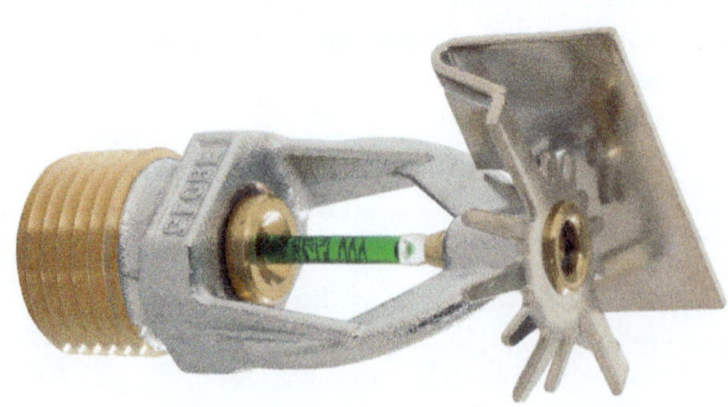

Figure 8 Sidewall sprinkler.

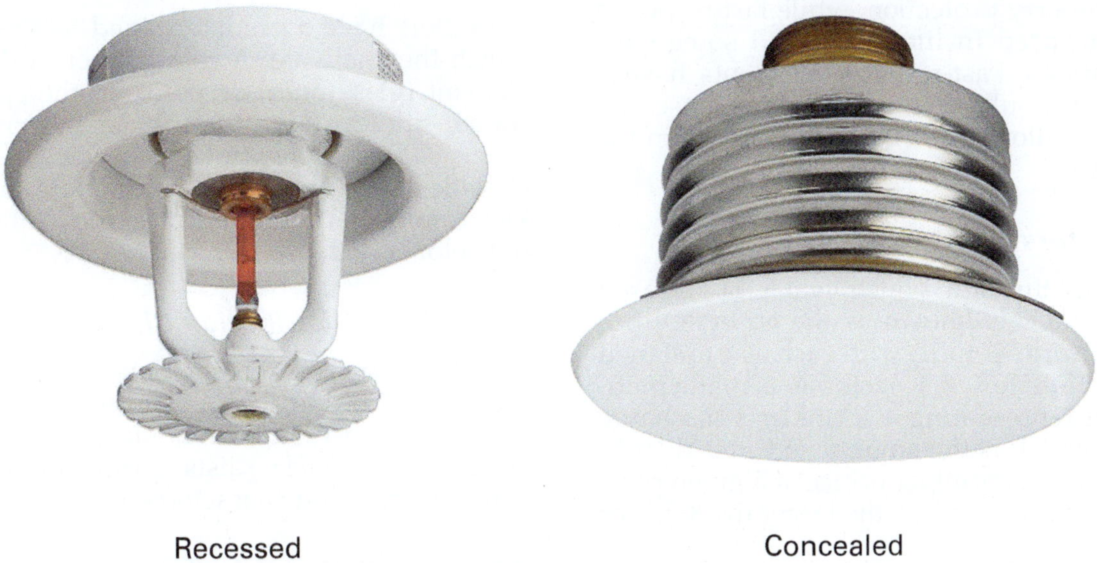

Figure 9 Recessed and concealed sprinklers.

Figure 10 Conventional sprinkler.

3.2.1 Thermal Sensitivity

Thermal sensitivity measures how quickly a sprinkler's thermal element operates. It is stated in terms of a sprinkler element's Response Time Index (RTI), as measured in (meter-seconds)$^{1/2}$. NFPA 13 recognizes two thermal sensitivity categories: fast-response elements with an RTI of 50 (meter-seconds)$^{1/2}$ or less, and standard-response elements with an RTI of 80 (meter-seconds)$^{1/2}$ or more.

Fusible Link Sprinklers

A fusible link sprinkler contains a two-part thermal element connected by a heat-sensitive alloy. When ambient conditions reach the rated temperature, the alloy melts and the thermal element components separate, causing the seat to open and activate the sprinkler.

Standard-response elements are primarily used for property protection, while fast-response elements are used in life safety and some storage applications. Fast-response elements have a narrower glass bulb than standard-response elements, which allows for a quicker response in the event of a fire.

3.2.2 K-Factors

The opening through which water is released from a sprinkler is known as the *orifice*. Orifice sizes range from ¼" to 1", with each size assigned a different **K-factor**. A K-factor is a mathematical constant representing a sprinkler's discharge coefficient, which is the amount of water distributed through the sprinkler orifice at a given pressure. The larger the orifice, the larger the K-factor (see *Table 1*).

As the K-factor increases, the rated discharge from the sprinkler also increases, with K-5.6 set as the 100 percent nominal discharge. Therefore, a sprinkler rated at a K-factor of 1.4 produces much less water (25 percent of nominal discharge) than a sprinkler rated at a K-factor of 11.2 (200 percent of nominal discharge).

Did You Know?

A Listed strainer must be provided on the supply side of sprinklers with nominal K-factors of less than 2.8.

3.2.3 Temperature Ratings

Sprinklers have a predetermined temperature at which they activate. A color, either in the fusing element (glass bulb) or paint on the sprinkler's frame, indicates the sprinkler's temperature range. *Table 2* lists sprinkler temperature ratings, classifications, and color codes. Note that sprinkler frames with painted or chrome finishes are not color coded.

3.2.4 Water Distribution Characteristics

Sprinklers can also be characterized by the way they distribute water, including the coverage area, speed of distribution, size of the droplets, and direction. *Table 3* lists sprinklers defined by water distribution characteristics.

3.2.5 Installation Orientation

Sprinklers can also be characterized by the way they are designed to be installed. *Table 4* lists sprinklers defined by installation orientation.

3.2.6 Special Service Conditions

Table 5 lists sprinklers defined by special service conditions. These devices are intended for use in specific environments. They are either covered with a special coating or designed for a specific function. Only the sprinkler manufacturer can apply coatings. For additional information on sprinkler types, refer to the NFPA 13 sprinkler definitions.

Table 1 Sprinkler Discharge Characteristics Identification (Data from NFPA 13, *Table 7.2.2.1*)

Nominal K-Factor [gpm/(psi)$^{1/2}$]	K-Factor Range [gpm/(psi)$^{1/2}$]	Percent of Nominal K-5.6 Discharge	Thread Type
1.4	1.3–1.5	25	½" NPT
1.9	1.8–2.0	33.3	½" NPT
2.8	2.6–2.9	50	½" NPT
4.2	4.0–4.4	75	½" NPT
5.6	5.3–5.8	100	½" NPT
8.0	7.4–8.2	140	¾" NPT or ½" NPT
11.2	10.7–11.7	200	½" NPT or ¾" NPT
14.0	13.5–14.5	250	¾" NPT
16.8	16.0–17.6	300	¾" NPT
19.6	18.6–20.6	350	1" NPT
22.4	21.3–23.5	400	1" NPT
25.2	23.9–26.5	450	1" NPT
28.0	26.6–29.4	500	1" NPT

Note: The nominal K-factors for dry-type sprinklers are used for sprinkler selection. See 27.2.4.10.3 for use of adjusted dry-type sprinkler K-factors for hydraulic calculation purposes.

Table 2 Sprinkler Temperature Ratings, Classifications, and Color Codings (Data from NFPA 13, *Table 7.2.4.1*)

Max. Ceiling Temperature (°F)	Temperature Rating(°F)	Temperature Classification	Frame Color Code	Glass Bulb Colors
100	135–170	Ordinary	Uncolored or black	Orange or red
150	175–225	Intermediate	White	Yellow or green
225	250–300	High	Blue	Blue
300	325–375	Extra high	Red	Purple
375	400–475	Very extra high	Green	Black
475	500–575	Ultra high	Orange	Black
625	650	Ultra high	Orange	Black

Table 3 Water Distribution Characteristics

Sprinkler Type	Description
Control Mode Specific Application (CMSA) Sprinkler	Specific-application control mode sprinkler capable of producing large water droplets; can provide fire control of specific high-challenge fire hazards.
Control Mode Density/Area (CMDA) Sprinkler	Spray sprinkler intended to provide fire control using density/area criteria.
Early Suppression Fast-Response (ESFR) Sprinkler	Has a thermal element with an RTI of 50 (meter-seconds)$^{1/2}$ or less. Provides fast-response fire suppression of specific high-challenge fire hazards.
Extended Coverage Sprinkler	Spray sprinkler with maximum coverage areas as specified in NFPA 13 and the individual Listing.
Nozzles	Used in applications requiring special water discharge patterns, spray direction, or other unusual discharge patterns.
Old-Style Conventional Sprinkler	Directs 40 percent to 60 percent of total water initially downward. Designed to be installed with deflector either upright or pendent.
Open Sprinkler	Does not have actuators or heat-responsive elements.
Quick-Response Extended Coverage (QREC) Sprinkler	Listed as having a thermal element with an RTI of 50 (meter-seconds)$^{1/2}$ or less and complies with extended protection areas as defined in NFPA 13, *Chapter 11*.
Quick-Response (QR) Sprinkler	Listed as a quick-response sprinkler. Has a thermal element with an RTI of 50 (meter-seconds)$^{1/2}$ or less.
Residential Sprinkler	Fast-response type; has a thermal element with an RTI of 50 (meter-seconds)$^{1/2}$ or less. Investigated for its ability to enhance survivability in the room of fire origin. Listed for protection of dwelling units.
Special Sprinkler	Tested and prescribed for use by its individual Listing.
Spray Sprinkler	Listed for its capability to provide fire control for a wide range of fire hazards. Maximum coverage areas specified in individual Listing.
Standard Spray Sprinkler	A spray sprinkler with maximum coverage area as specified in *Sections 10.2 and 10.3* of NFPA 13.

Did You Know?

George Manby invented the portable fire extinguisher in the early 1800s. His idea was that a small amount of water applied early and in the right place would accomplish what even large amounts of water could not once the fire had spread. The idea of putting out a fire in the earliest stages led to the invention of the automatic fire sprinkler in 1864 by Henry S. Parmalee, who needed to protect his piano factory.

Table 4 Sprinklers Defined by Installation Orientation

Sprinkler Type	Description
Concealed Sprinkler	A recessed sprinkler with a cover plate.
Flush Sprinkler	All or part of the body, including the shank thread, is mounted above the lower plane of the ceiling.
Pendent Sprinkler	Designed to be installed so that the water stream is directed downward against the deflector.
Recessed Sprinkler	All or part of the body, other than the shank thread, is mounted within the recessed housing.
Sidewall Sprinkler	Has special deflectors that are designed to discharge most of the water away from the nearby wall in a pattern equal to roughly half of a standard circular spray pattern, with a small portion of the discharge directed at the wall behind the sprinkler.
Upright Sprinkler	Designed to be installed so the water spray is directed upward against the deflector.

Table 5 Sprinklers Defined by Special Service Conditions

Sprinkler Type	Description
Corrosion-Resistant Sprinkler	Fabricated with corrosion-resistant material, or with special coatings or platings, to be used in a corrosive atmosphere.
Dry Sprinkler	Secured in an extension nipple with a seal at the inlet end that prevents water from entering the nipple until the sprinkler operates.
Intermediate Level Sprinkler/Rack Storage Sprinkler	Equipped with integral shields to protect its operating elements from the discharge of sprinklers installed at higher elevations.
Ornamental/Decorative Sprinkler	Has been painted or plated by the manufacturer.
Institutional Sprinkler	A sprinkler specifically designed for resistance to load-bearing purposes and with components not readily converted for use as weapons.
Pilot Line Detector	Standard spray sprinkler or thermostatic fixed-temperature release device used as a detector to release the main valve pneumatically or hydraulically, controlling the flow of water into a fire protection system.

3.0.0 Section Review

1. A sprinkler that directs 40 percent of its water flow upward and 60 percent downward is a(n) _____.
   a. pendent sprinkler
   b. conventional sprinkler
   c. recessed sprinkler
   d. upright sprinkler

2. A sprinkler with a temperature rating of 150°F will have a glass bulb color that is _____.
   a. black
   b. purple
   c. yellow or green
   d. orange or red

SECTION FOUR

4.0.0 PIPING, VALVES, AND OTHER SYSTEM COMPONENTS

Objective

Identify different types of piping, valves, and fire pumps.

a. Describe the installation requirements for various piping systems.
b. Identify sprinkler system valves and waterflow switches.
c. Describe the purpose of fire pumps.

Trade Terms

C-factor: Refers to the roughness of the inside of the sprinkler pipe. The higher the C-factor, the smoother the pipe.

Check valve: A valve that allows water to flow in only one direction.

Control valve: Controls the water supply to a sprinkler system.

Electrolysis: The decomposition of a substance by the passage of electricity through it.

Galvanized: Dipped in zinc during the manufacturing process.

Inside diameter (ID): The distance between the inner walls of a pipe. Used as the standard measure for tubing used in heating and plumbing applications.

Outside diameter (OD): The distance between the outer walls of a pipe. Used as the standard measure for air conditioning and refrigeration (ACR) tubing.

Sway bracing: Bracing of sprinkler pipe to minimize pipe movement during seismic activity.

Test and drain valve: A type of valve that includes both a test function and a drain function. The test function is used to simulate sprinkler activation to verify the operation of the waterflow alarm. The drain function can be used to drain all or part of a sprinkler system.

Water hammer: Shock wave in piping caused by a sudden change in water flow.

Sprinklers are connected to a system of piping, valves, pumps, and control panels.

4.1.0 Piping Systems

Piping systems in sprinkler systems can be divided into two broad categories: aboveground pipe and tubing, and underground pipe. Each has specific Listing requirements and application considerations.

When pipe sizes are determined for sprinkler systems, one of the things that must be determined is the **C-factor** of the pipe. The C-factor refers to the roughness of the inside of the pipe. The higher the C-factor, the smoother the pipe and the more easily the water flows. This translates into smaller pipe sizes, which in turn saves money. Typically, black steel has a C-factor of 120; copper and chlorinated polyvinyl chloride (CPVC) have a C-factor of 150.

4.1.1 Aboveground Pipe and Tube

The most common type of aboveground pipe used in sprinkler systems is black steel. Stainless steel, copper tubing, and CPVC piping may also be used. Steel pipe and copper tubing do not have to be Listed if they meet the applicable American Society for Testing and Materials International (ASTM) standards. Otherwise, they must be Listed. All plastic pipe must be Listed for use in sprinkler systems.

Steel Pipe

The most common form of steel pipe used in sprinkler systems is black steel. Black steel can also be **galvanized** (dipped in zinc during the manufacturing process) for corrosion resistance. Stainless steel pipe is rarely used because it is costly and difficult to thread. Steel pipe can be joined by various means, including the following:

- *Threading* — For use with threaded fittings.
- *Grooving* — For use with grooved fittings.
- *Drilling* — For use with mechanical tees or welded outlets.
- *Welding* — Either end-to-end or with welded, grooved, or threaded outlets.

Steel pipe is manufactured in many sizes. The most common sizes in the industry range from $\frac{1}{2}$" to 10". Steel pipe is manufactured with different wall thicknesses known as *schedules*. A given pipe size may be manufactured in different schedules. For example, a 2" pipe may be manufactured as Schedule 40 or even Schedule 10. With pipe, the **outside diameter (OD)** of a given pipe size is the same no matter what the schedule is. Thus, the thinner the wall, the larger the **inside diameter (ID)** of the pipe.

Copper Tube

Like steel pipe, copper tube is manufactured in many sizes. The most common sizes in the sprinkler industry range from $\frac{3}{4}$" to 2". Copper tube is not used as extensively as steel pipe.

Copper tube differs from steel pipe in that it is not manufactured in schedules. Rather, it is identified by types. The type refers to its hardness or flexibility. The three types used in sprinkler systems are K, L, and M. Types K, L, and M copper tube are allowed by NFPA 13 for use in sprinkler systems. Types K and L are flexible and come in straight lengths and rolls. Type K is normally used in commercial plumbing systems and Type L is normally used in residential plumbing systems. Type M is hard-drawn copper and is available in straight sections only. Hard-drawn copper will kink before it will bend, and care must be taken to maintain the quality of the tube during installation. The cost of copper tube is directly related to wall thickness and the thinnest wall permitted is normally used. The actual internal diameter of copper tube is very close to the nominal tube size. Copper specifications are often referred to as *CTS*, or *copper tube size*, as differentiated from *IPS*, which originally stood for *iron pipe size*.

There are other types of copper tube available, such as drain, waste, and vent (DWV) and air conditioning and refrigeration (ACR). These are not permitted by NFPA 13 for use in sprinkler systems. Common methods of joining copper tube are soldering, compression, and brazing.

Chlorinated Polyvinyl Chloride (CPVC) Pipe

CPVC pipe must be Listed for aboveground sprinkler applications. It is typically identified by its bright orange color. The most common industry sizes of CPVC pipe are $\frac{3}{4}$" to 3". CPVC pipe can be joined by the solvent-cement method using CPVC fittings or with mechanical couplings.

Fittings

Fittings used with steel pipe or copper tube do not have to be Listed but must meet or exceed other nationally recognized standards. Fittings used with steel pipe in aboveground systems must be manufactured of cast, malleable, or ductile iron. Fittings used with CPVC must be Listed. *Figure 11* shows examples of various pipe fittings.

4.1.2 Underground Pipe

Underground piping must comply with the standards in NFPA 24 or NFPA 13, as applicable. Pipe must be Listed or comply with American Water Works Association (AWWA) or ASTM standards.

Piping Materials

Typical underground pipe materials are high-density polyethylene (HDPE), polyvinyl chloride (PVC), chlorinated polyvinyl chloride (CPVC), ductile iron, and copper. Steel pipe is not typically used for underground service. When it is, such as between the **check valve** and the Fire Department Connection (FDC), it must be externally coated and wrapped and internally galvanized. Pipe must be designed to a working pressure of not less than 150 psi.

The type and class of pipe for a particular underground installation are determined by the following factors:

- Maximum system working pressure
- Depth at which the pipe is to be installed
- Soil conditions
- Corrosion
- Susceptibility of the pipe to other external loads, including earth loads, installation beneath buildings, and traffic or vehicle loads

Fittings

Fittings for underground pipe must meet the standards of NFPA 24 or NFPA 13, as applicable. Listed fittings are permitted for the system pressures as specified in their Listing, but not less than 150 psi.

Depth of Cover

The depth of cover depends on the maximum depth of frost penetration. The top of the pipe must be buried at least 1' below the frost line for the locality. Where frost is not a factor, the depth of cover should not be less than $2\frac{1}{2}$' to prevent mechanical damage. Where pipe is installed below driveways and railroad tracks, the minimum depth to bury the pipe is 3' and 4', respectively.

Straight Threaded Tee

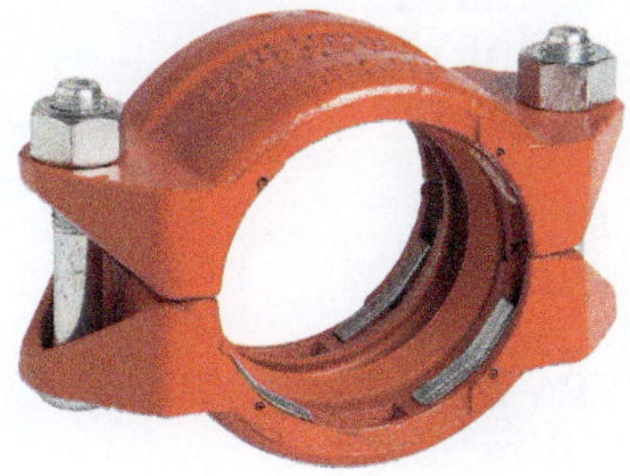

Plain End Coupling with Grippers

Grooved Tee with Threaded Outlet

Threaded Mechanical Tee

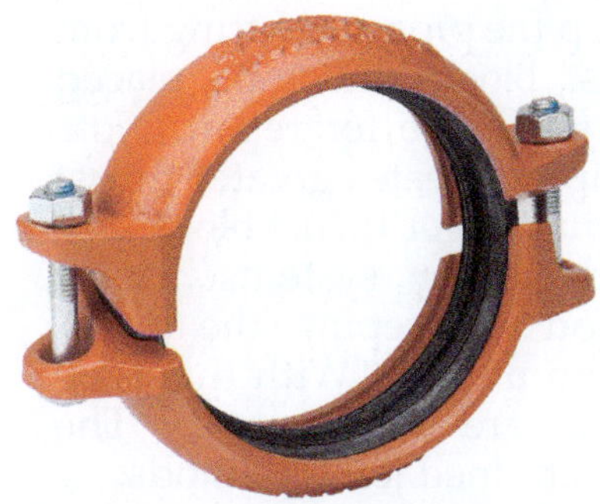

Rigid Coupling

Figure 11 Pipe fittings.

> **NOTE**
>
> Where pipe cannot be buried, it can be laid aboveground, provided that the pipe is protected against freezing and mechanical damage.

Protection Against Damage

Ideally, pipe should not be run under buildings. However, where pipe must be run under buildings, special precautions must be taken, including the following:

- Arching the foundation walls over the pipe
- Running pipe in covered trenches
- Providing valves to isolate sections of pipe under buildings

Fire service mains may enter the building adjacent to the foundation. Where adjacent structures or physical conditions make it impractical to locate risers immediately inside an exterior wall, such risers can be located as close as practical to exterior walls to minimize underground piping under the building.

Where a riser is located close to building foundations, underground fittings of the proper design and type must be used to avoid locating pipe joints in or under the foundations. To prevent **electrolysis** when joining pipes of dissimilar metal, the joint must be insulated against the passage of an electric current using an approved method.

Requirement for Laying Pipe

Care must be taken to not drop or damage pipe, valves, and fittings. Small cracks and fractures can occur that may go unnoticed until the pipe has been installed and the pressure checked.

Bolted joints must be properly torqued. Pipes, valves, hydrants, and fittings must be clean inside and at their mating surfaces. Care must be taken to prevent stones and foreign materials from entering pipes, valves, and fittings.

Carefully lower all pipe, fittings, valves, and hydrants into the trench using appropriate equipment. Carefully examine each component for cracks or other defects while the pipe is suspended over the trench.

Joint Restraint

Large poured-in-place concrete blocks, called *thrust blocks*, are used to restrain movement of underground piping systems using push-on or mechanical-joint fittings.

Underground piping systems that use fused, threaded, grooved, or welded joints should not require additional restraining, provided that such joints can pass the hydrostatic test of NFPA 24 or NFPA 13, as applicable, without the shifting of piping or leakage in excess of permitted amounts.

Thrust blocks are limited to use where the soil is suitable. Thrust blocks are installed between undisturbed earth and the fitting to be restrained and must be able to keep the pipe and fitting from moving. Ideally, thrust blocks must be placed so that the joints are accessible for repair. Note that larger diameter piping creates greater thrust under pressure, requiring larger thrust blocks.

Some underground piping systems use a restrained joint method of keeping the pipes, valves, and fittings from moving. With this type of system, thrust blocks are not necessary. The following are typical restrained joint methods:

- Locking mechanical or push-on joints
- Mechanical joints using setscrew retainer glands
- Bolted flange joints
- Heat-fused or welded joints
- Pipe clamps and tie rods
- Other approved methods or devices

4.1.3 Hangers, Bracing, and Restraints

Because of **water hammer** (a shock wave in the piping caused by a sudden change in water flow) and vibration, sprinkler systems can move vertically and horizontally. Hangers, bracing, and restraints are installed at specified intervals to control this movement.

Hangers, bracing, and restraints are chosen by sprinkler contractors and installed by sprinkler fitters according to sprinkler system shop drawings. Correctly installed hangers, bracing, and restraints allow pipes to drain, expand, contract, and vibrate without damaging the sprinkler system. The pieces of a hanger assembly that attach directly to the pipe or the building structure must be Listed. The other components do not have to be Listed.

Hangers (*Figure 12*) attach sprinkler piping to the building structure. The type of hanger is determined by the size of the pipe and the material of the structure (concrete, steel, or wood).

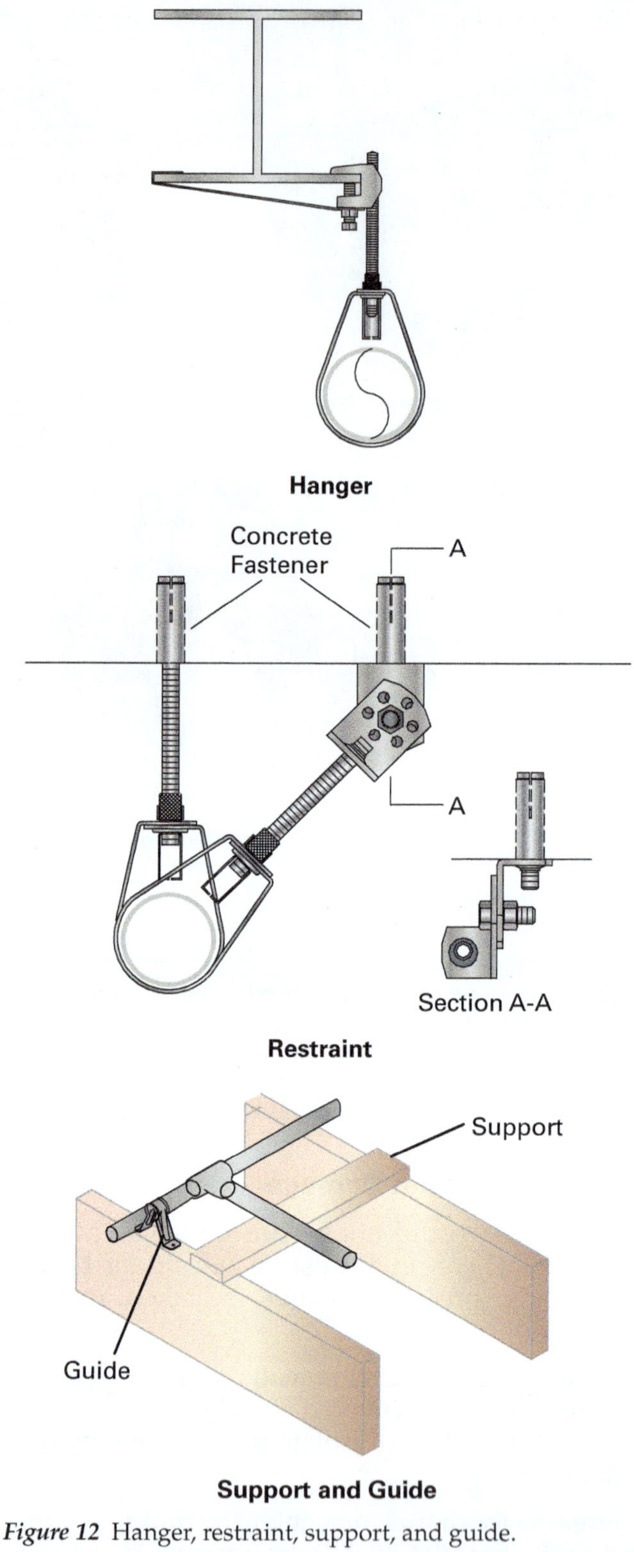

Figure 12 Hanger, restraint, support, and guide.

Hangers must be capable of supporting five times the weight of the water-filled pipe plus 250 pounds at each point of piping support. When it is determined that a single point of connection cannot meet this requirement, a trapeze hanger (*Figure 13*) must be used.

A trapeze hanger is designed to distribute the load of water-filled pipe over at least two points of the structure. Make sure the pipe hangers used on a trapeze unit can support the pipe load in accordance with NFPA 13 requirements. Typical hanger assembly components are shown in *Figure 14*.

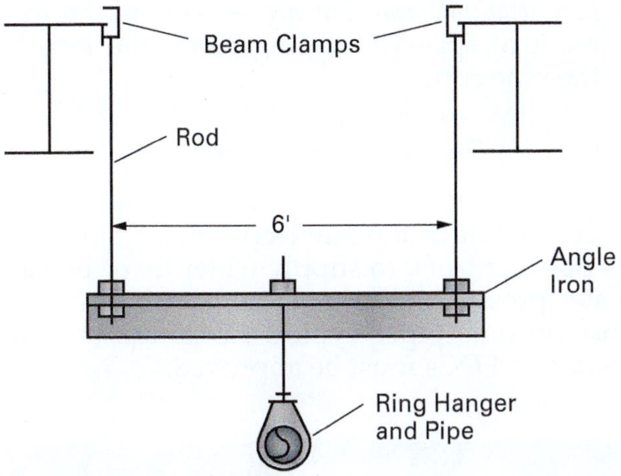

Figure 13 Trapeze hanger.

Certain areas of the country require that sprinkler systems be designed and installed to provide protection against seismic activity or earthquakes. During seismic activity, piping systems tend to move in different directions from the structure to which they are attached; therefore, flexibility of the pipe and clearance from the building structure must be considered. This is accomplished using flexible grooved fittings and proper clearance around piping where it passes through walls and floors (*Figure 15*). The four-way brace should be attached above the upper flexible coupling required for the riser, and preferably to the roof structure, if the structure is suitable. The brace should not be attached directly to a plywood or metal deck. As with hangers, attachment points to the structure are very important. Always follow the shop drawings.

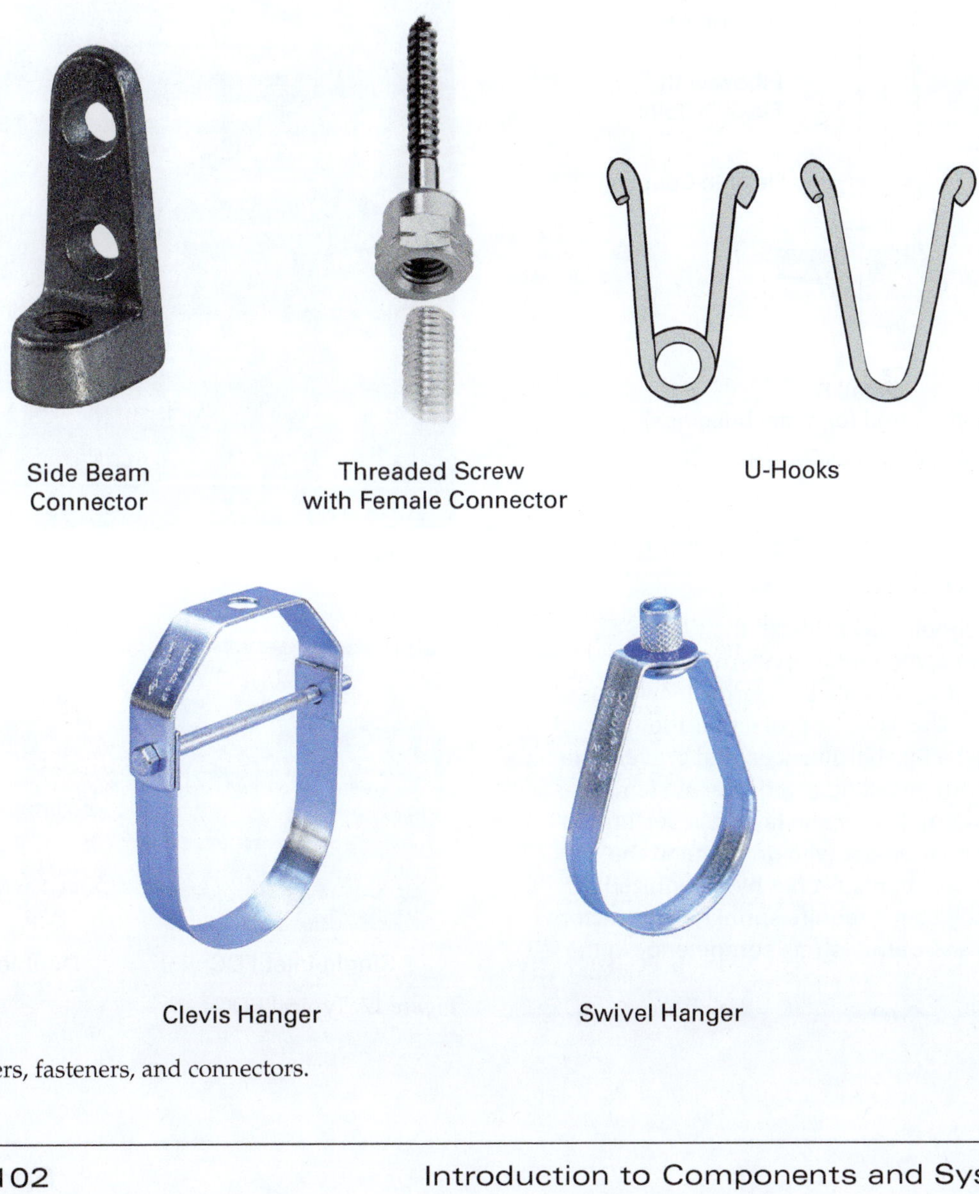

Figure 14 Hangers, fasteners, and connectors.

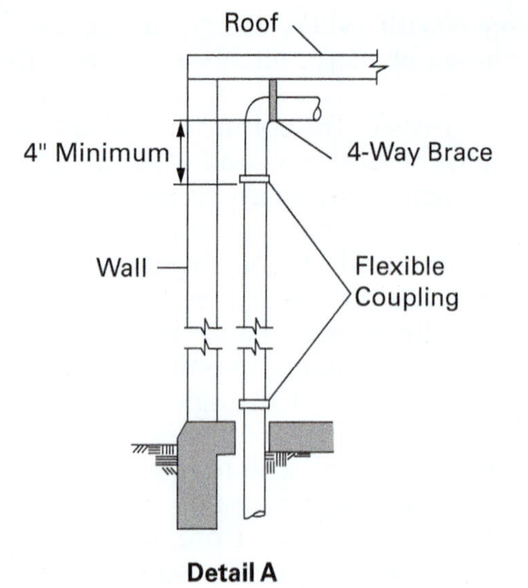

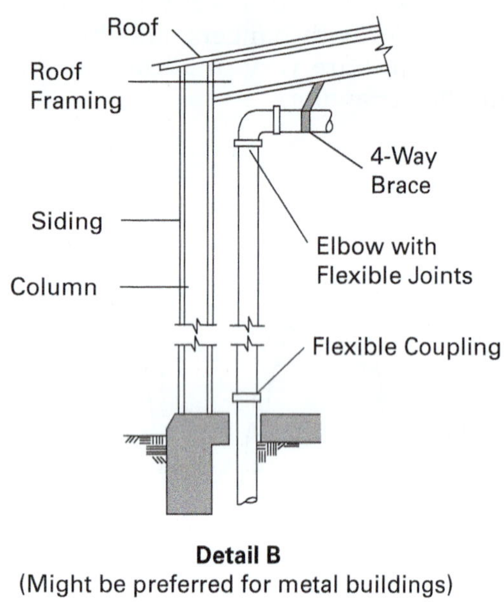

Figure 15 Pipe bracing and clearance details.

Did You Know?

Homeowners should never install their own sprinkler systems. A sprinkler system must be installed to meet applicable standards, local codes, and ordinances. This is not a project for the novice, and there may be local ordinances that prevent the homeowner from installing a sprinkler system.

Sprinkler system design and layout is performed by specialized contractors who understand the installation requirements set out by recognized standards. Many states require sprinkler contractors to be licensed and demonstrate competency in the trade.

Sway bracing (*Figure 16*) of sprinkler pipe minimizes movement during seismic activity. Currently, there are two types of sway bracing:

- *Lateral sway bracing* — Resists the movement of sprinkler piping from side to side.
- *Longitudinal sway bracing* — Resists the movement of sprinkler piping along the length of the pipe run.

4.1.4 Fire Department Connection

A Fire Department Connection, or FDC, is an inlet into which the Fire Department can connect its pumper truck to supply water to, or boost the water pressure of, an operating sprinkler system and/or standpipe. Typical FDCs are shown in *Figure 17*. FDCs must be approved.

Figure 16 Sway bracing.

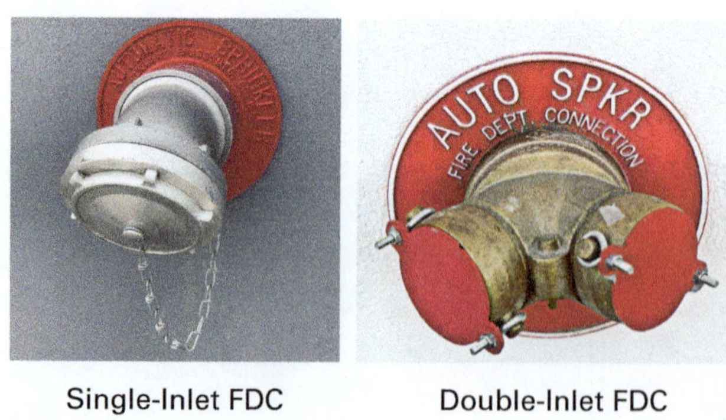

Figure 17 Typical FDCs.

4.2.0 Valves and Waterflow Switches

Valves are devices used to control the flow of water, gas, or other fluid through a pipe. Common sprinkler valves include control valves, check valves, and test and drain valves. Like a valve, a waterflow switch is connected inline with the system piping. Unlike a valve, it is used to monitor rather than control the flow through the system.

4.2.1 Control Valves

A **control valve** controls the water supply to a sprinkler system. Control valves must be Listed and require a visual indicator that shows whether the valve is open or closed. Typical control valves include gate valves, post indicator valves (PIV), and butterfly valves (*Figure 18*). A PIV is basically a post over a stemless gate valve with a window that indicates whether the valve is open or closed.

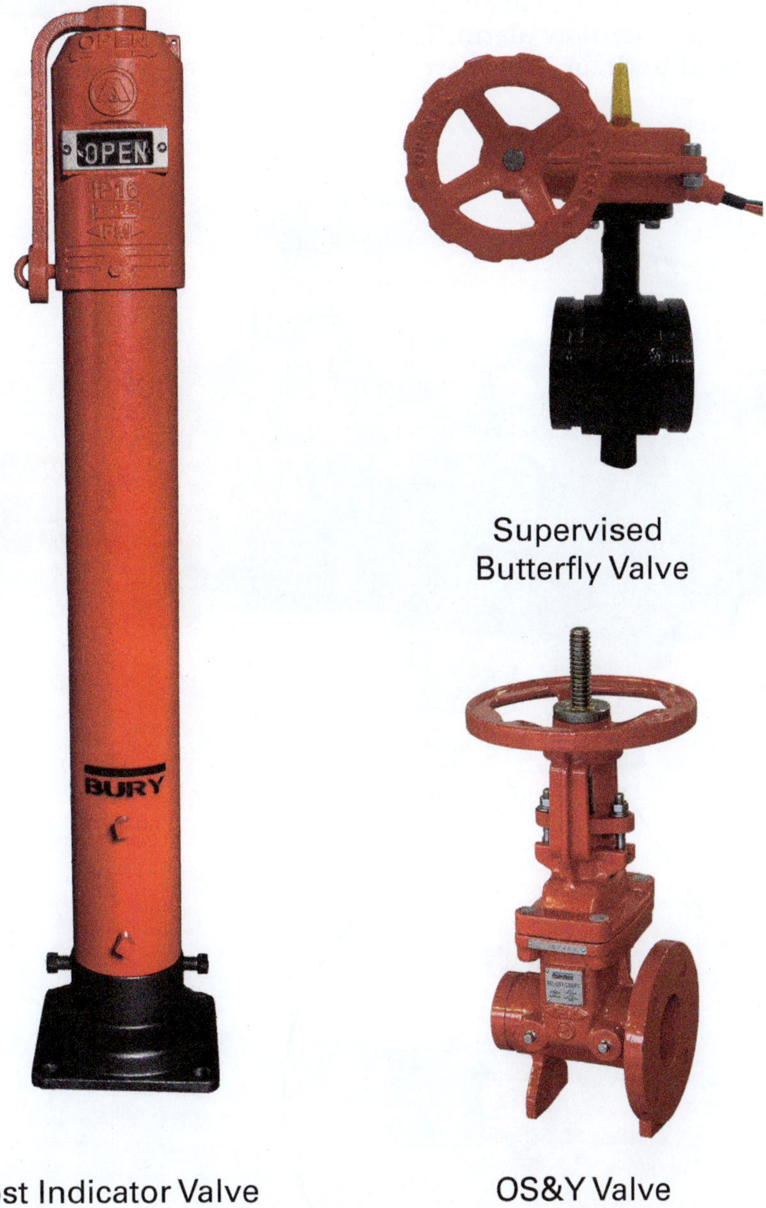

Figure 18 Control valves.

4.2.2 Check Valves

Check valves (*Figure 19*) allow water to flow into the sprinkler system, but not back out. Check valves must be Listed. Check valves must be installed in the correct direction and orientation. Some check valves are marked with an arrow to indicate the direction of water flow.

4.2.3 Test and Drain Valves

A **test and drain valve** (*Figure 20*) includes both a test function and a drain function. The test function is used to simulate sprinkler activation to verify the operation of the waterflow alarm. The drain function can be used to drain all or part of a sprinkler system. Test and drain valves are usually 1", $1\frac{1}{4}$", or 2" in size. They must be approved but need not be Listed.

Figure 19 Flow through a check valve.

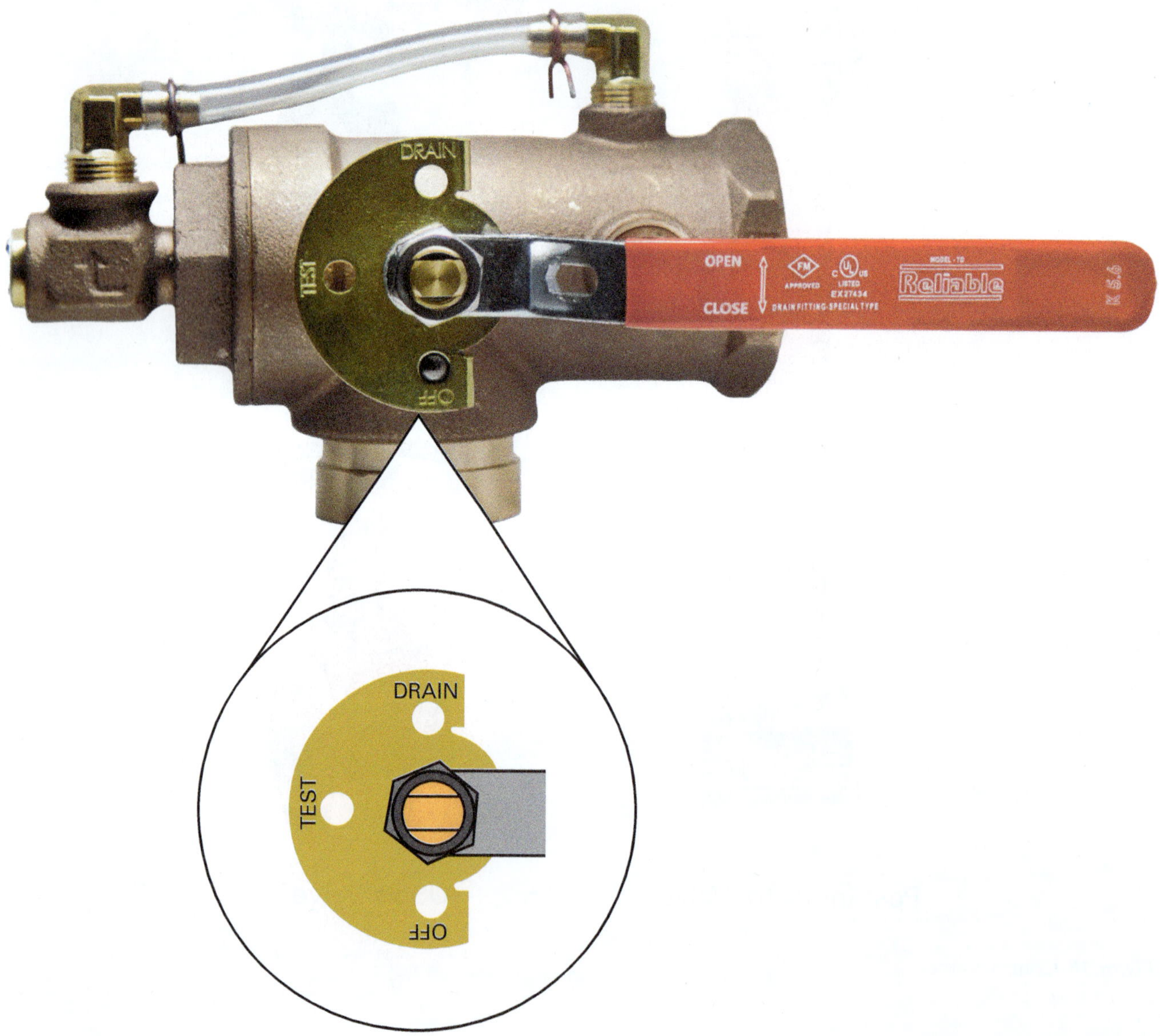

Figure 20 Test and drain valve.

4.2.4 Waterflow Switches

A waterflow switch (*Figure 21*) is designed to send an alarm signal to a local alarm bell, fire alarm panel, and/or the Fire Department when an activated sprinkler causes continuous water flow through the system piping. Waterflow switches must be Listed. They may be electrically operated, such as a flow switch or pressure switch, or mechanically operated upon the flow of water into the system, such as a water motor gong. If water flow greater than the specified rate (e.g., 10 gallons per minute) is detected, the vane or paddle triggers a switch that sends a signal to an alarm panel and bell (*Figure 22*). Sprinkler systems are required to have an audible alarm on the premises.

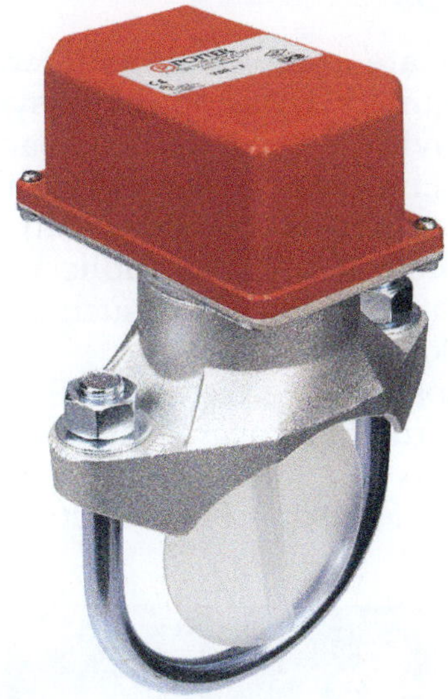

Figure 21 Waterflow switch.

Commercial Fire Alarm System Control Panel

Residential Fire Pump Control Panel

Figure 22 Control panels.

4.3.0 Pumps

Fire pumps are used to provide water at the pressure required by the sprinkler system. Some systems provide a secondary or backup pump (*Figure 23*). Fire pumps can be powered by electric, diesel, or steam (less common). The pump intake is connected to the public water supply and/or a static water source, such as a tank, reservoir, or lake.

Fire pumps must be Listed for fire service. The main standard that governs fire pump installations is NFPA 20, *Standard for the Installation of Stationary Fire Pumps for Fire Protection*.

System Monitoring

The flow of water in a sprinkler system is typically shut off for maintenance and testing, and it is essential that it be turned back on when the work has been completed. Failure to do so will result in the loss of sprinkler protection in the event of a fire. For this reason, NFPA 13 requires monitoring of any control valves directing the flow of water to most types of sprinkler systems.

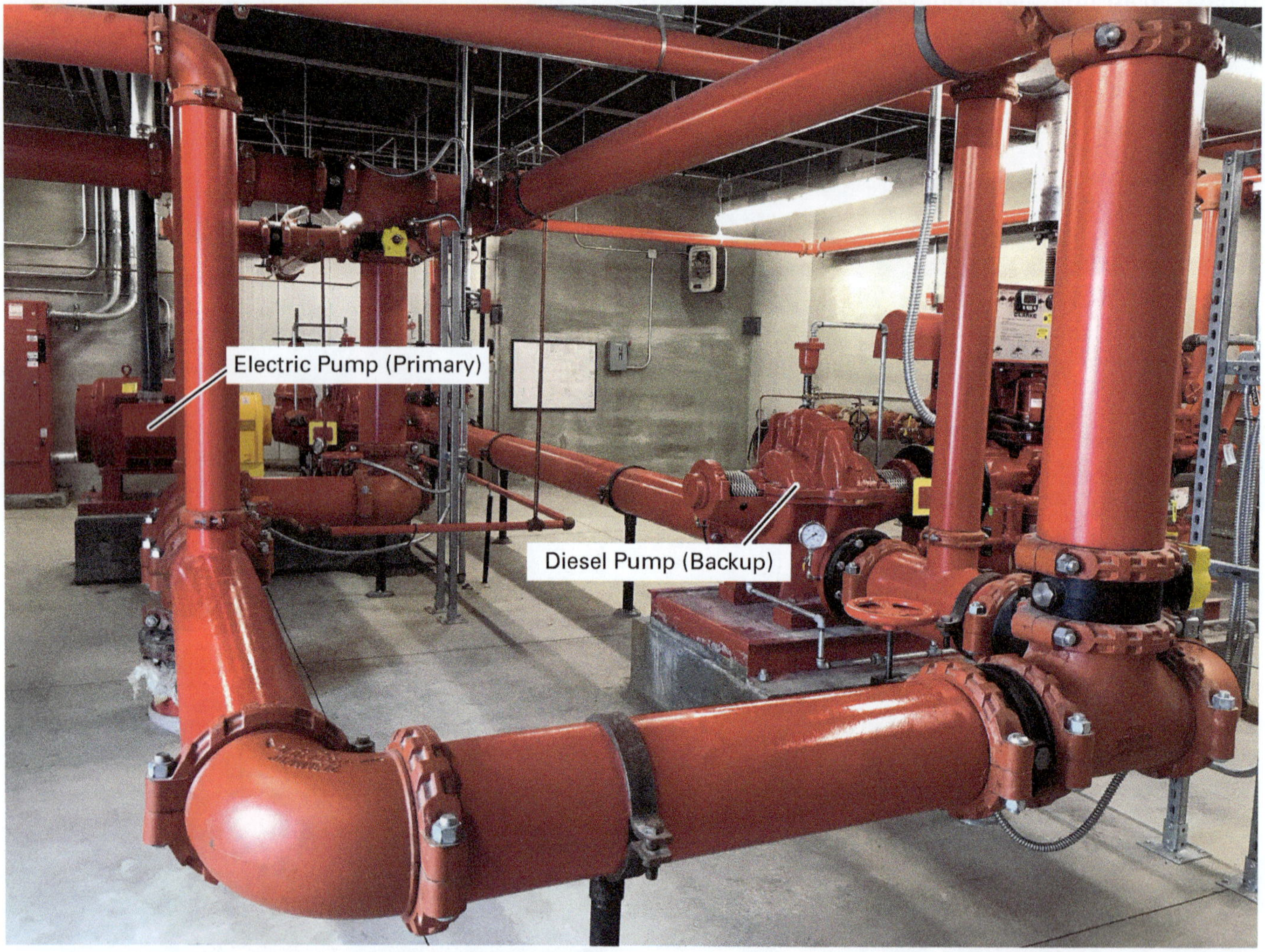

Figure 23 Fire pumps.

Preassembled Manifolds

Sprinkler system valves are also available as part of preassembled manifolds to save labor during system assembly. The commercial riser system manifold shown here is a factory assembled unit that includes a waterflow alarm switch, pressure gauge, main drain, and pressure relief kit.

Figure Credit: Reliable Automatic Sprinkler Co., Inc.

Did You Know?

You can often tell whether a valve is open or closed by the position of the valve handle. For valves with straight handles, if the valve handle is parallel to or in line with the pipe, the valve is open. If it is perpendicular, the valve is closed. For an outside screw and yoke (OS&Y) valve, if the valve stem is out, it is open. If it is in, it is closed.

Ball Valve (Closed)

OS&Y Valve (Open)

4.0.0 Section Review

1. A shock wave in sprinkler system piping caused by a sudden change in water flow is known as _____.
   a. subsidence
   b. water hammer
   c. seismic activity
   d. thermal expansion

2. The device used to signal an alarm when a sprinkler is activated is a _____.
   a. temperature sensor
   b. control valve
   c. check valve
   d. waterflow switch

3. The main code that governs fire pump installations is _____.
   a. NFPA 1
   b. NFPA 13
   c. NFPA 20
   d. NFPA 24

Review Questions

1. Which of the following is *true* regarding a deluge sprinkler system?
   a. The deluge valve triggers immediate water flow through all the sprinklers at one time.
   b. Deluge systems are often used in museums.
   c. Deluge systems use automatic sprinklers that activate independently.
   d. Deluge systems use normally closed sprinklers.

2. Underwriters Laboratories® (UL) is a type of _____.
   a. sprinkler association
   b. fire protection organization that writes regulations for the NFPA
   c. organization that evaluates products to determine compliance with safety standards
   d. Authority Having Jurisdiction over sprinkler systems

3. The UL product directory contains the names of companies that _____.
   a. donate funds to UL
   b. buy only Listed equipment
   c. are qualified to use a UL Mark on products tested by UL
   d. have the most UL product Listings

4. The frame arms must be parallel to the branch line in a(n) _____.
   a. upright sprinkler
   b. pendent sprinkler
   c. recessed sprinkler
   d. concealed sprinkler

5. The sprinkler shown in *Figure RQ01* is an example of a(n) _____.

Figure RQ01

   a. upright sprinkler
   b. sidewall sprinkler
   c. recessed sprinkler
   d. concealed sprinkler

6. A sprinkler with a nominal K-factor of 5.6 would require a *minimum* thread size of _____.
   a. ¼" NPT
   b. ½" NPT
   c. ¾" NPT
   d. 1" NPT

7. A sprinkler with a temperature rating of 175°F will have a glass bulb color that is _____.
   a. black
   b. purple
   c. yellow or green
   d. orange or red

8. A sprinkler color coded with blue paint means that the sprinkler has a temperature classification of _____.

   a. ordinary
   b. intermediate
   c. high
   d. extra high

9. The roughness of the inside of a pipe is represented by the _____.

   a. K-factor
   b. color code
   c. inside diameter
   d. C-factor

10. Steel piping may be used for general underground service if it is _____.

    a. Listed for that service
    b. stainless
    c. installed in dry soil
    d. used in a residential application

11. The hanger shown in *Figure RQ02* is an example of a _____.

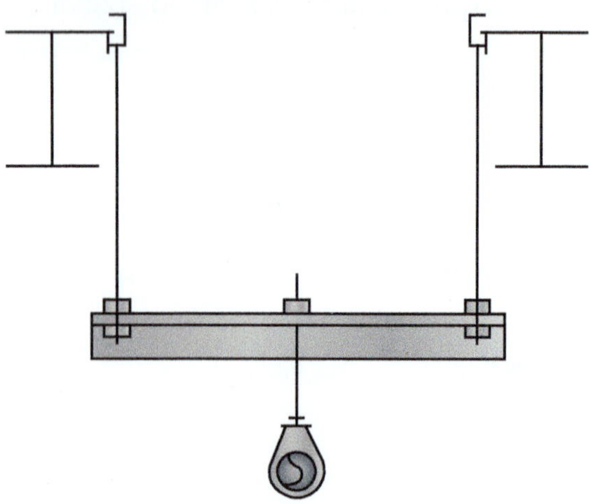

Figure RQ02

    a. trapeze hanger
    b. single point hanger
    c. sway control brace
    d. swing support

12. Techniques used to protect piping in a building during seismic activity include adequate clearance, four-way bracing, and _____.

    a. trapeze hangers
    b. Schedule 10 pipe
    c. thrust blocks
    d. flexible couplings

13. Sway bracing is used primarily to *minimize* pipe _____.

    a. movement due to subsidence
    b. damage due to inferior installation
    c. movement during seismic activity
    d. damage due to freeze/thaw conditions

14. The valve shown in *Figure RQ03* is a type of _____.

Figure RQ03

    a. control valve
    b. check valve
    c. gate valve
    d. test and drain valve

15. Sprinkler systems are required to be equipped with _____.

    a. a flashing warning light
    b. an audible alarm
    c. video monitoring
    d. remote metering

Trade Terms Quiz

Fill in the blank with the correct term learned from your study of this module.

1. ______________________________ and ______________ are Listing agencies that certify new products for service in particular locations and for specific applications.
2. The water supply to a sprinkler system is controlled by a(n) ______________.
3. A component that is tested by a recognized agency and found suitable for use in a sprinkler system for a specific purpose is said to be ______________.
4. A ceiling that is not part of the structural framework of the building is called a(n) ______________.
5. ______________ is a mathematical constant that represents a sprinkler's discharge coefficient.
6. The roughness inside the sprinkler pipe is referred to as the ______________.
7. ______________ causes a shock wave in the pipe and is due to a sudden change in water flow.
8. A product dipped in zinc during the manufacturing process is said to be ______________.
9. Equipment or materials to which has been attached a label, symbol, or other identifying mark of an organization is said to be ______________.
10. A(n) ______________ allows water to flow in one direction only.
11. A(n) ______________ can be used to simulate sprinkler activation to verify the operation of the waterflow alarm.
12. Movement of sprinkler pipe during seismic activity is minimized by ______________.
13. The decomposition of a substance caused by the passage of electricity through it is ______________.
14. The thinner the wall, the larger the ______________ of the pipe.
15. The distance between the outer walls of a pipe is known as its ______________.

Trade Terms

C-factor
Check valve
Control valve
Drop/suspended ceiling
Electrolysis
FM Approvals®
Galvanized
Inside diameter (ID)
K-factor
Labeled
Listed
Outside diameter (OD)
Sway bracing
Test and drain valve
Underwriters Laboratories® (UL)
Water hammer

Cornerstone of Craftsmanship

Chuck Bamford
Senior Engineering Technician
Wolfe Fire Protection, Inc.

How did you choose a career in the industry?

One summer when I was 17, I was asked what I wanted to do for work, and all I could think was no more mowing the lawn! My father had a plumbing mechanical company, and I worked three months each summer for six years installing and fabricating fire sprinkler systems.

Who inspired you to enter the industry?

Mike (Sly) Prussa was my foreman for a few summers. And my father, Jim Bamford, also encouraged me.

What types of training have you been through?

Those first six summers installing and fabricating really got me started in the industry. From there, I took wood shop, metal shop, and other industrial arts courses in my school district. Then, I went to University of Nebraska's Lincoln Construction Management in the College of Engineering. I also took an AFSA course in 1986 at Azusa Pacific.

How important is education and training in construction?

My junior high and high school industrial arts teachers combined with on-the-job training really got me started. I see some trainees come in with GEDs and others who go all the way through college. Just jump in and see if you like the occupation. Then, take as many classes as you can and pursue more education.

How important are NCCER credentials to your career?

I've helped with training over the last 30 years using NCCER and AFSA. The books really help in our state for them to earn that certification. We prepare them as much as we can but passing that exam with those trainings is what really sets them apart.

How has training/construction impacted your life?

It impacts me in every way. In '86 I took a three-week course, and since then I've received so many opportunities in the industry, like earning my NICET Level 4 (layout technician) and serving on committees with the NFPA 13 from 2001 to 2013.

What kinds of work have you done in your career?

I've done everything with fire sprinklers from fabrication, installation, and estimating to design work.

Tell us about your present job.

I'm currently a senior engineering technician. I help wherever I can with estimating, installation, and design.

What do you enjoy most about your job?

I really enjoy when I get to help installers advance in their career. A few guys just passed their tests and it's cool to see them keep moving on. Four years ago, they were just the helper, and now they are the foreman. It's awesome to see that growth.

What factors have contributed most to your success?

I credit that to my family. The support of my wife, sons, mom, and dad has helped me succeed.

Would you suggest construction as a career to others? Why?

Yes, but you should like to use a tape measure. Honestly, you just have to dive in to see if you like it. There are a lot of opportunities if you give it a try.

What advice would you give to those new to the field?

Just keep learning and continuing your education whenever possible. Codes and standards are constantly changing, and you have to be ready to keep up.

Interesting career-related fact or accomplishment:

Well, I started in the summer at 17, and now I've spent almost 40 years in the industry.

How do you define craftsmanship?

Repeat customers. It's all about building those relationships, producing a high quality of work, and finding that balance that keeps them coming back.

Trade Terms Introduced in This Module

C-factor: Refers to the roughness of the inside of the sprinkler pipe. The higher the C-factor, the smoother the pipe.

Check valve: A valve that allows water to flow in only one direction.

Control valve: Controls the water supply to a sprinkler system.

Drop/suspended ceiling: A ceiling that is not part of the structural framework of a building. This ceiling is installed by suspending it from the floor above, or from the roof, and is commonly used to provide space for services such as cables, recessed lighting, and piping.

Electrolysis: The decomposition of a substance by the passage of electricity through it.

FM Approvals®: A testing laboratory and Listing agency that certifies new products for service in particular locations and for specific applications.

Galvanized: Dipped in zinc during the manufacturing process.

Inside diameter (ID): The distance between the inner walls of a pipe. Used as the standard measure for tubing used in heating and plumbing applications.

K-factor: A mathematical constant that represents a sprinkler's discharge coefficient, which is the amount of water distributed through the sprinkler orifice at a given pressure. The larger the orifice, the larger the K-factor.

Labeled: Equipment or materials to which has been attached a label, symbol, or other identifying mark of an organization that is acceptable to the Authority Having Jurisdiction (AHJ).

Listed: Included in a list of products or services published by a recognized Listing agency and acceptable to the AHJ for use in sprinkler systems for a specified purpose.

Outside diameter (OD): The distance between the outer walls of a pipe. Used as the standard measure for air conditioning and refrigeration (ACR) tubing.

Sway bracing: Bracing of sprinkler pipe to minimize pipe movement during seismic activity.

Test and drain valve: A type of valve that includes both a test function and a drain function. The test function is used to simulate sprinkler activation to verify the operation of the waterflow alarm. The drain function can be used to drain all or part of a sprinkler system.

Underwriters Laboratories® (UL): A testing laboratory and Listing agency that certifies new products for service in particular locations and for specific applications.

Water hammer: Shock wave in piping caused by a sudden change in water flow.

Additional Resources

This module presents thorough resources for task training. The following reference material is suggested for further study.

NFPA 13, *Standard for the Installation of Sprinkler Systems*, Latest Edition. Quincy, MA: National Fire Protection Association.

The Pipefitters Blue Book, 2010. Graves. Webster, TX: W.V. Graves Publishing Company.

Figure Credits

Courtesy of Viking Group, Inc., Figures 1–6

Reliable Automatic Sprinkler Co., Inc., Figures 7, 10, 18 (Supervised Butterfly Valve, OS&Y Valve), 20

Courtesy of Victaulic Company, Figures 8, 11 (Plain End Coupling with Grippers, Grooved Tee with Threaded Outlet, Rigid Coupling), 19

Used with permission of Johnson Controls, Figures 9, 11 (Threaded Mechanical Tee), 21

Anvil International & Smith-Cooper International, Figure 11 (Straight Threaded Tee)

Courtesy of ITW Buildex, Figure 14 (Threaded Screw with Female Connector)

Courtesy of nVent Electrical and Fastening Solutions, Figures 14 (Clevis Hanger, Swivel Hanger), 16

iStock@4kodiak, Figure 17 (Double-Inlet FDC)

Courtesy of AMERICAN Flow Control, Figure 18 (Post Indicator Valve)

Reproduced with permission of NFPA from NFPA 13, *Standard for the Installation of Sprinkler Systems*, 2019 edition. Copyright© 2018, National Fire Protection Association. For a full copy of NFPA 13, please go to www.nfpa.org, Tables 1 and 2

Section Review Answer Key

Section 1.0.0

Answer	Section Reference	Objective
1. a	1.1.0	1a
2. b	1.2.0	1b
3. c	1.3.0	1c
4. d	1.4.0	1d

Section 2.0.0

Answer	Section Reference	Objective
1. d	2.1.0	2a
2. a	2.2.2	2b

Section 3.0.0

Answer	Section Reference	Objective
1. b	3.1.5	3a
2. d	3.2.3; *Table 2*	3b

Section 4.0.0

Answer	Section Reference	Objective
1. b	4.1.3	4a
2. d	4.2.4	4b
3. c	4.3.0	4c

NCCER CURRICULA — USER UPDATE

NCCER makes every effort to keep its textbooks up-to-date and free of technical errors. We appreciate your help in this process. If you find an error, a typographical mistake, or an inaccuracy in NCCER's curricula, please fill out this form (or a photocopy), or complete the online form at **www.nccer.org/olf**. Be sure to include the exact module ID number, page number, a detailed description, and your recommended correction. Your input will be brought to the attention of the Authoring Team. Thank you for your assistance.

Instructors – If you have an idea for improving this textbook, or have found that additional materials were necessary to teach this module effectively, please let us know so that we may present your suggestions to the Authoring Team.

NCCER Product Development and Revision
13614 Progress Blvd., Alachua, FL 32615

Email: curriculum@nccer.org
Online: www.nccer.org/olf

❑ Trainee Guide ❑ Lesson Plans ❑ Exam ❑ PowerPoints Other ______________________

Craft / Level: Copyright Date:

Module ID Number / Title:

Section Number(s):

Description:

Recommended Correction:

Your Name:

Address:

Email: Phone:

Steel Pipe and Fittings

Overview

This module describes the steel pipe used in sprinkler systems, including pipe types, schedules, and sizes. It also covers common methods of end preparation, including cutting, reaming, grooving, and threading pipe. Various types of grooved, threaded, flanged, and plain-end fittings are described, along with the installation methods and applications of each type.

Module 18103

Trainees with successful module completions may be eligible for credentialing through the NCCER Registry. To learn more, go to **www.nccer.org** or contact us at 1.888.622.3720. Our website, **www.nccer.org**, has information on the latest product releases and training.

Your feedback is welcome. You may email your comments to **curriculum@nccer.org**, send general comments and inquiries to **info@nccer.org**, or fill in the User Update form at the back of this module.

This information is general in nature and intended for training purposes only. Actual performance of activities described in this manual requires compliance with all applicable operating, service, maintenance, and safety procedures under the direction of qualified personnel. References in this manual to patented or proprietary devices do not constitute a recommendation of their use.

18103 V4.0

18103
Steel Pipe and Fittings

Objectives

Successful completion of this module prepares you to do the following:

1. Size and select steel pipe.
   a. Identify pipe sizes.
   b. Identify schedules and wall thicknesses.
   c. List pipe manufacturing processes.
   d. Identify ASTM International piping standards.
2. Prepare pipe for fittings.
   a. Recognize the tools used when cutting and reaming pipe.
   b. Cut and ream pipe.
3. Groove pipe and install grooved fittings.
   a. Identify types of grooved connections.
   b. Recognize the tools used when grooving pipe.
   c. Identify and install grooved fittings.
4. Thread pipe and install threaded fittings.
   a. Identify types of threads.
   b. Recognize the tools used when threading pipe.
   c. Thread pipe.
   d. Identify and install threaded fittings.
5. Select and install flanged fittings.
   a. Identify flanged fittings and their applications.
   b. Install flanged pipe fittings.
6. Identify the applications of plain-end pipe fittings.
   a. Identify plain-end fittings with grippers and their applications.
   b. Identify press-fit fittings and their applications.

Performance Tasks

Under supervision, you should be able to do the following:

1. Identify the correct pipe size and schedule for a given application.
2. Cut and ream pipe.
3. Groove pipe and install grooved fittings.
4. Thread pipe and install threaded fittings.
5. Install flanged fittings using the proper bolt tightening sequence.

Trade Terms

Burr
Bushing
Cast iron
Close nipple
Companion flange
Concentric reducing coupling
Coupon
Cut groove
Ductile iron
Eccentric reducing coupling
Elbow
Fin
Flexible sprinkler hose fittings
Gasket
Makeup
Malleable iron
National Pipe Thread (NPT)
Nipple
Pipe schedule
Plain-end pipe
Roll groove
Standpipe
Tee
Teflon® tape

Industry Recognized Credentials

If you are training through an NCCER-accredited sponsor, you may be eligible for credentials from NCCER's Registry. The ID number for this module is 18103. Note that this module may have been used in other NCCER curricula and may apply to other level completions. Contact NCCER's Registry at 1.888.622.3720 or go to **www.nccer.org** for more information.

You can also show off your industry-recognized credentials online with NCCER's digital badges. Transform your knowledge, skills, and achievements into badges that you can share across social media platforms, send to your network, and add to your resume. For more information, visit **www.nccer.org**.

NOTE

Unless otherwise specified, references in parentheses following figure and table numbers refer to NFPA 13.

Contents

Contents (continued)

Figures and Tables

Figures and Tables (continued)

Section One

1.0.0 Sizing and Selecting Steel Pipe

Objective

Size and select steel pipe.

a. Identify pipe sizes.
b. Identify schedules and wall thicknesses.
c. List pipe manufacturing processes.
d. Identify ASTM International piping standards.

Performance Task

1. Identify the correct pipe size and schedule for a given application.

Trade Term

Pipe schedule: A pipe labeling system based on pipe wall thickness. As the schedule increases, the wall thickness also increases.

Steel pipe is the most common type of pipe used in sprinkler systems. It is durable, machinable, and less expensive than most other types of pipe. Like other piping materials, steel is classified according to the diameter of the pipe, the **pipe schedule** or wall thickness of the pipe, and the material of construction.

Sprinkler systems are typically constructed using black steel pipe. This pipe usually has a rust-delaying varnish or lacquer coat on the outside. However, this lacquer coat does not classify the pipe as corrosion-resistant. Hot-dipped galvanized steel pipe is black steel pipe that is submerged in molten zinc that coats the inside and outside of the pipe. Galvanized steel pipe is more expensive than black steel pipe and is not used except in systems where external corrosion is an issue or when specifications or shop drawings require it.

Stainless steel can also be used in threaded systems; however, it is generally avoided as it is difficult to cut the threads correctly and create leakproof joints. It is also expensive, although it is good for special applications that require maximum resistance to corrosion.

1.1.0 Pipe Sizes

Pipe is classified by its nominal size in inches. For pipe sizes up to and including 12", nominal size approximates the inside diameter. From 14" and larger, nominal size reflects the outside diameter of the pipe. *Figure 1* shows the inside and outside diameters of pipe.

1.2.0 Schedules and Wall Thicknesses

Wall thickness can be defined using two classification systems: pipe schedules and manufacturer weights.

1.2.1 Pipe Schedules

The schedule refers to the thickness of the pipe wall. The larger the schedule number, the thicker the pipe wall. Pipe with a higher schedule number is stronger and can withstand more pressure. Common steel pipe schedules for sprinkler systems include Schedule 5, Schedule 10, Schedule 30, and Schedule 40.

Table 1 shows the schedule system used in the fire sprinkler industry. Note that the schedule numbers refer only to the wall thickness of a pipe of a given nominal size. For example, a ¾" Schedule 40 pipe does not have the same thickness as a 1" Schedule 40 pipe.

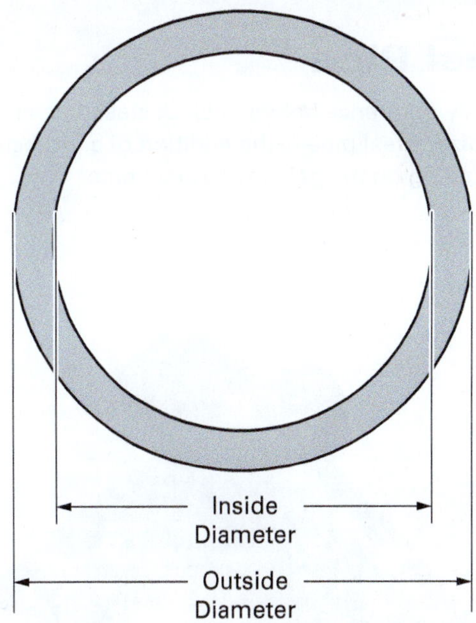

Figure 1 Inside and outside diameters of pipe.

Table 1 Steel Pipe Dimensions (Data from NFPA 13, *Table A.16.3.2*)

Nominal Pipe Size	Outside Diameter	Schedule 5		Schedule 10[a]		Schedule 30		Schedule 40	
		Inside Diameter	Wall Thickness	Inside Diameter	Wall Thickness	Inside Diameter	Wall Thickness	Inside Diameter	Wall Thickness
$\frac{1}{2}$[b]	0.840	—	—	0.674	0.083	—	—	0.622	0.109
$\frac{3}{4}$[b]	1.050	—	—	0.884	0.083	—	—	0.824	0.113
1	1.315	1.185	0.065	1.097	0.109	—	—	1.049	0.133
$1\frac{1}{4}$	1.660	1.530	0.065	1.442	0.109	—	—	1.380	0.140
$1\frac{1}{2}$	1.900	1.770	0.065	1.682	0.109	—	—	1.610	0.145
2	2.375	2.245	0.065	2.157	0.109	—	—	2.067	0.154
$2\frac{1}{2}$	2.875	2.709	0.083	2.635	0.120	—	—	2.469	0.203
3	3.500	3.334	0.083	3.260	0.120	—	—	3.068	0.216
$3\frac{1}{2}$	4.000	3.834	0.083	3.760	0.120	—	—	3.548	0.226
4	4.500	4.334	0.083	4.260	0.120	—	—	4.026	0.237
5	5.563	—	—	5.295	0.134	—	—	5.047	0.258
6	6.625	6.407	0.109	6.357	0.134[c]	—	—	6.065	0.280
8	8.625	—	—	8.249	0.188[c]	8.071	0.277[d]	7.981	0.322
10	10.750	—	—	10.370	0.188[c]	10.140	0.307[d]	10.020	0.365
12	12.750	—	—	—	—	12.090	0.330[c]	11.938	0.406

Notes:
All dimensions in inches.
[a]Schedule 10 defined to 5 in. nominal pipe size by ASTM A135, *Standard Specification for Electric-Resistance-Welded Steel Pipe*.
[b]These values applicable when used in conjunction with ***Section 29.4*** and ***Section 29.5***.
[c]Wall thickness specified in ***16.3.2***.
[d]Wall thickness specified in ***16.3.3***.

Steel Pipe

The only difference between black steel pipe and galvanized steel pipe is the addition of a protective zinc coating on the galvanized steel pipe.

NOTE

Both $3\frac{1}{2}$" and 5" pipe are available but are rarely used in modern systems. They may be seen in older or retrofit installations.

As shown in *Table 1*, the outside diameter remains the same for all schedules, although the wall thicknesses and inside diameters change. Therefore, the same fittings can be used on the pipe regardless of the schedule. *Figure 2* shows a comparison of wall thicknesses for $1\frac{1}{2}$" pipe.

1.2.2 Manufacturer Weights

From smallest to largest, there are four manufacturer weight classifications in use today:

- Thin wall
- Standard wall (STD)
- Extra-strong wall (XS)
- Double extra-strong wall (XXS)

Years ago, almost all sprinkler systems were installed using Schedule 40 and Schedule 30 black steel pipe. All sizes up to 6" were Schedule 40. Sizes 8" and greater were Schedule 30. Schedule 30 and Schedule 40 pipe therefore became known as *standard (STD) pipe*.

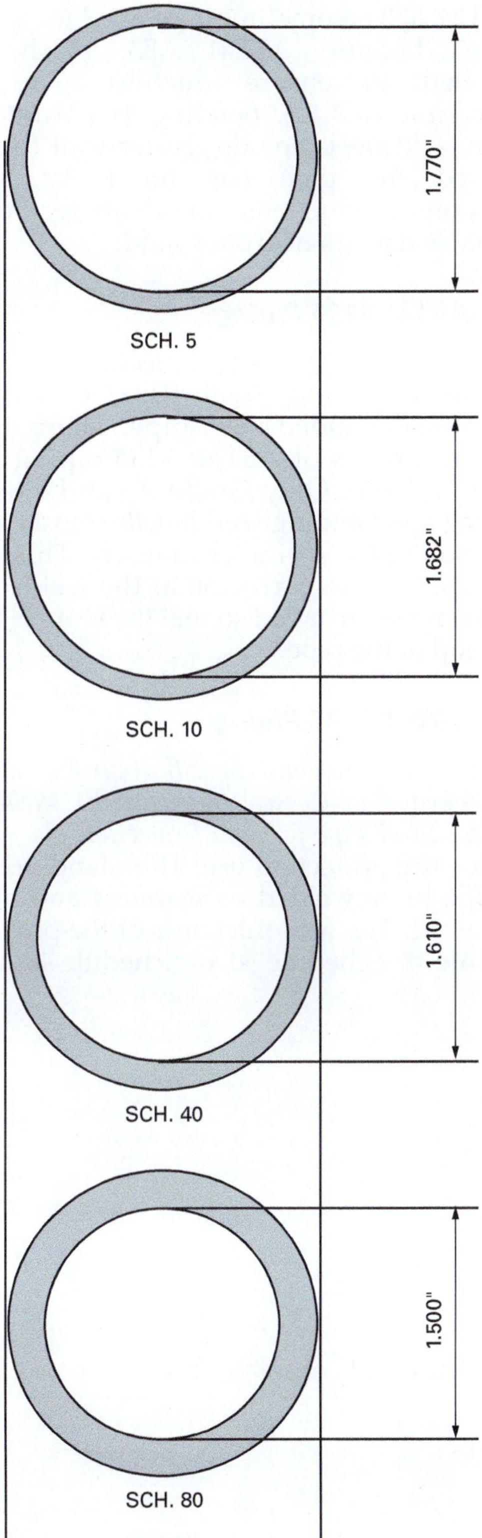

Figure 2 Visual comparison of wall thicknesses for 1½" pipe.

According to NFPA 13, standard pipe is usable at pressures up to 300 pounds per square inch (psi). Extra-strong pipe (Schedule 80) and double extra-strong pipe (Schedule 160) are also available but are rarely used in fire sprinkler work. Thin-wall pipe (Schedule 5 or Schedule 10) can be used in sprinkler work, but only with special fittings Listed for use with this type of pipe.

1.3.0 Pipe Manufacturing Processes

Three different methods are used to manufacture pipe. The piping industry has typically supplied steel pipe in lengths of 21' and 25'. Some specifications require the use of domestic-made pipe as opposed to foreign-made pipe. This is a contractual requirement based on national economics. As long as the pipe is made to specifications, there should be no difference in quality between foreign and domestic material.

1.3.1 Continuous-Weld Pipe

Continuous-weld pipe is manufactured from skelp, which is a coil of flat steel. In a furnace, the skelp is drawn through a series of forming rollers. The edges are heated to welding temperature. Rollers form the round section and squeeze the edges together, producing the weld. Sizing rollers adjust the outside diameter, and saws cut the pipe into the desired lengths.

1.3.2 Seamless Pipe

Seamless pipe is made from round steel billets called *tube rounds*, which are heated and then pierced. A mandrel, which is a cylindrical bar of a precise diameter, is inserted into the shell or pierced tube round. The size of the mandrel determines the interior diameter of the pipe. Rollers form the outside to the proper diameter and wall thickness. Seamless tubing is more expensive than other kinds of pipe or tubing and is not frequently used in the sprinkler industry.

1.3.3 Electric-Resistance-Welded (ERW) Pipe

Electric-resistance-welded (ERW) pipe is formed by drawing the skelp through a series of rollers. The rollers gradually form the material into a round shape. It then passes through two copper welding wheels. One wheel has a positive charge and the other has a negative charge. The skelp edges are heated to welding condition by the resistance to electrical current in the joint. After the rollers pass, the material cools in a bonded condition.

Did You Know?

Most of the steel produced today uses a method known as the *basic oxygen process.* Molten iron from a blast furnace (a giant oven lined with firebrick) is poured into a basic oxygen furnace. Pure oxygen is blown onto the metal through a tube called an *oxygen lance.* The oxygen combines with the impurities in the iron to form oxides, leaving molten steel. After the liquid steel has been created in the oxygen furnace, it must be further processed and cast into solid form before manufacturers can use it.

1.4.0 ASTM International Pipe Standards

There are three primary ASTM International standards referenced in NFPA 13 that relate to steel pipe. The applicable ASTM standard is marked on the pipe. Always install the type of pipe specified in the shop drawings.

1.4.1 ASTM A53 Piping

ASTM A53, *Standard Specification for Pipe, Steel, Black and Hot-Dipped Zinc-Coated (Galvanized) Welded and Seamless,* is premium grade piping. The standard identifies the following types of ASTM A53:

- *Type F* — Furnace butt-welded, continuous-weld
- *Type E* — Electric resistance welded, Grades A and B
- *Type S* — Seamless, Grades A and B

ASTM A53 is sometimes specified for sprinkler systems because ASTM A53 is chemically controlled to ensure ductility. Per NFPA 13, ***Section 16.3.8.1***, bending is permitted for Schedule 10 steel pipe or greater wall thickness, provided the pipe has no kinks, ripples, distortions, reductions in diameter, or any noticeable deviations from round.

1.4.2 ASTM A135 Piping

ASTM A135, *Standard Specification for Electric-Resistance-Welded Steel Pipe,* describes a general-use, welded steel pipe. There are two different grades of ASTM A135 piping, Grade A and Grade B. Only Grade A can be used for flanging and bending, and it is the most common type used in the sprinkler industry. This type of pipe is subject to corrosion at the welded seam and should be installed so that the seam is located at the top of the pipe.

1.4.3 ASTM A795 Piping

ASTM A795, *Standard Specification for Black and Hot-Dipped Zinc-Coated (Galvanized) Welded and Seamless Steel Pipe for Fire Protection Use,* is steel pipe for fire protection use. This standard allows the pipe to be welded or seamless and black or galvanized. The wall thickness of the pipe can be Schedule 10, Schedule 30, or Schedule 40.

1.0.0 Section Review

1. For a 2" pipe, the nominal pipe size reflects the _____.
   a. outside diameter
   b. inside diameter
   c. wall thickness
   d. weight

2. A type of thin-wall pipe is _____.
   a. Schedule 10
   b. Schedule 30
   c. Schedule 40
   d. Schedule 80

3. The type of steel formed from round steel billets is _____.
   a. continuous-weld pipe
   b. seamless pipe
   c. electric-resistance-welded pipe
   d. cast iron pipe

4. The ASTM standard that is considered premium grade piping is _____.
   a. ASTM A53
   b. ASTM A135
   c. ASTM A220
   d. ASTM A795

SECTION TWO

2.0.0 CUTTING AND REAMING PIPE

Objective

Prepare pipe for fittings.

a. Recognize the tools used when cutting and reaming pipe.
b. Cut and ream pipe.

Performance Task

2. Cut and ream pipe.

Trade Terms

Burr: A sharp, raised edge or other metal protrusion produced on a pipe end when cutting a pipe.

Fin: A sharp, flared metal protrusion produced on a pipe end when cutting it. A fin juts out at an angle to the pipe.

Before pipe is installed, it must be cut square and reamed to remove any burrs or raised edges.

2.1.0 Pipe Cutters and Reamers

Steel pipe is cut and prepared for installation using a variety of tools, including cutters and reamers.

2.1.1 Steel Pipe Cutters

Steel pipe cutters are used for cutting steel pipe up to 8" in diameter. The standard steel pipe cutter is available with one, three, or four cutting wheels. *Figure 3* shows examples of steel pipe cutters.

Steel pipe cutters operate like a vise clamped on a piece of pipe (*Figure 4*). The wheels are lined up on the cutting mark and tightened against the pipe. As the cutter is rotated around the pipe, the adjusting handle is turned to maintain uniform tightness on the pipe.

When using a steel pipe cutter, ensure that the pipe is clean. Dirt and grit act as abrasives on the cutting wheels and cause the cutters to wear rapidly.

Misalignment of the pipe cutter against the pipe causes the cutting wheels to cut on different lines. This causes the cutting wheels to crawl along the pipe and prevents the cutter from working properly.

2.1.2 Hinged Pipe Cutters

Hinged cutters can be used to cut 1" to 12" diameter steel pipe. They work well in close quarters and are safer than an abrasive saw. *Figure 5* shows a hinged cutter designed for 2½" to 4" steel pipe.

2.1.3 Reamers

The cutting process may leave a **burr** (raised edge) or a **fin** (flared protrusion) on the cut end of a pipe. Reaming removes these rough materials from the end of the pipe. NFPA 13, *Section 7.5.6.1*, requires that burrs and fins be removed whether the pipe is to be threaded, grooved, or welded. Reaming helps ensure free flow, as well as a leakproof, strong joint.

Pipe sizes up to 2" can be reamed with a hand reamer (*Figure 6*). Hand-reaming requires the use of a ratchet ring and a ratchet handle to turn the reamer. On larger pipes, it is more economical to use a power reamer because power reamers are faster and more uniform than hand reamers. When reamers are not available, a half-round file can be used to ream pipe.

Reaming must be done carefully to avoid oversized holes, bell-mouthed holes, or a poor finish. This is especially true with power reamers. If the reamer is out of alignment, the pipe end will be tapered, producing a mechanically weak, bell-mouth opening. To ensure a smooth finish, set the speed of the power drive according to the manufacturer's instructions.

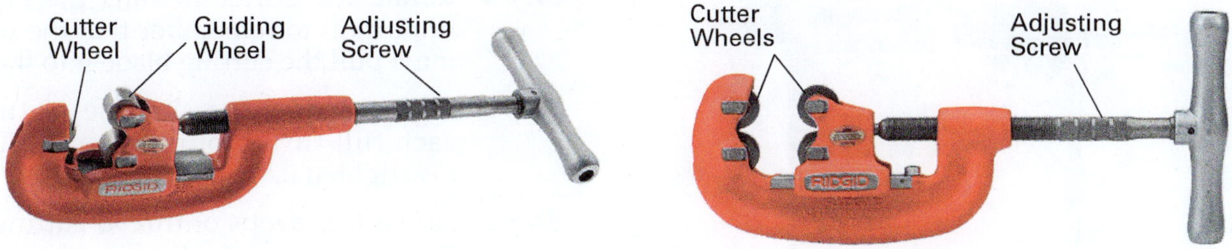

Conventional **Four-Wheeled**

Figure 3 Steel pipe cutters.

Figure 4 Using steel pipe cutters.

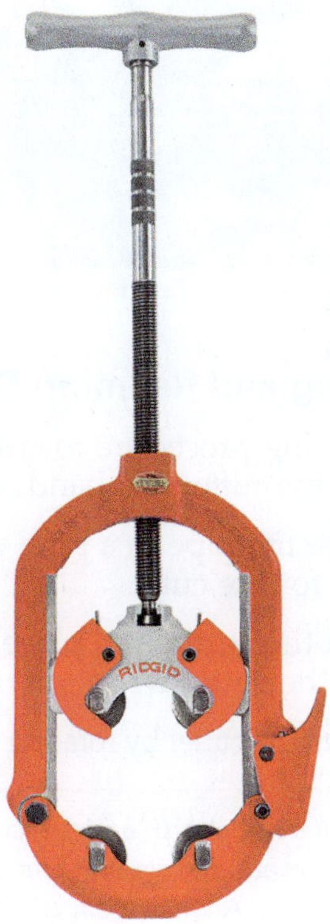

Figure 5 Hinged pipe cutter.

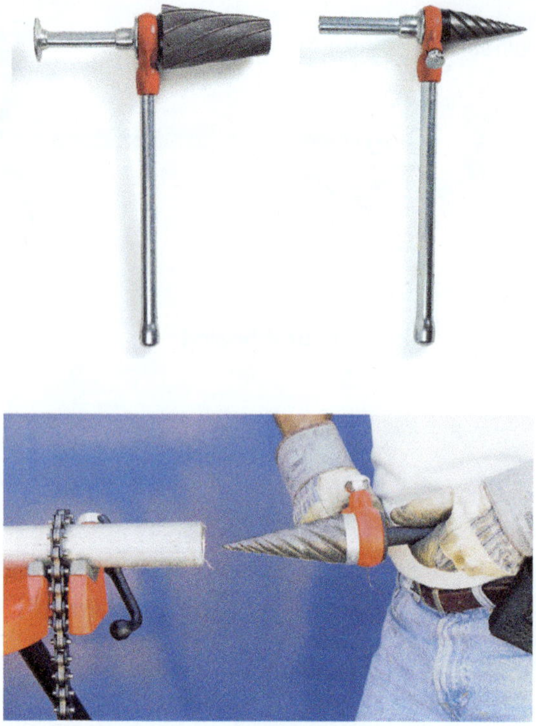

Figure 6 Reamers with ratchet housing attached.

2.2.0 Cutting and Reaming Steel Pipe

Use the following procedure to cut and ream steel pipe using a pipe cutter and hand reamer:

Step 1 Secure the pipe in a pipe vise and mark a place for the cut.

Step 2 Open the cutter and place it over the pipe with the cutter wheel on the mark.

Step 3 Tighten the cutter by rotating the screw handle.

WARNING!

Pipe edges are sharp, and the cutting process produces metal debris that can cut skin and damage eyes. Always wear appropriate personal protective equipment when cutting pipe.

Step 4 Rotate the cutter to start the cut. It is important to remember that the rotation must pull the cutting blade into the pipe.

Step 5 Tighten the cutter handle $\frac{1}{4}$ turn for each full turn around the pipe. Do not overtighten it.

Step 6 Add a few drops of thread cutting oil to the groove and continue cutting. Avoid skin contact with the oil.

Step 7 When the cut is almost finished, stop cutting and snap the pipe to finish the cut. This reduces the ridge that can form on the inside of the pipe.

Step 8 Select the appropriate reamer for the pipe size and insert the reamer tip in the pipe.

Step 9 Apply light forward pressure and start rotating the reamer in a clockwise direction.

CAUTION

You could damage the reamer if you rotate it in the wrong direction. The reamer should bite as soon as pressure is applied.

Step 10 Remove the reamer by pulling back on it while continuing to rotate it. Check the progress and then reinsert the reamer. Rotate the reamer until the inside edge is smooth. Stop when all burrs have been removed.

Step 11 Clean the pipe and tools using a shop towel.

After the pipe has been cut and reamed, it will be connected using grooved, threaded, flanged, or plain-end fittings. Pipe fittings are attached to pipe to change the direction of fluid flow, to connect a branch line to a main line, to close off the end of a line, or to join two pipes of the same or different sizes. The type of fitting to be used is specified on the shop drawings.

2.0.0 Section Review

1. The tool used to smooth a steel pipe edge after cutting is a _____.
   a. crimper
   b. plane
   c. reamer
   d. pocketknife

2. When using a pipe cutter, tighten the cutter handle _____.
   a. $\frac{1}{4}$ turn for each full turn around the pipe
   b. $\frac{1}{2}$ turn for each full turn around the pipe
   c. $\frac{3}{4}$ turn for each full turn around the pipe
   d. a full turn for each full turn around the pipe

SECTION THREE

3.0.0 GROOVED PIPE AND FITTINGS

Objective

Groove pipe and install grooved fittings.

a. Identify types of grooved connections.
b. Recognize the tools used when grooving pipe.
c. Identify and install grooved fittings.

Performance Task

3. Groove pipe and install grooved fittings.

Trade Terms

Coupon: The metal disc removed from a steel pipe when using a hole cutter.

Cut groove: A groove that is formed around pipe by cutting away the pipe material.

Elbow: A fitting that makes an angle between adjacent pipes. An elbow is always 90 degrees unless another angle is stated on the fitting or in the drawing specifications. Also known as an *ELL*.

Flexible sprinkler hose fittings: Flexible fittings designed to span the distance from the branch line to the sprinkler.

Gasket: A flat sheet or ring of rubber or other material used to seal a joint between metal surfaces to prevent gas or liquid from entering or escaping.

Roll groove: A groove that is formed by pressing the metal around the pipe.

Standpipe: A horizontal or vertical supply main running through buildings and having hose valves on each floor.

Tee: A fitting that has one side outlet 90 degrees to the run.

Grooved piping systems provide an alternative to welded, flanged, or threaded joints. Each joint in a grooved piping system serves as a union, allowing easy access to any part of the piping system for cleaning or servicing. Grooved piping systems have a wide range of applications and can be used with many types of piping materials. Grooved pipe and fittings are rapidly becoming the most popular form of steel pipe installation due to the speed of installation and ability to create tight seals.

The grooved piping system offers varied mechanical benefits, including the option of rigid or flexible couplings. Rigid and flexible couplings can be incorporated as needed into any system to take full advantage of the characteristics of each type. Rigid couplings create a rigid joint that is useful for risers, mechanical rooms, and other areas where positive clamping with no flexibility within the joints is desired. Flexible couplings allow for controlled pipe movement that occurs with expansion, contraction, and deflection. Rigid and flexible couplings have specific installation requirements for seismic bracing as referenced in NFPA 13. Another advantage of grooved piping is the ability to join thin-wall pipe.

3.1.0 Types of Grooved Connections

Grooved pipe consists of a rubber gasket and two housing halves that are bolted together. The housing halves are tightened in accordance with the manufacturer's instructions. This area must be clean and smooth for proper sealing to take place. The design of the gasket allows the internal pressure of the pipe to press on the back side of the gasket, forcing the gasket down on the pipe and forming the seal. This allows the pipe to flex in the grooved joint without leaking.

3.1.1 Cut Grooves Versus Roll Grooves

Grooves may be formed by rolling or cutting (*Figure 7*). Rolling uses a compression tool to *press* a groove into the pipe, while cutting uses a cutting tool to *remove* a groove of material from the pipe.

The original method of grooving a steel pipe was by cutting. The cutting method was designed for Schedule 30 or Schedule 40 pipe. Schedule 10 pipe does not have enough wall thickness to allow for a **cut groove**. Only a **roll groove** may be used on pipe whose walls are thinner than Schedule 40 (sizes less than 8") or Schedule 30 (8" and greater).

CAUTION

Never cut groove any pipe with a wall thinner than Schedule 30.

Cut grooving removes metal and thins the pipe wall. It is used less often than roll grooving in the fire sprinkler industry. A cut groove removes less metal than a standard thread; however, the strength of a threaded joint is greater for two reasons. First, in a threaded joint, the fitting reinforces the pipe where a groove would weaken it. Second, the edges of a cut groove introduce greater stress concentrations than threads. Roll grooves, on the other hand, can increase the strength of a pipe at the joint location. Cold working of the metal by displacing it increases the material's strength in a way that is similar to forging. However, there is a reduction of approximately 50 percent in available movement with a roll groove versus a cut groove. If flexibility in the joints is a specific issue, cut grooves may be necessary. Cut grooves also offer an advantage in dry and preaction systems in that water does not pool in the deformed grooved areas, thereby reducing the risk of corrosion.

3.1.2 Groove Specifications

Groove depth (diameter) is critical. If the groove is not deep enough, the coupling will not fit over the pipe. If the groove is too deep, the coupling will not grip the pipe. These depths may be gauged automatically by a stop on the grooving machine, or the sprinkler fitter may have to gauge each groove as it is prepared.

Did You Know?

Grooved pipe connections were developed in 1919 by Victaulic Corporation, then known as the *Victory Pipe Grooving Company*, to solve a military problem. The objective was to create a system that would allow troops to rapidly assemble systems for water and fuel distribution in the field.

Figure Credit: Courtesy of Victaulic Company

Did You Know?

Most valves from the factory are cut groove and most field piping is roll groove, but they all fit together as long as the dimensions match.

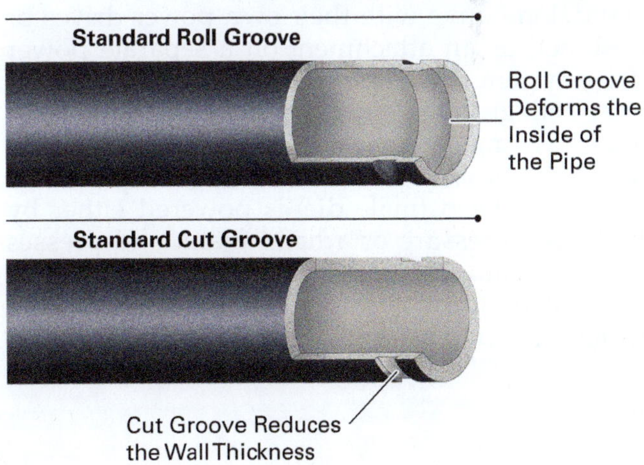

Figure 7 Roll groove and cut groove pipe.

The sprinkler fitter is responsible for quality control in the field. If the fitter makes a groove that leaks or does not make up, it is the fitter's responsibility. It only takes a moment to check the accuracy of each groove. It is easier to correct the groove on the ground in the assembly process than it is to disassemble an already installed portion of a system when it leaks or will not make up.

Misalignment of the groove rollers on a machine can cut the pipe in the process of making the groove. Whether using pipe that you have grooved, or pipe grooved in the shop, check the grooves before installing the pipe to be certain it is correct. Another possible malfunction is splitting at the weld joint on electric-resistance-welded (ERW) pipe.

3.2.0 Grooving Tools

Different types of tools are used for cut grooving and roll grooving. Schedule 40 pipe in 1" through 6" and Schedule 30 pipe in 8" and 10" can be cut grooved. Any pipe of lesser wall thickness must be roll grooved. However, Schedule 40 and Schedule 30 pipe also can be roll grooved.

3.2.1 Roll Groovers

Power roll groovers and hand-operated roll groovers are used to roll grooves at the ends of pipe to prepare the piping for grooved fittings and couplings.

Manual roll groovers are driven by a hand crank. Power roll grooving machines can be standalone units with their own power drives or designed as an attachment on a separate power drive system.

A roll grooving machine forms a groove in a pipe using upper and lower die heads. The lower die head (female die) rotates the pipe. The upper die head (male die) is powered either by hydraulic pressure or a hand crank and presses a groove into the pipe as the pipe is rotated. An adjustable stop controls the groove depth. Roll grooving removes no metal from the pipe but forms a groove by displacing the metal instead. Since the groove is cold formed, it has rounded edges that reduce pipe movement after the joint is made up.

Always follow the manufacturer's procedures for the proper setup and operation of any roll grooving tool, including roll selection for the material to be grooved. *Figure 8* shows a roll groover attachment for a power drive system. This tool can be used as a manual in-air groover or attached to a power drive system (*Figure 9*).

3.2.2 Cut Groovers

Cut grooving differs from roll grooving in that a groove is cut into the pipe instead of being cold formed. There are a variety of tools on the market for cut grooving pipe. The cut groover is a tool driven by an external power source. It is designed to be driven around a stationary pipe. This ensures that the groove is concentric with the outside diameter of the pipe. Most models have an adjustable pipe stop to ensure proper groove depth. The cut groover removes less than one-half of the pipe wall, which is less than the depth of thread cuts. Cut grooving is intended for standard weight or heavier pipe. *Figure 10* shows two types of cut groovers.

3.3.0 Grooved Fittings

Joining pipe using grooved fittings saves time as well as money. Grooved fittings are more expensive than threaded fittings, but fewer labor hours are needed to join pipe using grooved fittings. In large-diameter pipe, the labor savings in making joints with grooved fittings is greater than the material costs savings of using threaded fittings. Grooving pipe in the field is simple, whereas cutting threads on pipe larger than 2" usually requires a trip back to the shop. In addition, if the thin-wall pipe being used is not Listed for threading, one Listed means of connection in the field is with a grooved coupling.

Figure 11 shows the components in a grooved connection.

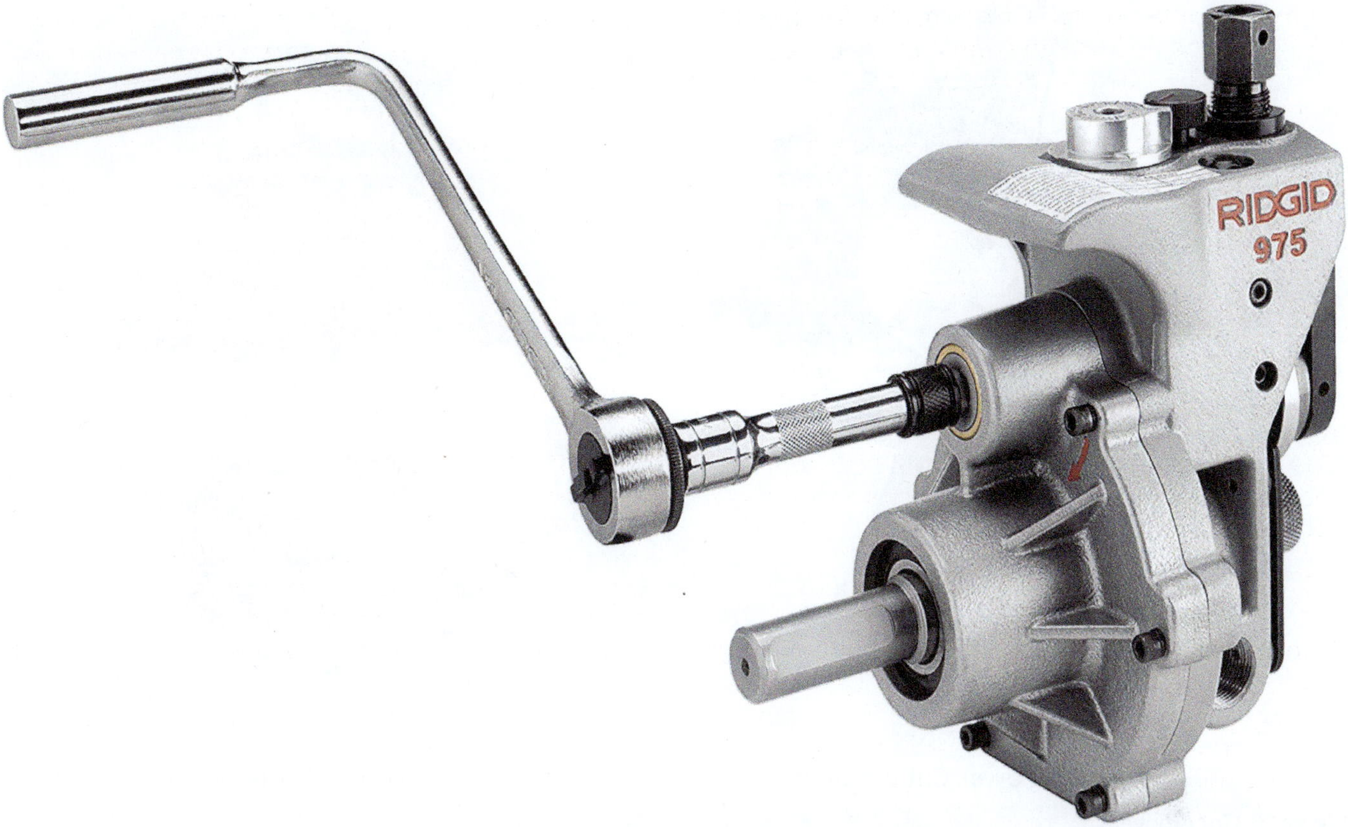

Figure 8 Roll groover attachment for power drive system.

Power Roll Groover
Attached to Power Drive System

Power Roll Groover
Used Manually in Air

Figure 9 Dual-use power roll groover applications.

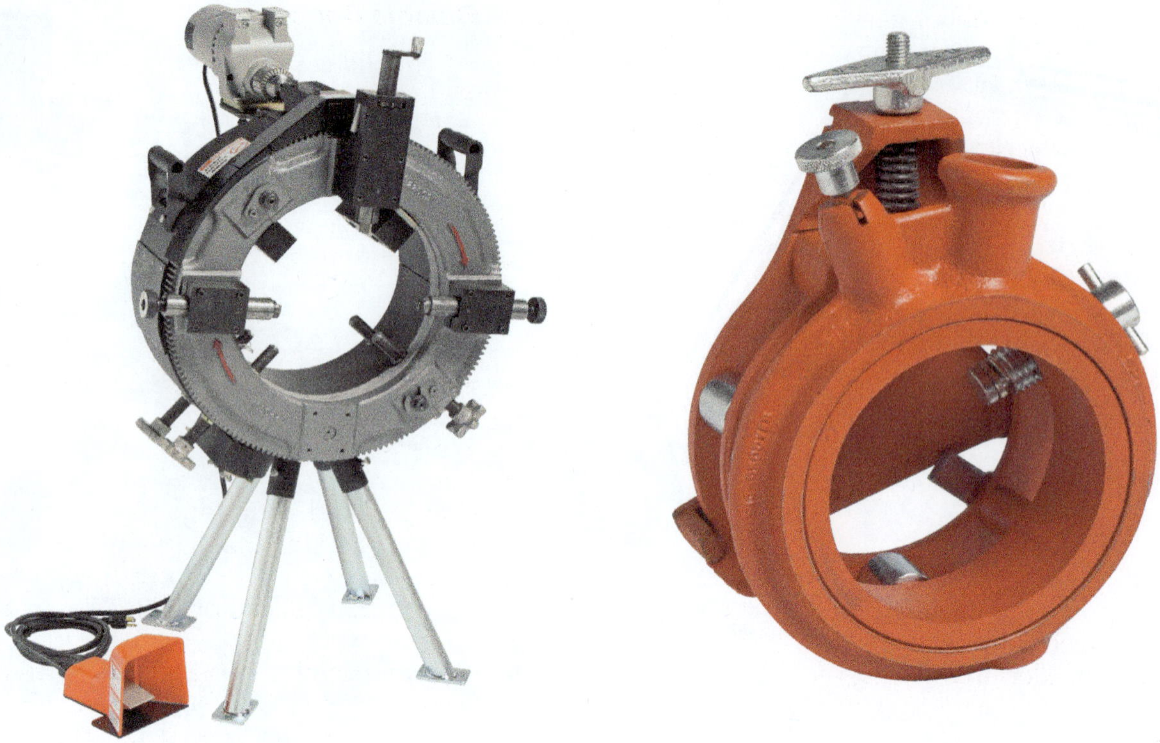

(A) Power Cut Groover

(B) Manual Cut Groover

Figure 10 Cut groovers.

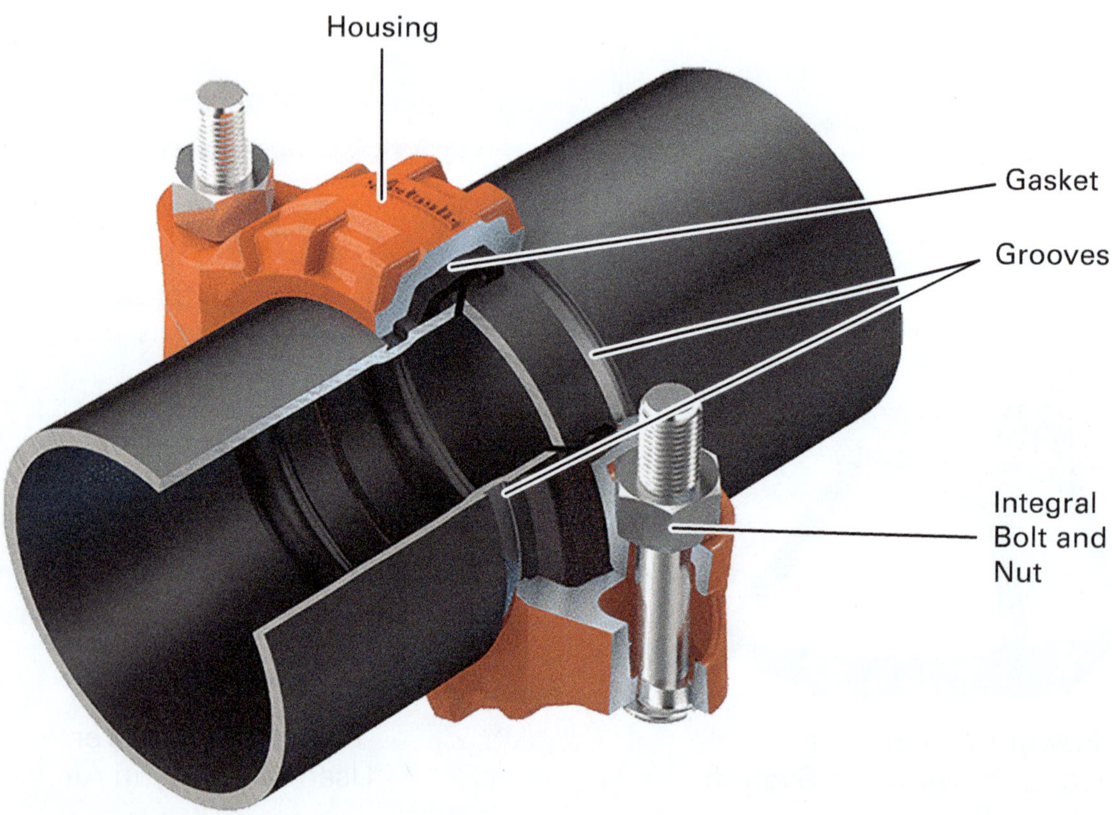

Figure 11 Typical grooved connection.

Lubrication of the rubber gaskets in accordance with manufacturer's instructions for these fittings is important. Use only a nonpetroleum-based lubricant or a pre-lubricated gasket recommended by the manufacturer. Petroleum products will attack the gasket and cause early failure. Removal of burrs is also very important. Failure to do either of these can result in cut gaskets and leaks. The manufacturer's instructions must be consulted to maintain the Listing and warranty status. The types of fittings and couplings used with grooved pipe include the following:

- Flexible couplings
- Rigid couplings
- Reducing couplings
- Elbows
- Tees
- Crosses
- Reducers
- Grooved mechanical branch tees
- Flexible sprinkler hose fittings
- Flanges
- Caps

Hydraulic Power Roll Groovers

Heavy-duty hydraulic power roll groovers are used with larger, heavier pipe. The power roll groover shown here can groove up to 12" Schedule 10 and 8" Schedule 40 standard wall pipe.

Figure Credit: Courtesy of RIDGID Tool Company

3.3.1 Flexible Couplings

Grooved flexible pipe couplings (*Figure 12*) are designed to allow for controlled linear and angular movement. The pipe coupling key in a flexible coupling is narrower than the pipe groove, allowing the coupling a certain amount of leeway to move within the groove. The flexible pipe coupling housing is also wider than normal, allowing for pipe-end separation while the internal gasket maintains a tight seal. Flexible grooved couplings can be a great advantage in areas where the pipe run needs to be run at a slight deviation from a straight line. They also have specific applications in areas prone to seismic activity.

Flexible grooved couplings are manufactured for standard, lightweight, and heavy-duty applications. They are rated according to working pressure. A rubber gasket provides the watertight seal.

There are several flexible coupling manufacturers. Each has its minor differences, and the gaskets from one manufacturer should not be used with couplings made by another. Some flexible couplings are approved for dry systems and some are not. It is important to determine which couplings work with the specific installation. Consult the manufacturer's literature to ensure that the application is suited for the use of flexible couplings.

3.3.2 Rigid Couplings

Unlike flexible couplings, rigid couplings do not allow for pipe movement. Rigid couplings have the same hanger and support requirements as threaded or flanged couplings. They are used in all applications unless flexibility is required. *Figure 13* shows an example of a rigid grooved coupling.

3.3.3 Reducing Couplings

Grooved reducing couplings save costs in both material and labor because they allow the pipe size to be reduced on the run with no takeout. *Figure 14* shows a reducing coupling.

3.3.4 Elbows

An **elbow**, also known as an *ELL*, is a fitting that makes an angle between adjacent pipes. A standard elbow is always 90 degrees unless another angle is stated on the fitting or in the shop drawings. Grooved elbows are available in 90-degree, 45-degree, $22\frac{1}{2}$-degree, and $11\frac{1}{4}$-degree models. In addition, there are

Figure 12 Flexible grooved coupling.

Figure 13 Rigid grooved coupling.

Figure 14 Grooved reducing coupling.

adapter elbows, which are grooved on one end and threaded on the other. These are available in 90-degree and 45-degree configurations. An additional selection is available in long-radius elbows. Models in the range of 90 degrees down to the shallowest angle of $11\frac{1}{4}$ degrees are available in radii of 3D, 5D, and 6D, where D is the diameter of the pipe. Another type of 90-degree elbow is an end-of-run elbow. An end-of-run fitting is a grooved cap with an outlet. It allows a sprinkler to be installed at the termination of a grooved piping run and functions as a cap. *Figure 15* shows types of elbows.

3.3.5 Tees

A **tee** is a fitting that has one side outlet 90 degrees to the run. There are many types of tees available for use in sprinkler systems, some of which are shown in *Figure 16*.

Grooved tees are standard or reducing and are available in straight grooved form or with a threaded side outlet.

The outlet coupling is a special kind of tee. It serves as both a coupling and a tee in one. The outlet is available either threaded or grooved.

One-Bolt Couplings

One-bolt couplings are also available. These couplings speed installation by eliminating the need for alternate bolt tightening.

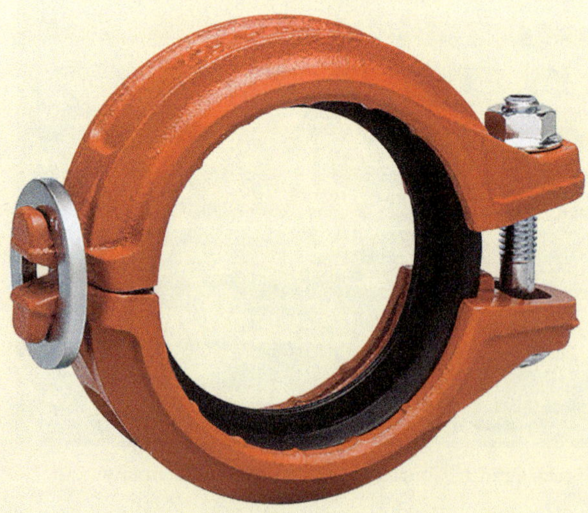

Figure Credit: Courtesy of Victaulic Company

Figure 15 Grooved elbows.

Elbows and Friction Loss

A 90-degree elbow creates an abrupt change in the direction of water flow that produces added resistance in the piping system. This resistance is referred to as *friction loss* and must be accounted for when designing the system. NFPA 13, **Section 27.2.4.8.1**, covers the friction losses contributed by various fittings. Piping systems should be designed to use as few fittings as possible.

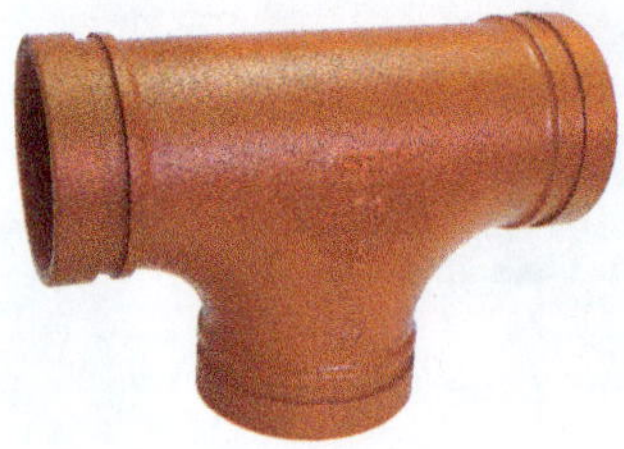

Straight Tee

Reducing Tee
with Grooved Outlet

Reducing Tee with
Male Threaded Outlet

Figure 16 Grooved tees.

Standpipe tees have a $2\frac{1}{2}$" threaded side outlet. They are used specifically for outlets on standpipes that use grooved connections instead of flanged connections. The bullhead tee has a large outlet on the side.

3.3.6 Crosses

Crosses are available in the standard pattern only; there are no reducing models. *Figure 17* shows a grooved cross.

3.3.7 Reducers

Grooved reducers are used to join two different sized grooved pipes available in the grooved configuration. There are concentric reducers and eccentric reducers (*Figure 18*).

3.3.8 Grooved Mechanical Branch Tees

Grooved mechanical branch tees provide a direct branch connection at any location where a hole can be cut in the pipe (*Figure 19*). They are installed using a hole cutter (*Figure 20*).

Hole-cutting tools are used to create openings in steel pipe for mechanical tees, other fittings, and devices. When using a hole cutter, make certain to remove the metal disc cutout, also known as a **coupon**. Failure to do so may result in corrosion and prevent the system from functioning properly in the event of a fire. NFPA 13, ***Figure 28.1***, *Contractor's Material and Test Certificate for Aboveground Piping*, requires a control feature to ensure that all cutouts have been removed from the pipe. To verify removal, some jurisdictions require that each coupon be attached to the outside of the pipe (*Figure 21*). Always follow the requirements of the location in which the system is to be installed.

Figure 17 Grooved cross.

Concentric Reducer

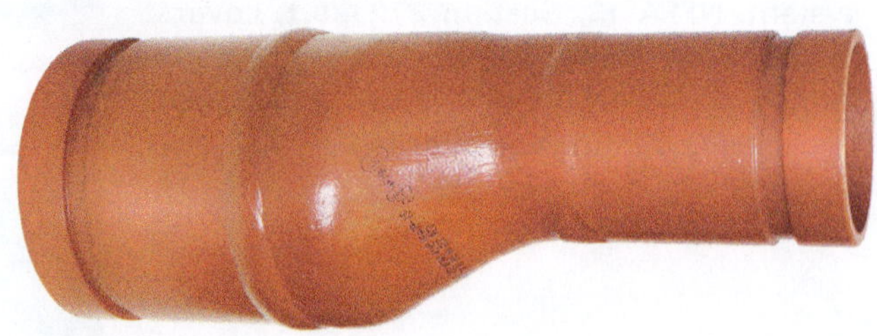

Eccentric Reducer

Figure 18 Grooved reducers.

3.3.9 Flexible Sprinkler Hose Fittings

Flexible fittings called **flexible sprinkler hose fittings** (*Figure 22*) are designed to span the distance from the branch line to the sprinkler.

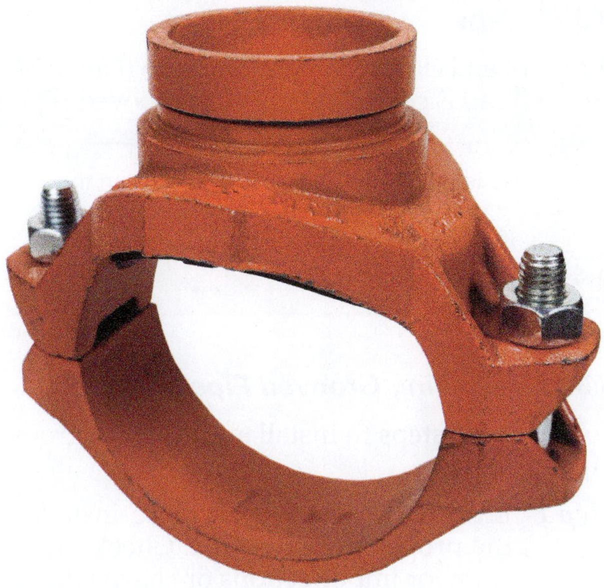

Figure 19 Grooved mechanical branch tee.

This fitting is usually made up of a grooved pipe coupling, a flexible connection that may be a corrugated tube or braided steel, a reducing fitting, and a mounting bracket that connects to the sprinkler and attaches to the T-bar or framing.

Flexible sprinkler hose fittings cost more than other types of fittings but also provide several advantages. Some benefits of using flexible sprinkler hose fittings include the following:

- The system can be tested with the sprinkler hose fitting and sprinkler in place.
- They are easy to install in tight spaces and less labor intensive than other types of fittings.
- They resist damage due to seismic activity.

Flexible sprinkler hose fittings typically vary in length from 24" to 72". Longer lengths create greater friction losses, which must be accounted for in the design of the system. In addition, all installations must follow NFPA requirements for the maximum number of bends and minimum bend radius.

Typical applications include intake or exhaust duct protection, hard ceiling and T-bars, and clean rooms. Flexible sprinkler hose fittings must be Listed for the application.

Figure 20 Hole cutter.

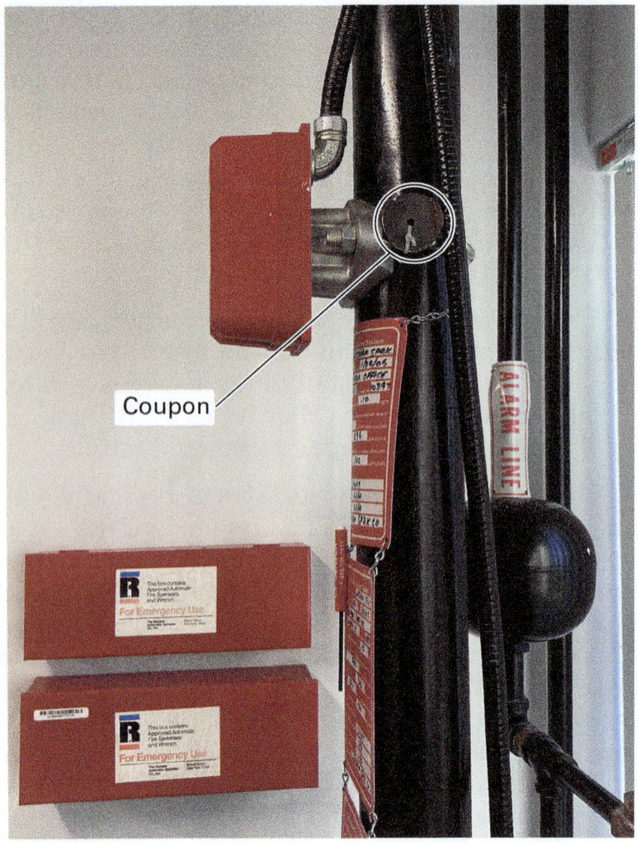

Figure 21 Pipe coupon.

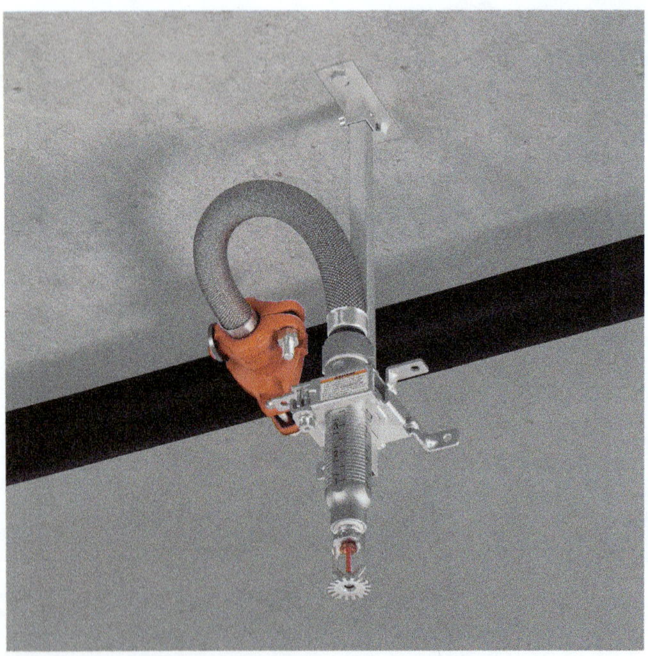

Figure 22 Flexible sprinkler hose fitting.

3.3.10 Flanges

Grooved flanges use the hinge point and the rubber gasket principle to attach a flange to grooved pipe. *Figure 23* shows a grooved flange adapter.

3.3.11 Caps

A type of end closure is a standard cap (*Figure 24*). A grooved cap is commonly used on grooved piping.

> **WARNING!**
> Release the system pressure and drain the system before removing a cap or any other fitting. Failure to do so can result in severe injury and flooding.

3.3.12 Installing Grooved Pipe Fittings

Follow these steps to install a traditional grooved pipe fitting:

Step 1 Ensure that the gasket to be installed is the proper size and type. Check the pipe ends for imperfections or sharp edges.

Step 2 Apply the approved lubricant per the manufacturer's instructions.

Step 3 Place the gasket over the pipe end, arranging it so that it does not hang over the pipe end.

Figure 23 Grooved flange adapter.

Figure 24 Grooved caps.

Step 4 Align and bring the two pipe ends together.

Step 5 Center the gasket between the grooves on each pipe.

Step 6 Ensure that no portion of the gasket extends into the groove on either pipe.

Step 7 Place housings over the gasket.

Step 8 Engage the housing keys into the grooves.

Step 9 Insert the bolts, then tighten the nuts finger-tight.

Step 10 Tighten the nuts with a wrench, alternately and equally until the manufacturer's torque specifications are met (as required).

CAUTION

Excessive nut tightening is not necessary. Uneven tightening may cause the gasket to pinch.

Some manufacturers offer fittings that do not require assembly. Follow these steps to install a ready-for-installation coupling:

Step 1 Insert the pipe into one end of the fitting until it bottoms out in the fitting (typically against the gasket).

Step 2 Insert the other end of the fitting until it bottoms out.

Step 3 Alternately tighten the coupling screws using an impact wrench or socket wrench per the manufacturer's instructions.

3.0.0 Section Review

1. Cut grooving is designed for _____.
   a. Schedule 5 pipe
   b. Schedule 10 pipe
   c. Schedule 20 pipe
   d. Schedule 40 pipe

2. The tool *most* likely to be used when making an in-place groove on Schedule 10 pipe would be a(n) _____.
   a. manual cut groover
   b. hydraulic roll groover
   c. in-air roll groover
   d. power cut groover

3. The type of fitting that might be used to install a sprinkler at a pipe termination is a(n) _____.
   a. bullhead tee
   b. end-of-run fitting
   c. cross fitting
   d. eccentric reducer

Section Four

4.0.0 Threaded Pipe and Fittings

Objective

Thread pipe and install threaded fittings.

a. Identify types of threads.
b. Recognize the tools used when threading pipe.
c. Thread pipe.
d. Identify and install threaded fittings.

Performance Task

4. Thread pipe and install threaded fittings.

Trade Terms

Bushing: A pipe fitting with both internal and external threads that can be used to connect a pipe or fitting with another pipe or fitting of a different size.

Cast iron: An alloy of iron and carbon that is cast in a mold. Cast iron fittings have a classification of either 125 or 250.

Close nipple: A nipple that is about twice the length of a standard thread that is threaded from each end with no shoulder.

Concentric reducing coupling: A reducing coupling that maintains the same center line between the two pipes that it joins.

Ductile iron: A type of cast iron in which magnesium is added to the molten gray iron to reduce brittleness.

Eccentric reducing coupling: A reducing coupling that displaces the center line of the smaller of the two joining pipes to one side.

Makeup: A term used to describe the amount of pipe that screws into a threaded fitting.

Malleable iron: A metallic material that is produced by slowly reheating and cooling cast iron. Malleable iron fittings have a classification of either 150 or 300.

National Pipe Thread (NPT): The US standard for pipe threads. NPT has a $\frac{1}{16}$" taper per inch from back to front.

Nipple: A piece of pipe that is threaded on both ends and is less than 12" in length. Any pipe over 12" is referred to as *cut pipe*.

Teflon® tape: Tape made of Teflon® that is wrapped around the male threads of a pipe before the pipe is screwed into a fitting. Teflon® tape serves as both a lubricant and a sealant.

There are several advantages to using threaded pipe. Threaded systems require no specialized in-shop fabrication and can be easily fabricated on the jobsite using pipe and various fittings. Portable threading equipment can be used almost anywhere on the jobsite. Threaded pipe can be cleaned on the inside before it is installed, reducing the possibility of metal particles becoming trapped and damaging valves or strainers.

Threaded pipe also has its disadvantages. Threaded pipe joints are more prone to leak than other types of joints, and the strength of the pipe is weakened by the threading process because the wall thickness of the pipe is reduced at the threads.

A pipe thread is an inclined plane wrapped around a cylinder. The thread angle is the angle formed by two inclined faces of the thread. The thread angle varies depending on the specific thread standard used. The thread pitch is the distance between the adjacent screw threads, measured from center line to center line. Pitch can also mean inch when referring to pitch as *threads per inch*. The crest of the thread is the outermost part of the thread, or the top of the thread when viewing it from the side. The root of the thread begins at the innermost part of the thread, or from the bottom if viewed from the side.

4.1.0 Types of Threads

The **National Pipe Thread (NPT)** is the standard thread for pipe connections. The NPT has a threaded angle of 60 degrees and a slight flat at the crest and the root. While NPT can be either straight or tapered, the type normally used has a tapered internal and external thread. The taper used is $\frac{1}{16}$" per inch of threads. *Figure 25* shows the taper used in cutting pipe threads.

In addition, there are several other standard pipe threads for specific applications.

All National Pipe Threads are designated by the nominal size of the pipe, the number of threads per inch, and the symbols for the thread series, respectively. For example, $\frac{3}{8}$–18–NPT is the designation for a $\frac{3}{8}$" nominal size pipe that has 18 threads per inch and uses National Pipe Threads.

4.1.1 Threads Per Inch

The proper number of threads on the end of a piece of pipe is specified in *Standard USAS (ASME) B1.20.1, Pipe Threads, General Purpose (Inch)*. From 1" pipe through 2" pipe, the number of threads per inch is $11\frac{1}{2}$. This is determined by

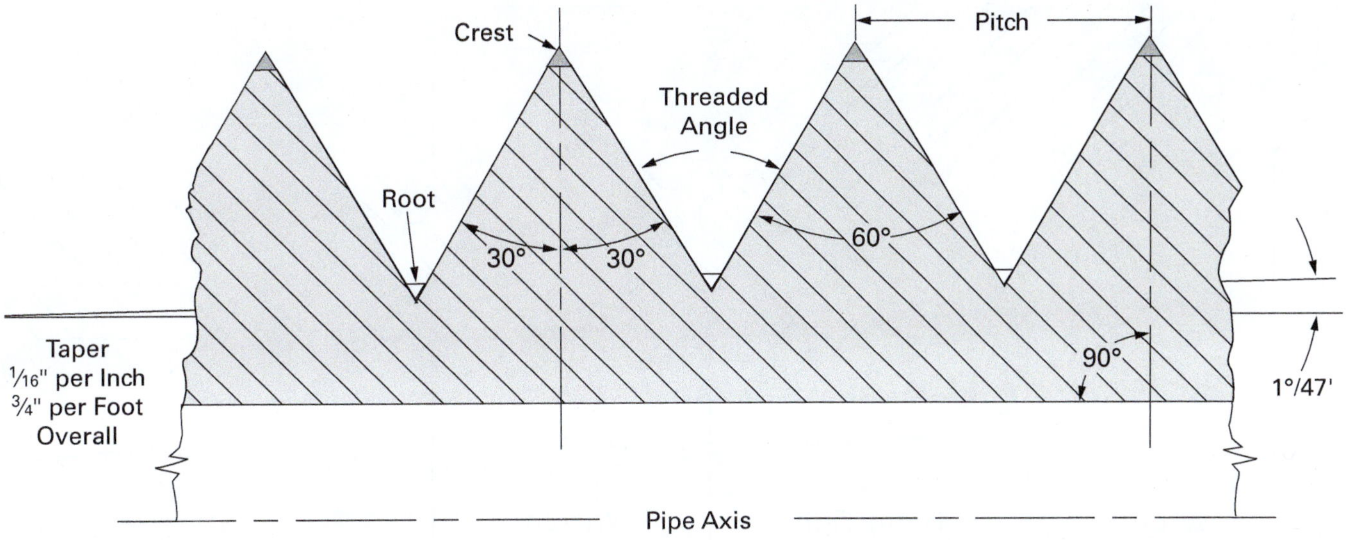

Figure 25 Taper used in cutting pipe threads.

the dies or die segments that are used in the pipe threader. For larger sizes of commonly used pipe, there are only eight threads per inch.

The wall thickness of the pipe and the setting of the dies affect how many threads are present. If the dies are set improperly, too many threads can be cut on Schedule 40 and Schedule 80 pipe. In this case, the die continues down the pipe making threads until you decide to stop, and the pipe will make up too short. On the other hand, improper settings can also result in too few threads and the pipe will make up long and have reduced joint strength.

4.1.2 Pipe Thread Identification

The number of threads per inch varies by pipe size. The number of threads is determined in trade practice by *ASME B1.20.1*. An abbreviated version is shown below. These threads are identified as NPT, with the following designation and meanings:

3⁄4–14–NPT

Where:

3⁄4 = Nominal pipe size in inches
14 = Number of threads per inch
N = National (United States)
P = Pipe
T = Taper

4.1.3 Taper Thread Connection

The joining, or **makeup**, of a taper-threaded pipe connection is performed in two distinct operations known as *hand engagement* and *wrench makeup* (*Figure 26*). Hand engagement is how far the fitting can be tightened by hand. Wrench makeup is the additional turning of the fitting, using a wrench to completely make up the joint. For an average size pipe, hand engagement on properly cut threads is normally three to four revolutions. Hand engagement is a good way to check the quality of threads after cutting. Wrench makeup is usually an additional three turns, for a total of approximately seven revolutions for complete makeup.

The standard for taper thread pipe makeup gives basic thread dimensions and the number of revolutions for hand engagement and wrench makeup. The taper thread makeup may vary, plus or minus one turn, from the standard thread makeup allowances.

The overall external thread length is divided into three sections. The first section, from left to right, identifies the number of threads that engage with hand tightening. The second group consists of those additional threads that engage when a wrench is properly applied. The third group consists of those threads, called *vanishing threads*, that are essential in the threading process but do not engage. If there are too many threads showing, the joint needs to be made up further. If there are too few threads showing, the joint is made up too tightly. In either case, leaks are likely, and the system will not fit the way it was designed.

4.2.0 Threading Tools

The life of any tool depends on proper lubrication. If the frame is steel, a light coating of oil is essential to prevent rusting. If there are moving parts, the bearing surfaces require lubrication.

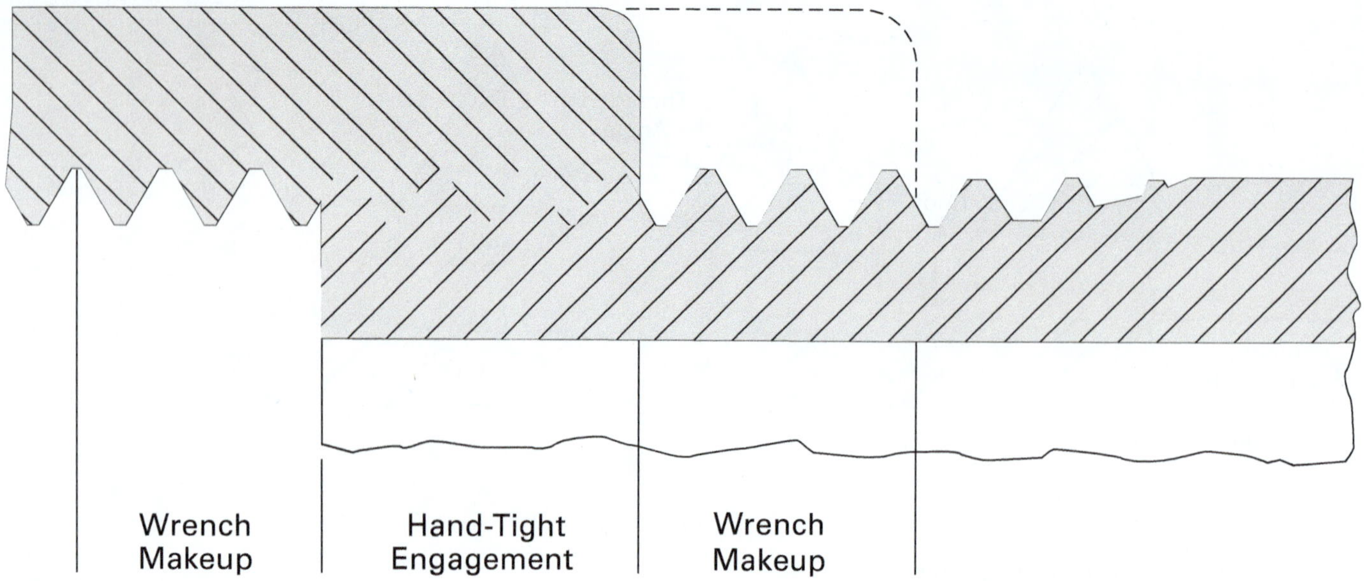

Figure 26 Hand engagement and wrench makeup.

Excess lubricants leave a trap for dust, dirt, grime, and abrasive action. Follow the manufacturer's specified lubrication requirements for all tools. Sprinkler fitters use a variety of hand tools for cutting, reaming, and threading pipe. The most common hand tools include steel pipe cutters, hinged cutters, reamers, and dies. In many cases, the tool selected must be the exact size required to perform the job, or personal injury and/or equipment damage can occur.

4.2.1 Hand Threaders

Occasionally, you may have to hand thread a piece of pipe or a hanger rod. A solid die with a holder is required for this procedure (*Figure 27*). Dies come in a wide range of sizes and thread patterns. Some holders are called *three-way pipe threaders* and hold up to three dies within a specified range. Each die set is locked in place and a guide keeps the threader straight for consistent, true threads. Most die sets include common die sizes as well as the holder.

Pipe Thread Standard

At its inception, this standard was known as the *Briggs Standard.* Robert Briggs originally formulated this basic thread schedule in England and published the Briggs Standard between 1820 and 1840. At that time, all manufacturers had their own standards and one manufacturer's threads would not fit those of another.

In 1886, manufacturers in the United States adopted the Briggs Standard jointly with the American Society of Mechanical Engineers (ASME). That standard resulted in the ability to interchange pipe and fittings made by various manufacturers and is the same basic standard that is in use today.

Threadable Thin-Wall Pipe

The threading process removes metal from the pipe wall, thereby reducing the wall thickness. If the pipe wall is thin to start with, the risk of failure is increased by threading. For that reason, NFPA 13 has traditionally limited the schedule of steel pipe that can be threaded. Over the years, however, pipe manufacturers have developed thin-wall steel pipe that is suitable for threading.

The annex of NFPA 13 states that threaded thin-wall pipe should be checked by the installer using working ring gauges conforming to the Basic Dimensions of Ring Gauges for USA (American) Standard Taper Pipe Threads, NPT as per ANSI/ASME. Not all Authorities Having Jurisdiction (AHJs) will accept threaded thin-wall pipe.

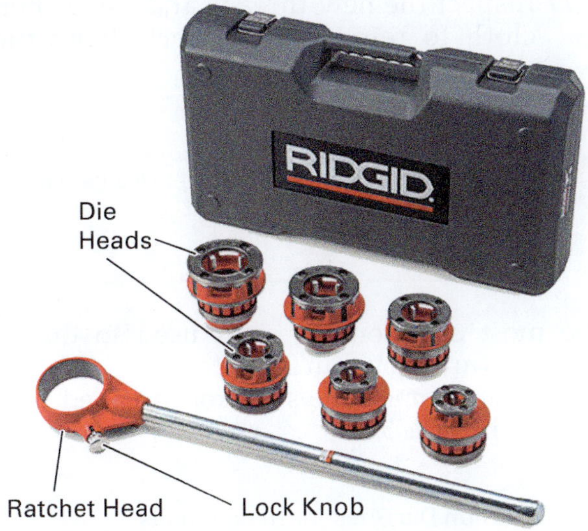

Figure 27 Hand threader.

Threading Pipe Using a Manual Pipe Threader

Follow these steps to thread pipe using a manual pipe threader:

Step 1 Identify the kind of material being threaded.

Step 2 Identify the diameter of the material being threaded.

Step 3 Select the die head best suited for the threading process.

Step 4 Insert the die head into the pipe threader handle.

Step 5 Inspect the die and die handle to ensure that all parts are clean and in good condition.

Step 6 Secure the pipe in a vise.

Step 7 Make sure that the end of the pipe is square, clean, and deburred.

Step 8 Position your body so that you are balanced, facing the work area, and within easy reach of the work area.

Step 9 Position the die flush on the end of the pipe.

Step 10 Apply thread cutting oil (*Figure 28*).

Step 11 Press the die against the end of the pipe and turn the die handle clockwise to start the threading process.

Step 12 Turn the die handle two full revolutions around the pipe.

Figure 28 Oiling the threader.

NOTE

Continue to apply thread cutting oil as necessary to keep a thin coat on the threads while cutting.

Step 13 Stop the work and back the dies off the pipe.

Step 14 Check the threads to ensure that they are properly started and clean the cutting threads from the die if necessary.

Step 15 Rethread the die onto the pipe and continue the process until the threading is complete. The threading is complete when one full thread extends past the back edge of the dies.

CAUTION

Do not thread the pipe too far. This can cause an improper fit between the fitting and the pipe, resulting in leaks.

Step 16 Back the die off the pipe.

Handheld Power Pipe Threaders

A handheld power pipe threader can be used in conjunction with a portable stand to thread pipe in the field. This battery-operated threader uses the same dies as an equivalent manual threader and can be used to thread black steel pipe up to 2" in diameter. When pipe is being threaded, a support arm is clamped to the pipe and the threader is rested against it to counteract the torque and hold the pipe in place.

Figure Credit: Courtesy of Milwaukee Electric Tool Corporation

Step 17 Inspect the new threads and use a shop cloth to remove any debris from the threads.

WARNING!

Freshly cut threads are very sharp. Use care in handling them to avoid cuts.

The most common die sizes used to thread hanger rods are shown in *Table 2*.

The most common die sizes for threaded pipe are shown in *Table 3*.

Table 2 Common Die Sizes for Hanger Rods

Hanger Rod Diameter	Die Size (threads/inch)
3⁄8"	16
1⁄2"	13
5⁄8"	11

Table 3 Common Die Sizes for Threaded Pipe

Nominal Pipe Size	Die Size (threads/inch)
1⁄16" to 1⁄8"	27
1⁄4" to 3⁄8"	18
1⁄2" to 3⁄4"	14
1" to 2"	11 1⁄2
2 1⁄2" and up	8

4.2.2 Portable Power Threading Machines

Portable power threading machines are used extensively in the sprinkler trade because they require less time than hand threading. The common portable models can thread pipe up to 2" in diameter. Always read the manufacturer's instructions for care, maintenance, and safety instructions before using a power threader. *Figure 29* shows the components of a compact threading machine. *Figure 30* shows the oil reservoir and pump assembly.

WARNING!

Never operate a power tool until you have been trained to use it. You can greatly reduce the chance of accidents by learning safety rules for each task that you perform. Wear appropriate personal protective equipment when using a tool or when working near someone who is using a tool.

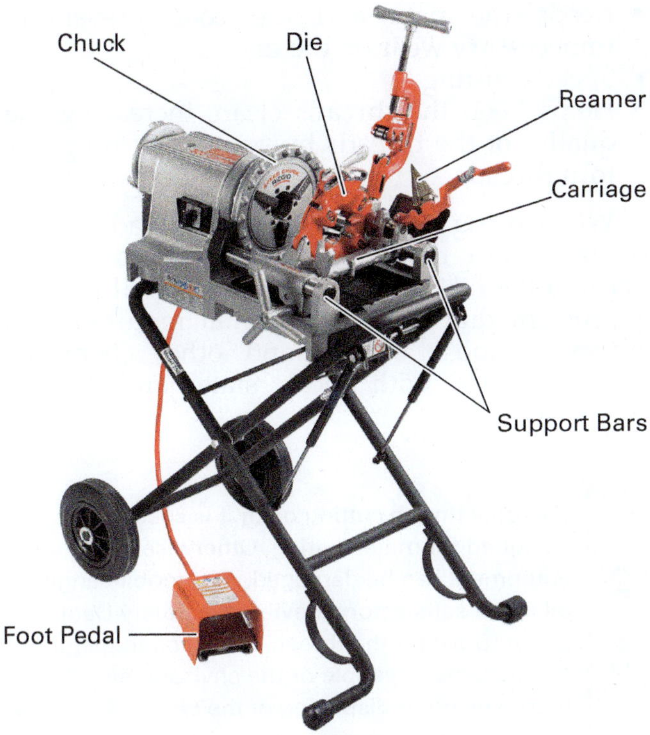

Figure 29 Compact threading machine.

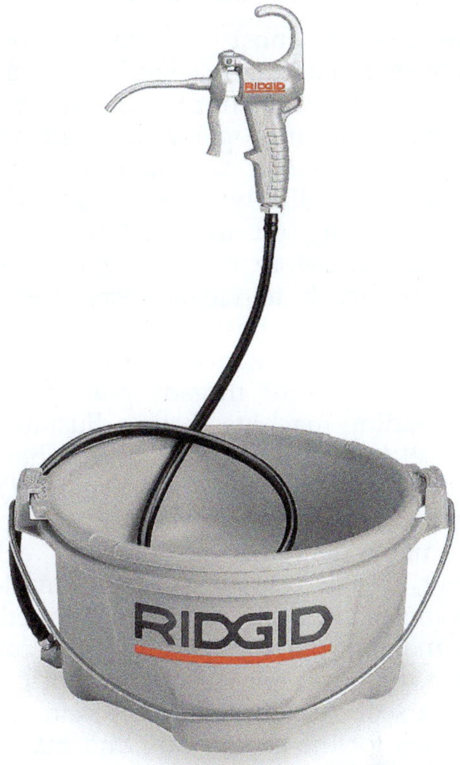

Figure 30 Oil reservoir and pump assembly.

As a safety precaution, portable power threading machines are equipped with a foot pedal for controlling the power. Lifting your foot off the pedal automatically turns the power off. Accessories for cutting and reaming pipe can also be added to the same machine for performing these tasks. The portable power threading machine's standard components include the following:

- Chuck
- Ways and support bars
- Carriage
- Dies for power threading machines

Chuck

The chuck on a power threading machine centers the pipe and holds it in place for threading. The dual-jaw chuck rotates with the pipe as it is being threaded. On most models, the rear jaw does the actual centering and must be tightened only enough to ensure that the jaws are touching the pipe. The front jaw clamps the pipe in place and must be tightened so that the pipe does not slip.

Ways and Support Bars

A carriage moves along a bedway or support bar and holds the die head in place. Horizontal bars that are supported on both ends are called *ways*. Horizontal bars that are supported on one end are called *support bars*. The ways or support bars can be either flat, round, or V-shaped. It is important to keep them free of dirt, grit, and metal shavings to allow free movement of the carriage. The ways or support bars align the die head with the pipe being threaded. Do not use the ways as a tool rack or handle. Exercise caution to ensure that the ways do not get dented or bent.

Carriage

Generally, there are two carriages on a power threader. One carriage carries the die head, and the other carriage carries the chuck. Both carriages must be kept free of dirt, grit, and metal shavings. If a carriage becomes excessively worn, there will be too much side play (side-to-side movement) and it must be replaced. Carriages on flat bedways will sometimes have gibs (collars) that can be adjusted to eliminate excessive side play.

Die Sets

It is very important to keep the die sets together. Do not mix dies of one set with dies from another set. Usually, each die is marked with a serial number and the four dies that are made together have the same number.

Dies for Power Threading Machines

The dies are the most important components of a power threading machine and can cause numerous problems if not maintained properly, from chipping to improperly sized threads. Dull dies can be sharpened or honed, and dies must be cleaned periodically using a manufacturer-approved cleaning solution and lightly oiled with lubricating oil. Before reinserting the die head, use a wire brush to clean away any foreign material that may have accumulated in the holding apparatus. Because 1" and 2" pipe have the same number of threads per square inch, some threading machines have adjustable heads that are set to size so the dies do not have to be changed for different pipe sizes (*Figure 31*).

Threaders are precision cutting devices and proper maintenance is essential to achieving consistent, reliable results. For example, torn threads can be caused by dies with chipped or dull teeth. A machine that is out of alignment can create off-center threads. Poor threads cannot provide a watertight seal. Always follow the manufacturer's instructions for the proper care and maintenance of threaders. Some manufacturers also provide troubleshooting tables that can be helpful when diagnosing thread issues.

Thread Cutting Oil

The use of thread cutting oil (*Figure 32*) is essential when threading pipe and rod by hand or using power tools. Thread cutting oil is available in a variety of types and sizes. Make sure the thread cutting oil selected is compatible with both the threader and the application. Thread cutting oil serves the following purposes:

CAUTION

If CPVC is being used on a system with steel piping, make sure the thread cutting oil is approved for use with CPVC. Incompatible thread cutting oil can damage CPVC.

- Keeps the pipe and die cool, preventing unnecessary wear on the die
- Speeds cutting
- Helps keep the threads clean, increasing the quality of the threads by preventing rough or torn threads

When using tools that require thread cutting oil, make sure that water or other liquids do not get into the oil. The disposal of used oil may be a problem due to environmental requirements. Always dispose of oils and other chemicals in accordance with local, state, and federal requirements.

CAUTION

Use only thread cutting oil that is specifically designed for metal cutting. Otherwise, the equipment can be damaged. Automobile engine oil is not satisfactory. Review the Safety Data Sheet (SDS) for the thread cutting oil to identify any hazards to people or the environment before using or disposing of the oil.

4.3.0 Threading Pipe

In order to create a strong, leakproof, threaded joint, the threads must be clean and smoothly cut. They must have the correct pitch, lead, taper, form, and size. If these conditions are not met, the connection will leak.

CAUTION

Steel pipe with wall thicknesses less than Schedule 30 in sizes 8" and larger or Schedule 40 in sizes smaller than 8" typically must not be joined by threaded fittings, although there are exceptions to this rule made for Listed thin-wall pipe that has been designated as threadable. Only if a threaded assembly is investigated for suitability in sprinkler installations and Listed for this service can it be used.

Figure 31 Universal die head.

Figure 32 Thread cutting oil.

4.3.1 Safety Precautions for Power Tools

Using power tools to cut, clean, ream, or deburr pipe requires specific safety precautions. Adhere to the following safety guidelines when using power tools:

- Check all cords for signs of excessive wear or fraying.
- Ensure that the plug has a proper grounding connection (three wires on 120-volt AC). If the grounding pin is missing or broken, do not use it.
- Ensure that there is a ground fault circuit interrupter (GFCI) on the line.
- Do not use power tools near water, while standing in water, or in wet, damp, or foggy weather.
- Ensure that extension cords are safe for the application. Discard cords that are cracked, frayed, or cut. Ensure that extension cords are properly grounded.
- Wear all appropriate personal protective equipment.

WARNING!

When working with a power threading machine, do not wear loose clothing or jewelry that could easily be caught in the rotating parts of the machine. Always wear appropriate personal protective equipment. Tie back long hair to prevent it from being entangled in the machine during operation. Use a piece of cloth or cardboard to absorb any oil that spills during the machine's operation, so the oil is not tracked around the jobsite.

4.3.2 Power Threading Machine Setup

It is important to read the manufacturer's instruction manual thoroughly before operating the equipment. Make sure you understand how the machine works, what its capabilities are, and what hazards are associated with using the equipment. Follow these steps to set up a power threading machine:

CAUTION

Never operate any tool or machine unless you have read the manufacturer's operating instructions and have been trained to use it safely. Always wear the appropriate personal protective equipment.

Step 1 Make sure the machine's power switch is in the OFF position before connecting the power.

Step 2 Make sure the machine is properly grounded. Insist on installation of a GFCI on the line or use a GFCI extension cord (*Figure 33*).

Step 3 Remove any tools that might be on the machine.

Step 4 Check the die heads, support bars, and carriage. They must be free of dirt, grime, shavings, or any other foreign material.

Step 5 Use a floor pipe stand or support for long lengths of pipe (*Figure 34*).

Step 6 Make sure the threading machine and pipe stand are level and aligned.

Power Threading Machines

Power threading machines operate at very high torques and have sharp metal teeth that can easily catch loose clothing or jewelry. Exercise caution and wear appropriate personal protective equipment when operating a power threader.

Figure Credit: Shutterstock.com/Valentinos71

Figure 33 Extension cord with GFCI.

Alignment

The components of a power threading machine must be properly aligned and adjusted to achieve quality threads. If the machine is out of alignment or adjustment, the threads will be defective. Misalignment may prevent the die from starting on the pipe, break the die, cause off-center threads, or cause spiraled threads. Follow these steps to check the alignment of a power threading machine:

Step 1 Ensure the power is off.

Step 2 Place a pipe in the chuck.

Step 3 Close the die head so that it touches the pipe.

Step 4 Ensure the pipe touches all the die segments simultaneously.

If the pipe does not touch all points of the die segments simultaneously, the machine is out of alignment. Follow the manufacturer's alignment procedure to align the machine before going ahead with the threading operation.

Leveling

It is also critical that the power threading machine be as level as possible and be set on a stable surface. In the field, this is the fitter's responsibility. Make sure that the machine is level and stable before the power is turned on.

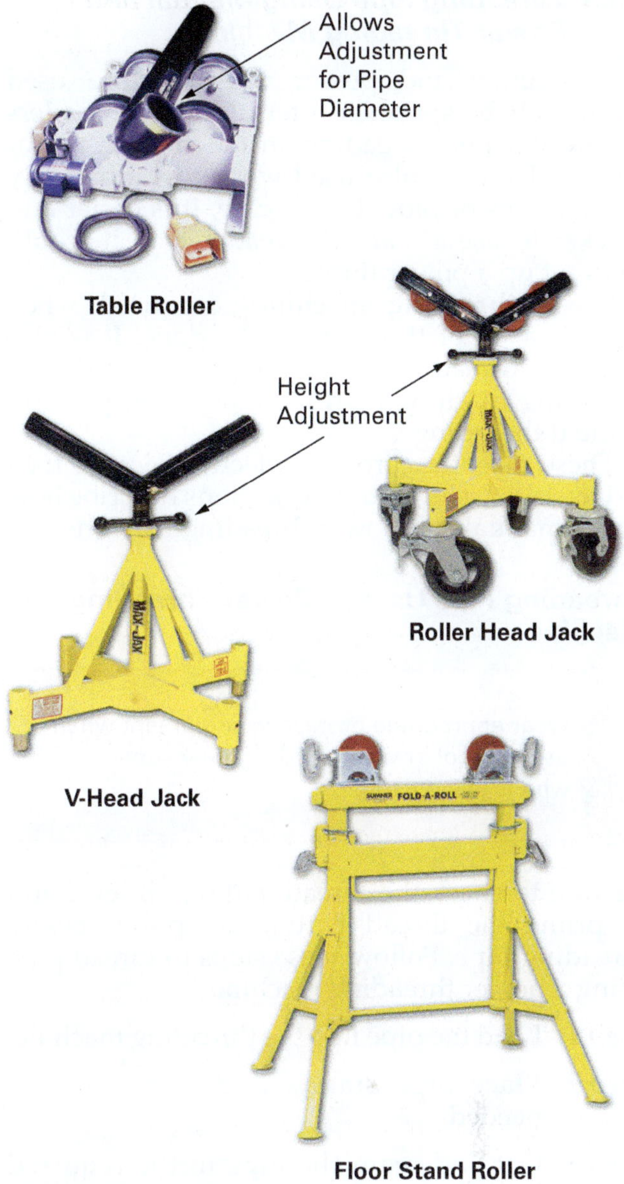

Figure 34 Pipe supports.

After ensuring that the threading machine is level, insert the pipe into the chuck and secure the other end of the pipe with a pipe stand or support. Some models allow the pipe to move backward and forward only; other models allow the pipe to rotate freely for operations such as cut grooving or roll grooving. A pipe support or stand keeps the pipe aligned with the machine.

> **NOTE**
> After leveling the machine, add a slight tilt to the machine so that thread cutting oil will not run up the pipe.

4.3.3 *Threading Pipe Using Manual and Power Threading Machines*

Both manual and power threaders are used extensively by sprinkler fitters. Manual threaders are used when threading installed pipe or for repairs. They are also used when threading only a few pieces of pipe. In this case, it is easier and quicker to use a manual threader than it would be to set up a power threader.

Power threading machines are used when there is a lot of threading to be done. It is also easier to thread larger size pipe with a power threading machine than it would be to perform manual threading.

These machines are cost effective because they reduce labor. The following sections describe how to use manual and power threading machines.

Threading Pipe Using a Power Threading Machine

> **WARNING!** Wear appropriate protective equipment when using a tool or when working near someone who is using a tool.

Be sure to check the thread cutting oil level and to prime the thread cutting oil pump before threading pipe. Follow these steps to thread pipe using a power threading machine:

Step 1 Load the pipe into the threading machine.

Step 2 Place pipe stands under the pipe as needed.

Step 3 Cut and ream the pipe to the required length.

Step 4 Make sure that the proper dies are in the die head.

Step 5 Loosen the clamp lever.

Step 6 Move the size bar to select the proper die setting for the size of pipe being threaded.

Step 7 Lock the clamp lever.

Step 8 Swing the die head down to the working position.

Step 9 Close the throw-out lever.

Step 10 Lower the lubrication arm and direct the oil supply onto the die or prime the hand pump.

Step 11 Turn the machine switch to the forward position.

Step 12 Step on the foot switch to start the machine.

Step 13 Turn the carriage to bring the die against the end of the pipe.

Step 14 Apply light pressure on the handwheel to start the die.

Step 15 Release the handwheel once the dies have started to thread the pipe. The dies feed onto the pipe automatically as they follow the newly cut threads.

> **CAUTION** Make sure that the die is continuously flooded with oil while cutting to prevent overheating the die and the pipe. Overheating can damage the die and the pipe threads.

Step 16 Open the throw-out lever as soon as one full thread extends from the back of the dies.

Step 17 Release the foot switch to stop the machine.

Step 18 Turn the carriage handwheel or arm to back the die off the pipe.

Step 19 Swing the die and the oil spout up and out of the way.

Step 20 Screw a fitting onto the end of the pipe to check the threads. If the threads are correct, you should be able to screw a fitting three to four revolutions by hand onto the new threads. If the threads are not deep enough, adjust the die with the clamp lever and repeat the threading operation.

Step 21 Turn the machine switch to the OFF position.

Step 22 Open the chuck and remove the pipe.

4.3.4 *Joint Lubricants*

A good grade of pipe thread compound (also referred to as *joint compound* or *pipe dope*) or polytetrafluoroethylene (PTFE) tape (also known as **Teflon® tape**) must be applied to the male threads to lubricate the joint in accordance with *ASME B1.20.1* (referenced in NFPA 13). This will prevent the threads from seizing up and ease the makeup of the joint. Always check the shop drawings before applying any type of lubricant. See *Figure 35*.

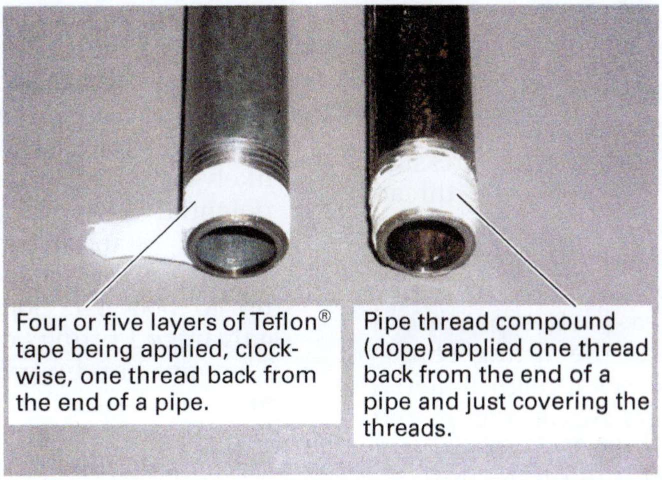

Figure 35 Teflon® tape and pipe thread compound.

Teflon® Tape

Teflon® tape is a special tape that is wrapped around threads to prevent metal-to-metal contact and to create a leakproof joint. Follow these steps to use Teflon® tape:

> **WARNING!**
>
> Never use Teflon® tape on pipe that is to be welded or pipe that carries steam or other high-temperature service. When heated, Teflon® tape emits a highly toxic gas that could be fatal.

Step 1 Remove all excess thread cutting oil from the threads to improve the grip of the tape on the threads.

Step 2 Start the tape at the end of the pipe, leaving the first full thread bare to prevent the tape from bunching up at the beginning of the thread.

Step 3 Wrap the tape around the pipe in the direction in which the joint is to be assembled. Wrap the tape around right-hand threads in a clockwise direction and around left-hand threads in a counterclockwise direction. This prevents the tape from coming off when the pipe and fitting are joined.

Step 4 Continue to wrap the tape around the joint, overlapping the edges of each wrap until all remaining threads have been covered.

Step 5 Press the tape against the threads to seal it to the threads and to prevent the tape from slipping off the threads once you start to make up the joint.

Did You Know?

Use ½" wide Teflon® tape for pipe that is ¾" nominal size and smaller. Use ¾" wide tape for pipe that is 1" nominal size or larger.

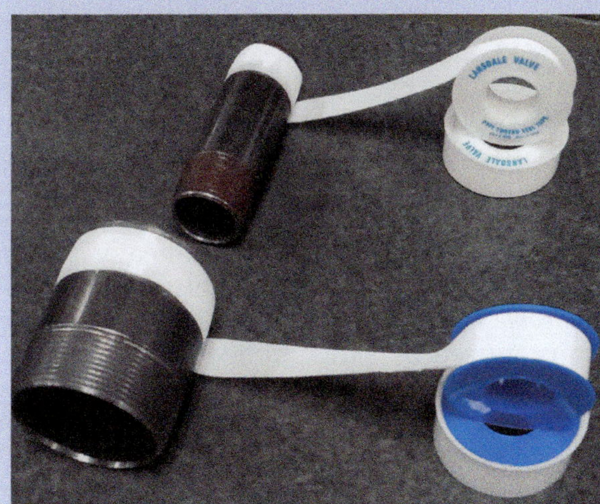

Figure Credit: Mac McGaffagan

Pipe Thread Compound

Pipe thread compound is used to fill the gap produced by the thread profile. Use caution when applying thread compound on dry systems. If the thread compound is too fluid, the air pressure of a dry system can literally push the thread compound out of the threads.

Apply thread compound properly and uniformly to the outside threads that screw into the fitting. If using a paste-type thread compound, placing it inside the fitting results in wasted sealant that becomes foreign material in the sprinkler system. Pipe thread compound is applied to the outside (male) threads using a small brush. It must be applied to fill the threads.

4.4.0 Types of Threaded Fittings

Threaded pipe fittings can be made of **ductile iron**, **malleable iron**, or **cast iron**.

- *Cast iron fittings* — Cast iron fittings are made of an alloy of iron and carbon that is cast in a mold. They are the most common type of threaded fitting. Cast iron fittings are manufactured in large quantities and are the least expensive type of fitting. They are also available in the widest variety of styles and sizes. One disadvantage of cast iron fittings is that they are brittle and break more easily than other fittings. Dropping one on the floor can crack the fitting and cause a leak. Under NFPA 13, the standard-weight pattern cast iron fittings can be used with working pressures up to 300 psi in sizes 2" or smaller. Above 2", the standard weight cast iron fitting is permitted up to only 175 psi. Beyond these ratings, the extra-heavy pattern (Class 250) must be used; however, there are fewer extra-heavy pattern fittings than standard weight fittings.
- *Malleable iron fittings* — Malleable iron fittings are produced by slowly reheating and cooling cast iron fittings. The result is a change in the crystalline structure of the carbon, giving the fitting increased strength, pliability, and shock resistance. Because of this, malleable fittings are usually more expensive than cast iron fittings. Although malleable fittings are more expensive, they weigh less than comparable cast iron fittings. In some instances, the higher cost of the fitting is more than offset by the reduction in shipping costs, which are based on weight. Instead of cracking, malleable fittings stretch. However, this stretching may result in leaks and unlike cast iron, malleable fittings that must be removed are often not reusable. On the other hand, there are times when the flexibility of a malleable fitting is important, such as in cold-temperature environments. If a malleable fitting accidentally freezes, it is more likely to stretch than it is to crack, reducing potential water damage. NFPA 13 allows the use of standard-weight pattern malleable fittings of 6" or smaller with working pressures up to 300 psi. Some manufacturers supply galvanized fittings only as the malleable type. They are hot-dip galvanized, and the threads are chased after the fitting is dipped. Therefore, the threads are untreated.
- *Ductile iron fittings* — Ductile iron fittings are similar in appearance, design, and function to those made from other types of iron. Ductile iron fittings are available in the same thickness classes, pressure ratings, linings, and joint designs as standard ductile iron pipe. The standard class numbers of the fittings must match those of the pipe being used.

Black Steel Pipe in Exposed Locations

Black steel pipe is run exposed in locations without ceilings, such as warehouses, gyms, industrial locations, and even residences. This residence features exposed black steel pipe as an architectural element that complements the black-tie beam connector plates.

Figure Credit: Metropolitan Fire Protection

Threaded fittings are cast and then machined. Sometimes the castings are imperfect and will leak. The most common problem is caused by sand holes. Whether a leak is created from a sand hole, a crack, or a distorted fitting, the fitting must be removed and replaced. Report all problems to your supervisor.

Many types of fittings are available in cast, malleable, and ductile iron. The fittings include the following:

- Elbows
- Tees
- Crosses
- Mechanical tees
- Couplings
- Nipples
- Extension fittings
- Plugs
- Bushings
- Caps

4.4.1 Elbows

The two most common types of elbows are standard elbows and reducing elbows (*Figure 36*). A standard elbow has internal (female) threads with the same nominal pipe size on both outlets. Standard elbows are available in 90-degree and 45-degree patterns in both black and galvanized finishes. A reducing elbow has internal threads with different nominal pipe sizes on the two ends. When specifying a reducing elbow, the largest opening is shown first (e.g., 2" × 1").

4.4.2 Tees

Tees allow three pieces of pipe to be joined (*Figure 37*). Tees can be straight tees with three outlets of the same nominal pipe size, or reducing tees, with outlets that are not the same size. The largest outlet on the straight-through waterway is read first, regardless of how the tee fits into the system. In the sprinkler industry, when a tee is installed with the largest outlet on the side, it is called a *bullhead tee*.

Local supply houses stock tee sizes that are frequently demanded. Others can be obtained from the foundry by special order. A lead time of as much as six months may be required to obtain some types of tees.

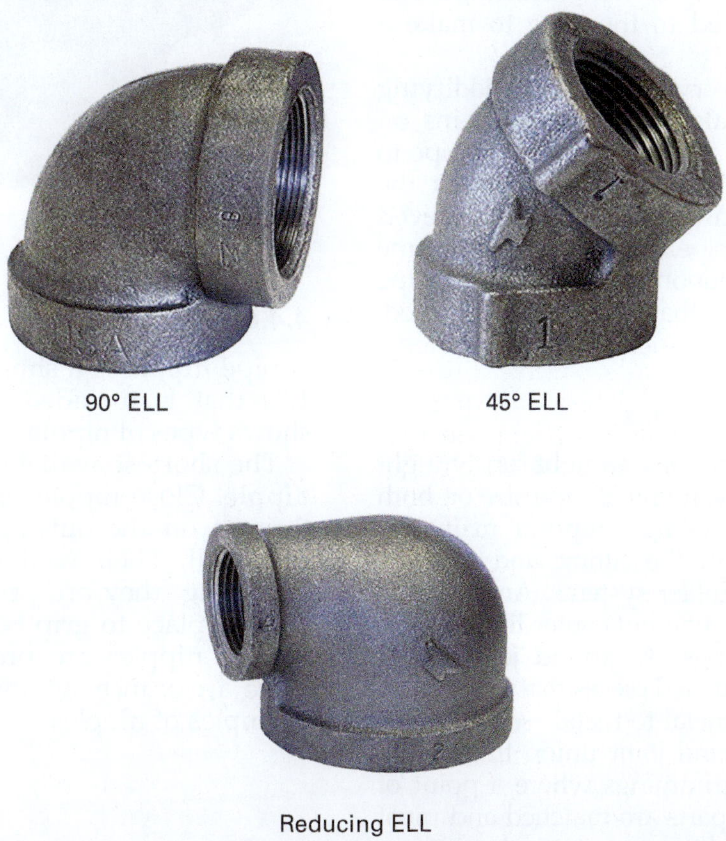

Figure 36 Threaded elbows.

Figure 37 Threaded tees.

4.4.3 Crosses

Crosses can be used to join four pieces of pipe at the same point. However, they are seldom used due to availability and cost (*Figure 38*).

4.4.4 Mechanical Tees

A threaded mechanical tee provides a side outlet without having to make a joint in the pipe. A hole of the proper size is cut into the pipe using a hole saw and the coupon retrieved. The mechanical tee is then clamped to the pipe so that the rubber gasket in the fitting aligns with the hole. The U-bolt is then tightened to the pipe to make a watertight connection.

Mechanical tees are very useful when modifying existing systems or making additional cut-ins on new systems without having to remove the pipe to be cut and threaded. After making a hole for the fitting, the pipe coupon (cutout) must be retrieved to avoid plugging the pipe. As stated earlier, some AHJs require that the coupon be attached to the pipe so the inspector can see that it has been removed. *Figure 39* shows a threaded mechanical branch tee.

4.4.5 Couplings and Unions

Couplings align two pipes on a straight run. Straight couplings have the same nominal pipe size on both ends. A **concentric reducing coupling** maintains one center line through the fitting and is more commonly used in sprinkler systems. An **eccentric reducing coupling** has different center lines for the inlet and outlet openings. A ground joint union fitting assembly has matched pieces that are ground together to form a metal-to-metal seal. In the sprinkler industry, ground joint union fittings are used mostly in valve trimmings where a point of joining is needed. The parts are matched and must be kept together. *Figure 40* shows a typical union.

Figure 38 Threaded cross.

4.4.6 Nipples

A pipe **nipple** is a short piece of pipe (less than 12") that is threaded on both ends. *Figure 41* shows types of nipples.

The shortest available nipple is called a **close nipple**. Close nipples consist almost entirely of threads on the outside, making them difficult to install. Their wall is weak, contributing to flattening, they are prone to leak, and there is no safe place to grip because of all the threads. Longer nipples are preferable as they provide more insurance against leaks. *Table 4* shows examples of nipples.

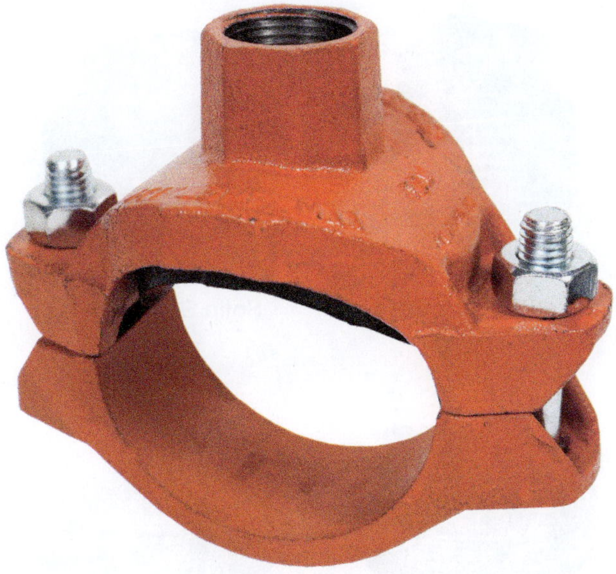

Figure 39 Threaded mechanical branch tee.

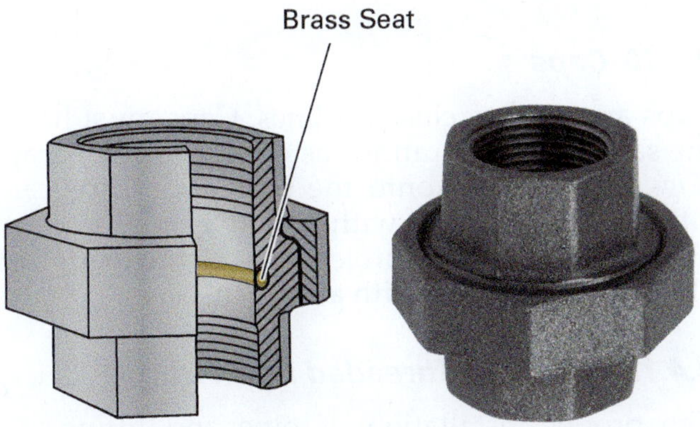

Figure 40 Malleable threaded union with brass seat.

4.4.7 Extension Fittings

Extension fittings are available and may be used to resolve discrepancies; however, they must be installed in accordance with NFPA, manufacturer, and Listing requirements. Extension fittings are covered in NFPA 13, ***Section 16.8.6***. *Figure 42* shows a typical extension fitting.

4.4.8 Plugs

Plugs are fittings that are used to close off lines and must have the same pressure ratings as other fittings. A plug fitting (*Figure 43*) screws into the fitting end. Plugs can either be solid square-head plugs or countersink plugs. The latter have a square depression where a square key is inserted for tightening.

Pipe Nipples

Pipe nipples are available from suppliers in a variety of short lengths. It generally makes more sense to buy commonly used nipples in bulk, especially close and shoulder nipples, rather than to cut and thread them in the field. Most threading equipment cannot be used to make close or shoulder nipples. However, both short and long nipples can be fabricated using pieces of scrap piping gathered from the field installations. This type of task can be accomplished in the shop during slow periods.

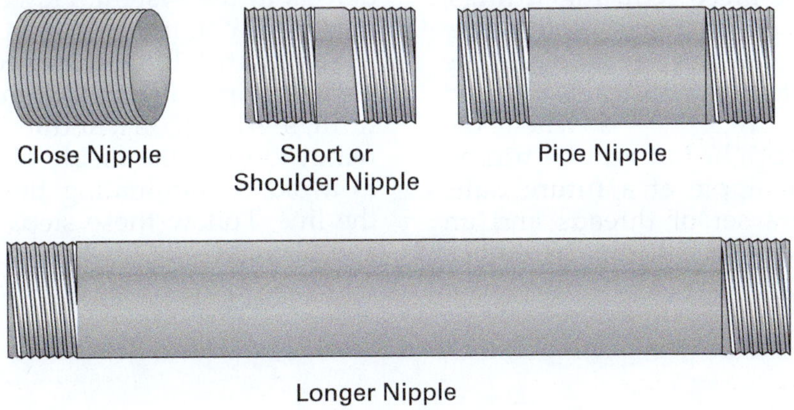

Figure 41 Types of nipples.

Table 4 Close and Next-to-Close Nipple Lengths

Nominal Pipe Size	Length of Close Nipple	Length of Shoulder Nipple
1	$1\frac{1}{2}$	2
$1\frac{1}{4}$	$1\frac{5}{8}$	2
$1\frac{1}{2}$	$1\frac{3}{4}$	2
2	2	$2\frac{1}{2}$
$2\frac{1}{2}$	$2\frac{1}{2}$	3

Note: All dimensions in inches.

Figure 42 Extension fitting.

4.4.9 Bushings

A **bushing** is occasionally used in fittings when a reducing fitting is not available. However, NFPA 13 calls for one-piece reducing fittings, prohibiting the use of bushings unless the required size fitting is not available. One exception for hex bushings specifically allowed in NFPA 13, *Section 16.3.12.2*, is when the sprinkler is initially upright but will be turned pendent with a drop nipple at a future date. Bushings add an extra set of threads and an extra potential for leaks to a fitting assembly. *Figure 44* shows types of threaded bushings.

Figure 43 Threaded plugs.

4.4.10 Caps

Caps are used to close off lines. Caps must have the same pressure ratings as other fittings. A cap (*Figure 45*) screws onto the pipe end. Caps can either be plain caps with round heads or caps that have a square projection on the body to allow for tightening with a wrench.

4.4.11 Installing Threaded Pipe Fittings

The proper installation of pipe and fittings is crucial to fabricating a quality sprinkler system. Poor fitting practices can result in leaks and other defects in the system. However, this skill can be mastered with training and practice.

Before installing threaded pipe and fittings, always wipe down and inspect the pipe and fittings. All oil, dirt, and metal shavings must be removed from the inside of the pipe and fittings to avoid contaminating the system or clogging the line. Follow these steps to fit threaded pipe and fittings:

Figure 44 Types of threaded bushings.

Figure 45 Threaded cap.

CAUTION

Never operate any tool or machine unless you have read the manufacturer's operating instructions and have been trained to use it safely. Always wear the appropriate personal protective equipment.

Step 1 Determine the type, size, and schedule of pipe being used in the system.

Step 2 Determine the length of pipe needed by measuring the center-to-center distance between fittings, then subtract the takeout on each side of the fitting. The takeout is simply a way to account for the pipe length that is threaded into the fitting. For example, if sprinklers are spaced 10' apart using a typical 1" threaded tee, the takeout is about 1" on both sides. Therefore, the distance between sprinklers would be 10'-0" – 1" – 1" = 9'-10". This is the center-to-center distance between fittings.

NOTE

Takeouts vary by manufacturer. Always consult the manufacturer for the fitting in use.

Step 3 Cut the pipe to the desired length.

Step 4 Ream the pipe to remove all burrs.

Step 5 Thread the pipe, taking care to cut the threads at the proper depth. Make sure you form at least seven threads to ensure proper fitting makeup. After cutting, wipe the pipe with a shop cloth to remove any oil or metal shavings.

Step 6 Select the proper size, shape, and type of fittings, as indicated on the shop drawings.

Step 7 Apply thread compound or Teflon® tape to the pipe threads.

Step 8 Screw the fitting on the pipe until it is hand tight (*Figure 46*).

Step 9 Tighten the fitting slowly using a pipe wrench.

CAUTION

Do not allow the threads to bottom out in the fitting. This can damage both the threads and the fitting and result in leaks.

Did You Know?

The monkey wrench is a general-purpose wrench with an adjustable jaw for turning nuts of various sizes. Charles Moncky invented this tool around 1858. Monkey wrench also has an informal meaning: to disrupt, as in "He threw a monkey wrench into our plans." Monkey comes from the last name of the wrench's inventor.

Figure 46 Screw the fitting on until it is hand tight.

4.0.0 Section Review

1. The National Pipe Thread (NPT) taper is _____.
   a. $\frac{1}{2}$"
   b. $\frac{1}{4}$"
   c. $\frac{1}{8}$"
   d. $\frac{1}{16}$"

2. The die size used with 1" pipe is _____.
   a. 8 threads per inch
   b. $11\frac{1}{2}$ threads per inch
   c. 14 threads per inch
   d. 18 threads per inch

3. Pipe thread compound is applied to the _____.
   a. inside of the fitting only
   b. inside of the fitting and outside threads
   c. outside threads only
   d. thread cutting die and inside of the fitting

4. The *most* brittle type of threaded fitting is made of _____.
   a. ductile iron
   b. cast iron
   c. malleable iron
   d. wrought iron

SECTION FIVE

5.0.0 FLANGED PIPE AND FITTINGS

Objective

Select and install flanged fittings.

a. Identify flanged fittings and their applications.
b. Install flanged pipe fittings.

Performance Task

5. Install flanged fittings using the proper bolt tightening sequence.

Trade Term

Companion flange: A flange that is used to connect a fitting or flange directly to threaded pipe.

A *flange* is a flat collar welded to the end of a pipe, valve, or fitting. It is aligned with a mating flange and connected using bolts. Flanged fittings may be used in a variety of applications, including riser assemblies, cold storage facilities, fire pump installations, and division boundaries between systems.

Flanges are specified by class designations, with each class representing the pressure rating at a specified temperature. Class 125 and Class 250 flanges are made of cast iron, while Class 150 and Class 300 flanges are made of ductile iron, steel, or stainless steel. As the class number increases, the rated pressure also increases, so a Class 250 flange is rated for a higher pressure/temperature combination than a Class 125 flange. The two most common class designations used in sprinkler fitting are Class 125 and Class 250.

NOTE

The class rating does not match the pressure rating at ambient temperatures. For example, Class 125 flanges are rated at 125 psi at 353°F, but up to 200 psi at ambient temperatures up to 150°F. Similarly, Class 250 flanges are rated at 250 psi at 450°F, but up to 500 psi at ambient temperatures.

5.1.0 Types of Flanged Pipe Fittings

Types of flanged fittings include the following:

- Elbows
- Tees
- Crosses
- Flanged reducers
- Companion flanges
- Blind flanges
- Floor flanges
- Extra-heavy flanges
- Grooved adapters

Unlike other types of fittings that are readily available from a supply house, flanged fittings must be ordered for each job. Be sure to allow enough lead time when ordering so they will be available when needed. Some common flanged fittings are shown in *Figure 47.*

5.1.1 Elbows

Elbows commonly used in Fire Department Connections (FDCs) are the 45-degree lateral and 90-degree flange-thread elbows. Some 90-degree straight elbows have flanges on both ends and others have flanges on only one end. Flanged-base elbows, although not often used, are especially useful in the horizontal run of riser assemblies where the end riser is fed by an elbow and support below the riser is needed. The cast-on base saves a lot of difficulty otherwise involved in supporting an elbow from below. The base may be round or square. Long-radius elbows also are available for fire sprinkler use. These are used to reduce friction losses or to make a connection in an installation that has special geometry.

5.1.2 Tees

In addition to the flanged-base and long-radius elbows discussed, there are two special drain fittings that are used in FDCs. One is the base-flanged tee which is used when support is required at a connection, and the other is the reducing standard flanged tee that is used when larger drain piping needs to connect with smaller drain piping.

5.1.3 Crosses

Flanged crosses are fittings used to connect four pieces of pipe. As with many of the flanged fittings, make sure to allow enough lead time when ordering crosses. Flanged crosses are rare and may have to be specially cast.

Figure 47 Flanged fittings.

5.1.4 Flanged Reducers

Flanged reducers are available in many sizes and ratings. Both concentric and eccentric versions are available. Concentric flanged reducers have the same center line. Eccentric flanged reducers do not have the same center line. Concentric and eccentric flanged reducers are used primarily in fire pump installations.

5.1.5 Companion Flanges

A **companion flange** is mated with a fitting, device, or flange of the same nominal pipe size and screwed to the end of a piece of pipe. *Figure 48* shows a standard companion flange. Reducing companion flanges are also available. *Table 5* shows cast iron flange dimensions. Mating flanges must be matched up in terms of these dimensions. Note that Class 250 flanges are configured in a different way and cannot be used in place of Class 125 flanges.

5.1.6 Blind Flanges

Blind flanges (*Figure 49*) are used to close off the ends of flanged pipes, valves, and pressure vessel openings. From the standpoint of internal pressure and bolt loading, blind flanges are the most highly stressed of all flange types since the maximum stresses in a blind flange are bending stresses at the center. A blind flange is used in place of a cap or plug to blank off a flanged fitting.

Did You Know?

A *skillet flange,* also called a *pie plate,* is a special type of blind flange with a flat handle that is used to slide the flange in between two companion flanges to isolate a section of piping, typically for the purpose of hydro static pressure testing. The handle provides a visual cue to remove the device after the testing has been completed.

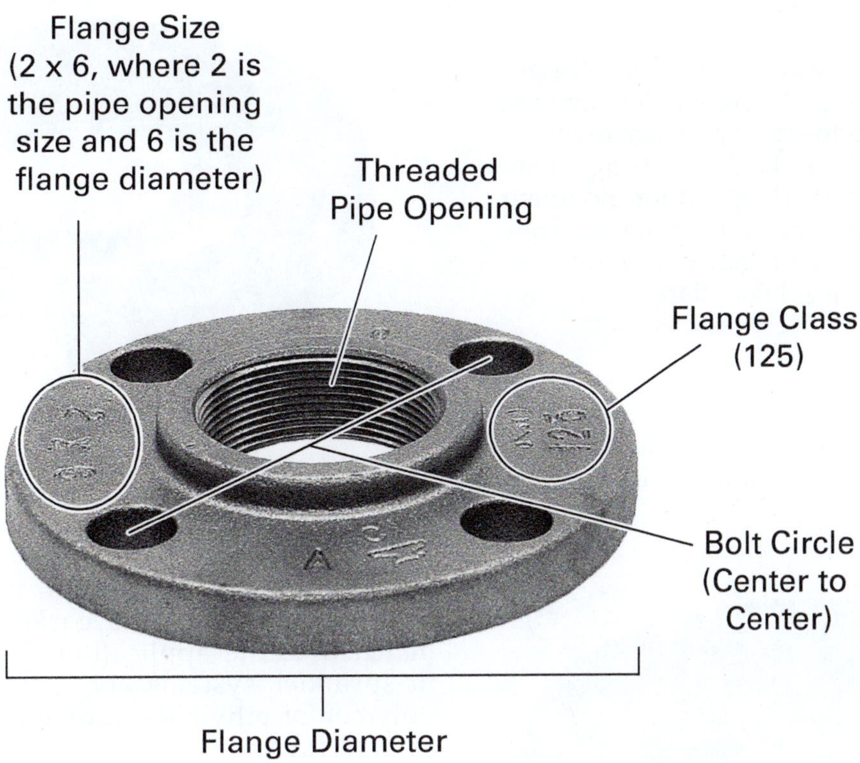

Figure 48 Companion flange.

Table 5 Cast Iron Flange Dimensions

Pipe Size	Outside Diameter	No. of Bolt Holes	Bolt Diameter	Bolt Hole Diameter	Bolt Circle
Class 125					
1	4¼	4	½	⅝	3⅛
1¼	4⅝	4	½	⅝	3½
1½	5	4	½	⅝	3⅞
2	6	4	⅝	¾	4¾
2½	7	4	⅝	¾	5½
3	7½	4	⅝	¾	6
4	9	8	⅝	¾	7½
6	11	8	¾	⅞	9½
8	13½	8	¾	⅞	11¾
Class 250					
Pipe Size	**Outside Diameter**	**No. of Bolt Holes**	**Bolt Diameter**	**Bolt Hole Diameter**	**Bolt Circle**
1	4⅞	4	⅝	¾	3½
1¼	5¼	4	⅝	¾	3⅞
1½	6⅛	4	¾	⅞	4½
2	6½	8	⅝	¾	5
2½	7½	8	¾	⅞	5⅞
3	8¼	8	¾	⅞	6⅝
4	10	8	¾	⅞	7⅞
6	12½	12	¾	⅞	10⅝
8	15	12	⅞	1	13

Note: All dimensions in inches.

5.1.7 Floor Flanges

Although not used in waterways, floor flanges are similar in appearance to companion flanges. They are typically used to create pipe stands. Floor flanges are less expensive and lighter in weight than companion flanges and are designed for mounting to a supporting surface using bolts or screws. They are available in coated and galvanized finishes. *Figure 50* shows a galvanized floor flange.

5.1.8 Grooved Flange Adapters

Grooved flange adapters are used for making connections between grooved pipe and flanged equipment. A grooved flange adapter is shown in *Figure 51*.

Figure 49 Blind flange.

Figure 50 Galvanized floor flange.

Figure 51 Grooved flange adapter.

5.2.0 Gaskets and Bolts

Gaskets are available in a variety of types and materials, and the gasket used must be matched to the application. The gaskets used in sprinkler systems are typically made of a polymer of ethylene-propylene rubber known as *EPM rubber*.

CAUTION

Some gasket materials are not compatible with CPVC and may damage the CPVC pipe or fitting over time. Always ensure that the gasket material selected is compatible with the piping system.

WARNING!

Some older gaskets may contain asbestos and require special Environmental Protection Agency (EPA) disposal and handling procedures.

Rubber gaskets are rated to match the class of the flange (typically Class 125 or Class 250). Both flat-ring gaskets and full-face gaskets are used in flanged piping systems.

CAUTION

Always check the class rating of the flange gasket against the shop drawings. Never use a Class 125 gasket in a Class 250 application. Use of an improperly rated gasket can result in system failure.

5.2.1 Flat-Ring Gaskets

Flat-ring gaskets are used on flanges with raised faces, and the outside diameter of the flat-ring gasket is slightly larger than the outside diameter of the raised face. Flat-ring gaskets sit inside of the flange bolt pattern. The raised-face flange dimensions and flange class rating determine the size of the flat-ring gasket. Bolts, nuts, and gaskets are usually combined in a package known as a *flange kit* or *flange pack*. *Figure 52* shows a flat-ring flange pack.

The following measurements are required when selecting a flat-ring gasket:

- Inside diameter (ID), which corresponds to the size of the pipe bore (opening)
- Outside diameter (OD), which is slightly larger than the OD of the raised face
- Class rating of the mating flange

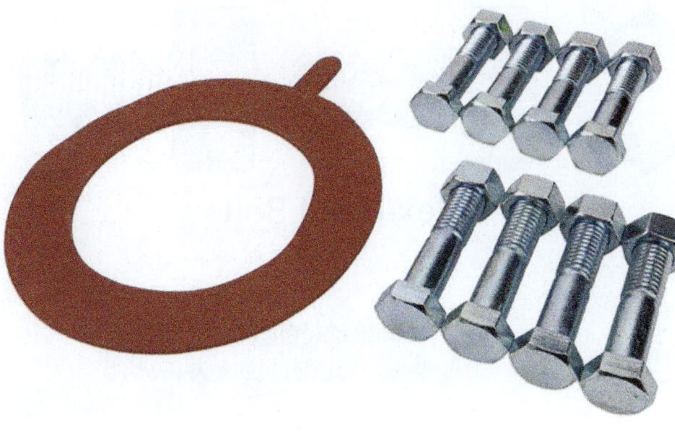

Figure 52 Flat-ring flange pack.

5.2.2 Full-Face Gaskets

Full-face gaskets are used with flat-face flanges and extend through the bolt holes to the outer edge of the flange. These gaskets have bolt holes that match the holes of the flanges with which they are used. *Figure 53* shows a full-face gasket.

The selection of a full-face gasket requires a few additional measurements than a flat-face gasket. The following measurements are required when selecting a full-face gasket:

- Inside diameter (ID), which corresponds to the size of the pipe bore (opening)
- Outside diameter (OD), which corresponds to the OD of the flange
- Number of bolt holes
- Bolt circle diameter
- Class rating of the mating flange

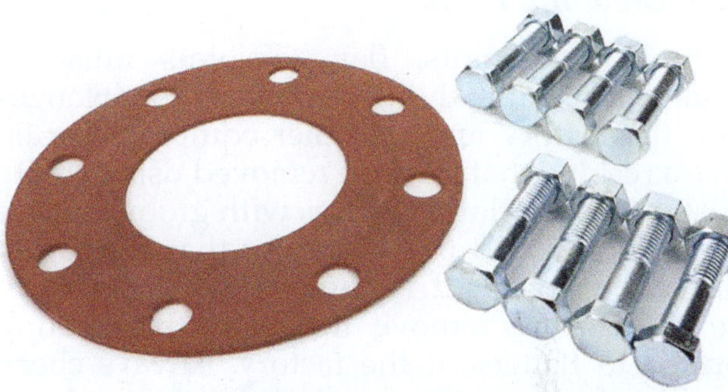

Figure 53 Full-face gasket pack.

5.2.3 Flange Bolts

Two types of bolts are used on flange connections: machine bolts and stud bolts. Machine bolts have square or hexagonal heads. Stud bolts are threaded on both ends or are threaded the entire length and cut to length in the field. There are some instances in which cap screws are used instead of bolts. *Figure 54* shows types of flange bolts.

5.3.0 Installing Pipe Flanges

ANSI standards require that flange bolts always straddle the horizontal and vertical center lines unless otherwise indicated. This straddling is known as the *two-hole method*. It is possible, though, for a job specification to require that the bolt holes line up with the vertical and horizontal lines. This method is known as the *one-hole method*. The general rule is to reference one center line only, depending on the type of drawing used. In elevation drawings, the usual reference is the vertical center line, and in plan drawings, the usual reference is the north-south center line. This rule ensures that all fabricators refer to the same center line.

The four steps in pipe flange installation are cleaning the parts, aligning the parts, installing the gaskets, and tightening the flange bolts.

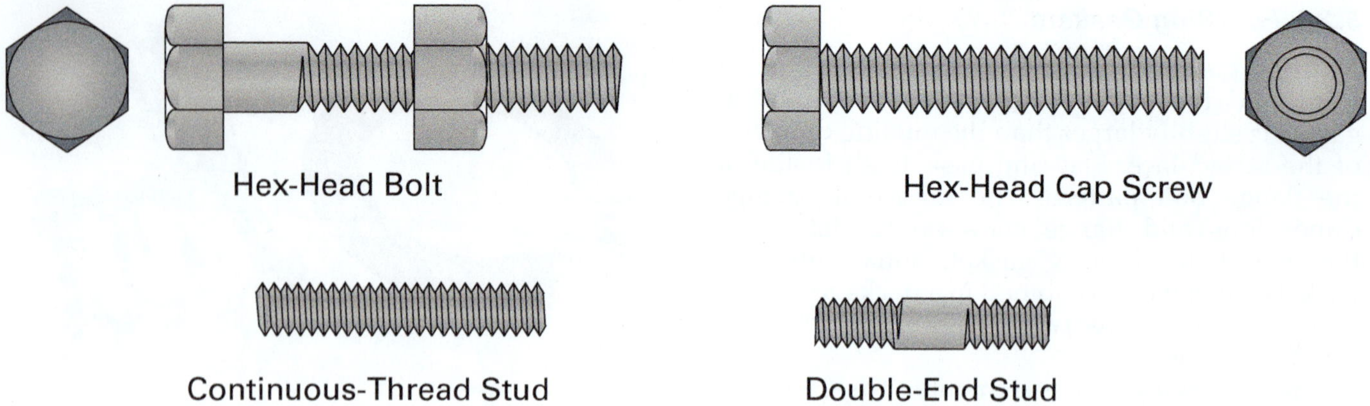

Figure 54 Flange bolts.

5.3.1 Cleaning Parts

Like all pipe joints, flanged joints must be cleaned to ensure the best possible fit. Flanges from the factory and on other equipment often have a residue that must be removed using a wire brush or emery cloth. Flanges with grooved faces must be clean and free of dirt so that the gasket will seal properly. It is sometimes necessary to use solvents to remove the protective coating applied to flanges at the factory. Always check the manufacturer's instructions before applying any type of solvent to the flanges.

Gaskets should also be cleaned before installation. Flanged joints are designed so that the face of each flange matches up squarely with a gasket to form a tight seal. Any type of obstruction or debris that could prevent the flanges from making up properly must be removed.

5.3.2 Flange Alignment

Aligning flanges properly is extremely important to producing a leak-free connection. After the pipe is cleaned, the pipe must be put in place and properly supported. Hangers and support stands are then adjusted until the flanges are precisely aligned. Some flanges include a drift pin that can be used to align the bolt holes. After the alignment is complete, you are ready to install the gasket. *Figure 55* shows proper flange alignment in a completed installation.

> **WARNING!**
>
> Never try to align the bolt holes with your fingers. Any slippage of the flange components can result in injury.

5.3.3 Installing Gaskets

Gaskets are selected to match the size and class rating of the flange. Some gasket manufacturers require the use of a special lubricant prior to installation. Follow the manufacturer's instructions for the gasket in use.

Follow these steps to install a full-face gasket between flat-face flanges:

Step 1 Check the job specifications to determine the type of gasket needed.

Step 2 Place the gasket between the flange faces.

Step 3 Hold the gasket in place using a drift pin and insert the bottom bolts into the flanges.

> **WARNING!**
>
> Never use your fingers to hold the gasket in place. Use a drift pin to avoid pinching your fingers.

Step 4 Insert all the bolts through the flange bolt holes.

Step 5 Place the nuts on the bolts and tighten the nuts finger-tight.

If raised-face flanges are used, install a flat-ring gasket. Follow these steps to install a flat-ring gasket between raised-face flanges:

Step 1 Check the job specifications to determine the type of gasket needed.

Step 2 Insert the bottom bolts into the flanges.

Step 3 Lubricate the gasket, if necessary.

Step 4 Place the gasket between the flange faces.

Step 5 Insert all the bolts through the flange bolt holes.

Step 6 Place the nuts on the bolts and tighten them finger-tight.

5.3.4 Tightening Flange Bolts

Flange bolts must be tightened carefully to avoid warping the flange. If the bolts are not tightened correctly, the joint may leak or crack. Always follow the flange manufacturer's torque requirements. The correct amount of torque must be applied in the proper tightening sequence to make a good seal. Tightening the bolts in an opposing sequence is critical. *Figure 56* shows the tightening sequence for flange bolts.

Figure 55 Proper flange alignment.

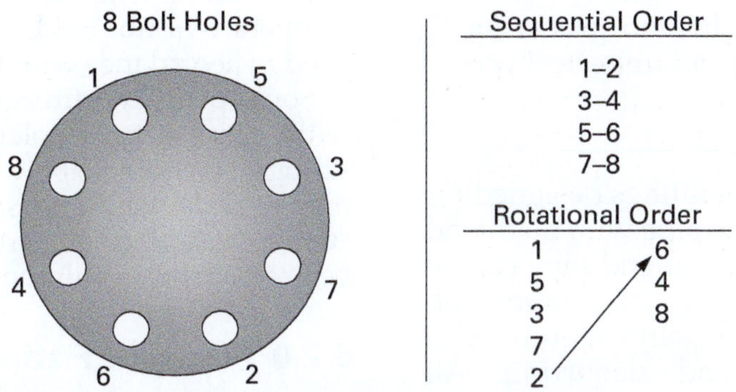

Figure 56 Proper tightening sequence.

5.0.0 Section Review

1. A blind flange can be used in place of a _____.
   a. cap
   b. cross
   c. tee
   d. reducer

2. The gaskets used in sprinkler systems are typically made of _____.
   a. silicone
   b. rubber
   c. plastic
   d. steel

3. Which of the following is *true* regarding flange bolt tightening?
   a. Bolts can be tightened in any sequence.
   b. Tightening the bolts in an opposing sequence is critical.
   c. It is only necessary to tighten the bolts finger-tight.
   d. It is not necessary to install all the bolts, only enough to hold the flange in place.

SECTION SIX

6.0.0 PLAIN-END PIPE AND FITTINGS

Objective

Identify the applications of plain-end pipe fittings.

a. Identify plain-end fittings with grippers and their applications.
b. Identify press-fit fittings and their applications.

Trade Term

Plain-end pipe: Pipe that has no groove or thread on the end and has all imperfections removed.

There are several types of fittings designed for **plain-end pipe**, which is pipe with no thread or groove on the end. Plain-end pipe can be connected using push-on fittings or compression (press-fit) fittings. These usually require end preparation by cleaning and deburring. All coatings, notches, bumps, weld beads, and score marks 1" to 1½" from the pipe end must be removed so the gasket will seat properly. Internal chamfers, normal abrasive wheel cuts that leave burrs, and dull wheel cutters will produce unusable surfaces. Special grinders are available that reliably prepare the pipe end.

Listings for plain-end fittings must be checked to ensure that the application is correct. Some of the older products Listed require significant peripheral penetration of the pipe wall with grippers.

Some plain-end fittings use a quarter-turn locking lug that has a small-diameter tooth pattern on the end of the lug that grips the pipe. Others have a hex-head locking bolt. These fittings come in various sizes and reducing combinations. Refer to the manufacturer's catalog to determine application limits.

6.1.0 Plain-End Fittings with Grippers

Plain-end fittings come in various sizes and reducing combinations. Check the manufacturer's data sheet for takeout dimensions.

Couplings with grippers have curved metal inserts, sometimes called *roustabouts*, that bite into the pipe to create a mechanical joint (*Figure 57*). This type of plain-end fitting is suitable for use on square-end or beveled-end pipe. Other than making a square cut and deburring the pipe, no special pipe preparation is necessary. Couplings with grippers are designed for Schedule 40 pipe. They are most useful when cutting into existing systems to extend or modify them. In areas where there is significant expansion and contraction of piping due to temperature changes, couplings with grippers will not hold. These fittings may only be used in accordance with their Listing.

Some plain-end fittings are designed to be used with gripper-type plain-end couplings. These fittings include elbows, tees, crosses, adapter nipples, adapter flanges, and swaged nipples. Be sure to reference the manufacturer's instructions before installing plain-end fittings.

6.2.0 Press-Fit Fittings

A press-fit piping system uses Schedule 10 thin-wall pipe and a steel fitting (*Figure 58*). The fitting has pre-lubricated gasket to seal against the outside of the pipe. (Because the fitting is pre-lubricated, some manufacturers recommend that they remain in their bags until the fitting is installed.) To install a press-fit fitting, the pipe is cut square, deburred, and then inserted into the fitting until it contacts the pipe stop. A special crimping tool is then applied, producing a permanent joint rated for up to 500 psi. Press-fit fittings are available as straight couplings, elbows, and tees. This system requires far less time to install than a welded connection and provides additional safety since no flame or arc is required.

NOTE

Prepared-end fittings were once popular but are no longer manufactured due to leakage issues. One type of prepared-end fitting used a special fitting and a U-shaped key to secure the piping. You may encounter these fittings in existing installations.

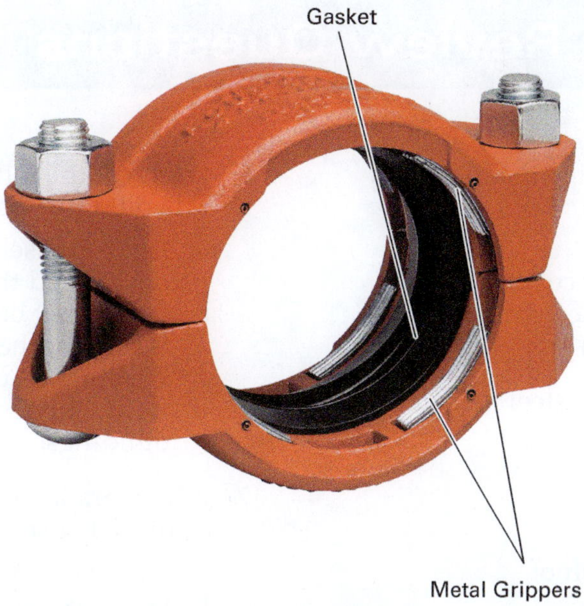

Figure 57 Plain-end fitting with grippers.

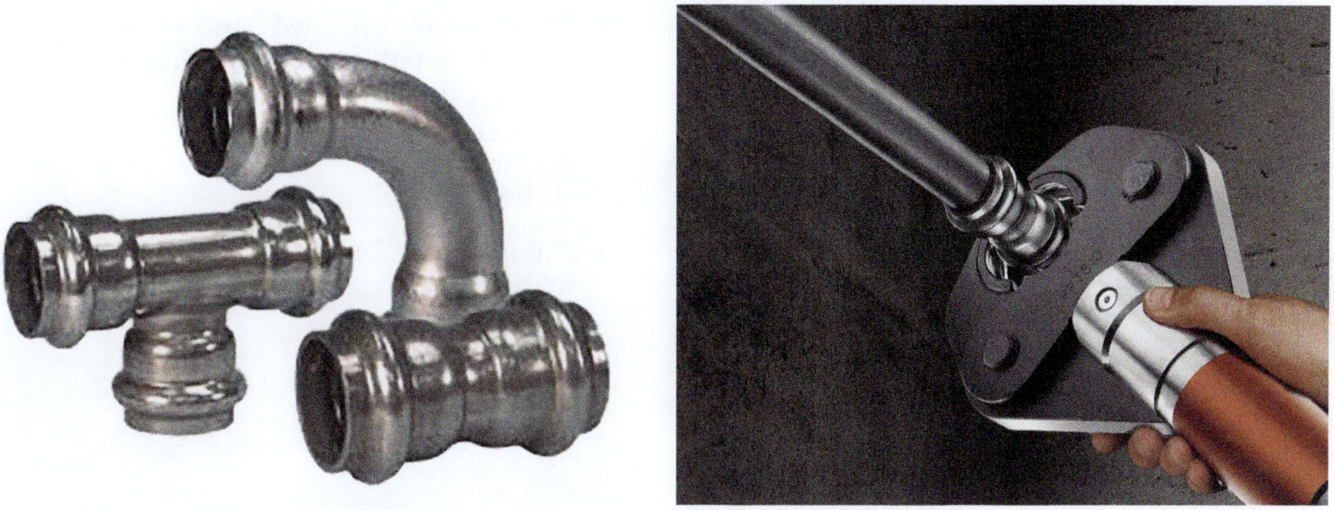

Figure 58 Press-fit fittings and crimping tool.

6.0.0 Section Review

1. The type of fitting that can be used on either square-end or beveled pipe is a(n) _____.
   a. press-fit fitting
   b. flanged fitting
   c. gripper-type fitting
   d. extension fitting

2. A plain-end pipe fitting that uses a special crimping tool to create a mechanical connection is a(n) _____.
   a. press-fit fitting
   b. flanged fitting
   c. gripper-type fitting
   d. extension fitting

Review Questions

1. Galvanized steel pipe is pipe that has been _____.
   a. painted silver
   b. sprayed with a lacquer coating
   c. dipped in a molten zinc mixture
   d. coated with aluminum
2. One advantage of stainless steel is that it is _____.
   a. inexpensive
   b. easy to thread
   c. resistant to corrosion
   d. difficult to make leakproof
3. The inside diameter of 2" Schedule 10 steel pipe is _____.
   a. 1.750"
   b. 2.067"
   c. 2.157"
   d. 2.625"
4. The outside diameter of 1" Schedule 40 steel pipe is _____.
   a. 1.315"
   b. 1.750"
   c. 2.067"
   d. 2.625"
5. The type of pipe with the *greatest* wall thickness is _____.
   a. Schedule 10
   b. Schedule 40
   c. Schedule 80
   d. Schedule 160
6. Which of the following is *true* regarding Schedule 10 pipe?
   a. It can be used in sprinkler work if it is designated thin-wall pipe by NFPA 13.
   b. It can only be used with special fittings Listed for the application.
   c. It cannot be used in sprinkler systems.
   d. It can only be used in single-story buildings.
7. Which grade of ASTM A135 pipe is *most* commonly used in the sprinkler industry?
   a. Grade A
   b. Grade B
   c. Grade C
   d. Grade D
8. The manually operated hand tool that can be used to cut pipe in tight quarters is a _____.
   a. hinged pipe cutter
   b. hacksaw
   c. tubing cutter
   d. bow saw
9. A raised edge formed when cutting pipe is known as a _____.
   a. burl
   b. burr
   c. roll
   d. crease

Figure RQ01

10. The standard ELL in *Figure RQ01* is labeled _____.
    a. A
    b. B
    c. E
    d. F

11. The fitting labeled B in *Figure RQ01* is a(n) _____.
    a. 90-degree elbow
    b. 45-degree elbow
    c. $22\frac{1}{2}$-degree elbow
    d. $11\frac{1}{4}$-degree elbow

12. The fitting labeled D in *Figure RQ01* is a(n) _____.
    a. 90-degree elbow
    b. 45-degree elbow
    c. $22\frac{1}{2}$-degree elbow
    d. $11\frac{1}{4}$-degree elbow

13. The fitting labeled F in *Figure RQ01* is a(n) _____.
    a. drain elbow
    b. 45-degree elbow
    c. 90-degree end-of-run
    d. eccentric reducer

14. The fitting with the *shallowest* angle in *Figure RQ01* is _____.
    a. B
    b. D
    c. E
    d. F

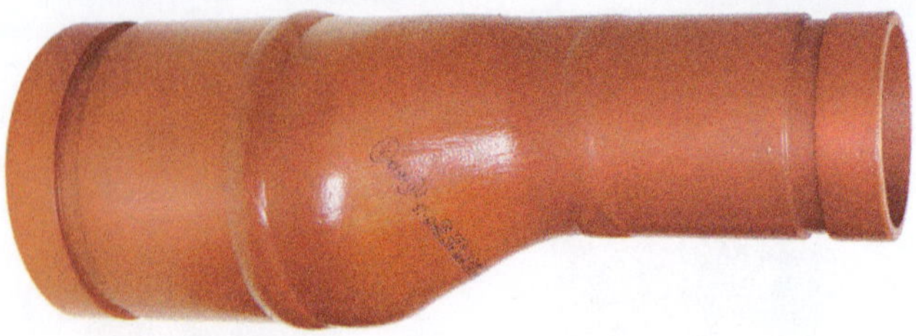

Figure RQ02

15. The fitting shown in *Figure RQ02* is a(n) _____.
    a. 11¼-degree elbow
    b. concentric reducer
    c. eccentric reducer
    d. bullhead tee
16. When completing the installation of a grooved fitting, it is important to tighten the nuts _____.
    a. alternately per the manufacturer's torque specifications
    b. only finger-tight without using a wrench
    c. as tightly as possible to ensure a strong connection
    d. to their maximum torque one at a time
17. In the thread designation 1–11½–NPT, the 11½ represents the _____.
    a. number of threads per inch
    b. year of manufacture
    c. bolt length in millimeters
    d. nominal pipe size
18. Given a properly threaded pipe, a threaded fitting should be able to be hand-threaded _____.
    a. one to two full revolutions
    b. two to three full revolutions
    c. three to four full revolutions
    d. four to five full revolutions
19. The die size commonly used to thread ½" hanger rods is _____.
    a. 11 threads/inch
    b. 13 threads/inch
    c. 16 threads/inch
    d. 18 threads/inch
20. The die size commonly used to thread 2" pipe is _____.
    a. 8 threads per inch
    b. 11½ threads per inch
    c. 14 threads per inch
    d. 18 threads per inch
21. Which of the following is *true* regarding thread cutting oil?
    a. There is only one type of oil, and it can be used in all applications.
    b. Thread cutting oil is necessary, but it tends to slow down the speed of cutting.
    c. Thread cutting oil keeps the pipe and die cool, preventing unnecessary wear on the die.
    d. All thread cutting oil is compatible with CPVC.
22. Pipe thread compound is also referred to as _____.
    a. thread-lock compound
    b. pipe dope
    c. joint compound
    d. caulk

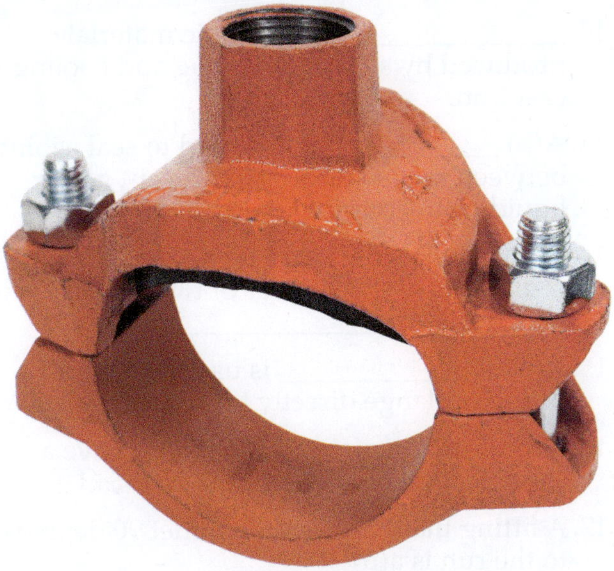

Figure RQ03

23. The fitting shown in *Figure RQ03* is a threaded mechanical _____.
    a. elbow
    b. branch tee
    c. nipple
    d. bushing

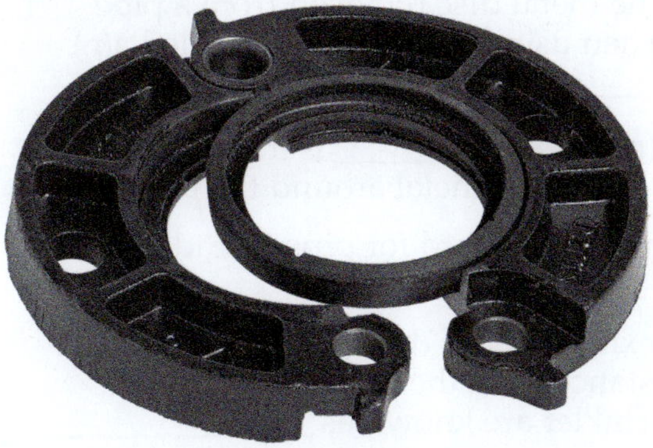

Figure RQ04

24. The component shown in *Figure RQ04* is a(n) _____.
    a. eccentric reducer
    b. reducing coupling
    c. blind flange
    d. grooved flange adapter

25. Press-fit fittings are used with _____.
    a. Schedule 10 pipe
    b. Schedule 30 pipe
    c. Schedule 40 pipe
    d. Schedule 80 pipe

Trade Terms Quiz

Fill in the blank with the correct term learned from your study of this module.

1. The metal disc removed from a pipe when using a hole cutter is called a(n) _______________.
2. A(n) _______________ is formed by pressing the metal around the pipe.
3. The US standard for pipe threads is _______________.
4. Flexible fittings designed to span the distance from the branch line to the sprinkler are known as _______________.
5. A reducing coupling that maintains the same center line between the two pipes that it joins is a(n) _______________.
6. A(n) _______________ makes an angle between adjacent pipes.
7. A piece of pipe about two times the length of a standard thread and threaded on both ends is a(n) _______________.
8. A(n) _______________ is a pipe fitting with internal and external threads that can be used to connect a pipe or fitting with another pipe or fitting of a different size.
9. _______________ is wrapped around the male threads of a pipe before that pipe is screwed into a fitting.
10. A horizontal or vertical supply main running through buildings and having hose valves on each floor is a(n) _______________.
11. A pipe that has no groove or threads on the end and has all imperfections removed is called a(n) _______________.
12. _______________ is a metallic material produced by slowly reheating and cooling cast iron.
13. A(n) _______________ is used to seal a joint between metal surfaces to prevent gas or liquid from entering or escaping.
14. A fitting that displaces to one side the center line of the smaller of two pipes being joined is a(n) _______________.
15. A(n) _______________ is used to connect a fitting or flange directly to threaded pipe.
16. Fittings made of _______________ have a classification of either 125 or 250.
17. A fitting that has one side outlet 90 degrees to the run is a(n) _______________.
18. A(n) _______________ is a piece of pipe threaded on both ends and less than 12" long.
19. _______________ is a term used to describe the amount of pipe that screws into a threaded fitting.
20. _______________ refers to a type of cast iron in which magnesium is added to the molten gray iron to reduce brittleness.
21. A(n) _______________ or a(n) _______________ can be removed by reaming the pipe after cutting it.
22. A(n) _______________ is formed around pipe by cutting away the pipe material.
23. A pipe labeling system based on pipe wall thickness is known as a(n) _______________.

Trade Terms

Burr
Bushing
Cast iron
Close nipple
Companion flange
Concentric reducing coupling
Coupon
Cut groove
Ductile iron
Eccentric reducing coupling
Elbow
Fin
Flexible sprinkler hose fittings
Gasket
Makeup
Malleable iron
National Pipe Thread (NPT)
Nipple
Pipe schedule
Plain-end pipe
Roll groove
Standpipe
Tee
Teflon® tape

Cornerstone of Craftsmanship

Kevin Kelly
Codes & Standards Specialist
Victaulic

How did you choose a career in the industry?

I got started in the fire protection industry as a volunteer firefighter at age 17. There was a senior firefighter there who prompted me to look into the Fire Protection Engineering program at the University of Maryland.

Who inspired you to enter the industry?

I'd say Professor Bryan at the University of Maryland inspired me to enter the industry.

What types of training have you been through?

My early training was as a firefighter. I was a Firefighter II, part of the FAST team responsible for responding to fallen firefighters, and I was also the safety officer for the fire department. After that I went on to pursue my degree in Fire Protection Engineering and obtained my professional engineering license.

How important is education and training in construction?

It's very important and necessary for good construction.

How important are NCCER credentials to your career?

NCCER provides an important role. The organization creates training, updates it, and recognizes new opportunities, all with a high degree of professionalism. I've gotten to know NCCER more with this project, and it reiterates all of that even more.

How has training/construction impacted your life?

I provide more training than I receive. I teach about fire protection systems for both Victaulic and NFPA, and I also have my own engineering firm that focuses a lot on training. I think it's important to return what I've learned back to the people who need it.

When I teach my students about construction, I tell them to ask themselves, "Will this last twenty years?" Good quality is important, and in some cases, that construction will last longer than your lifetime, so doing it safely and correctly is an incredibly important concept to teach.

What kinds of work have you done in your career?

Well, I started as a volunteer firefighter, but I'm also a third-generation farmer. I was in the Army Corps of Engineers (COE) for 13 years, and I also did some work as a consulting fire protection engineer and worked for Sutphen Corporation building fire trucks. After that, I worked at the National Fire Sprinkler Association as the manager of codes and then at a consulting firm for fire protection. I've been at Victaulic since 2014.

Tell us about your present job.

I monitor all codes and standards that impact our products across the globe, and it's larger than just fire protection. I'm making sure codes and standards allow for anything that connects pipe, from petroleum to steam and fire protection, can be installed.

What do you enjoy most about your job?

I enjoy learning; I'm always learning about new product lines and the part of the world they will be used in. I really enjoy that constant challenge.

What factors have contributed most to your success?

I think my ability to communicate and negotiate. I feel like I'm constantly in a gray area where the answers aren't always so clear cut with technical decisions. My ability to talk to people and figure out the best compromise on new situations, technology, or requirements is really helpful. I'm always willing to take on new situations and work through problems.

Would you suggest construction as a career to others? Why?

Definitely. It's a good career, there's good money, and it's in demand. There is absolutely a need for qualified construction workers.

What advice would you give to those new to the field?

You must be willing to learn new skills and understand new information. The industry changes on a regular basis, and you need to be able to adapt to those changes.

Interesting career-related fact or accomplishment?

I've dedicated my life to saving people and property from fire, and when I was manager of codes, I helped pass a residential sprinkler requirement for codes. But I didn't do it alone. There were 40 to 50 years of people before me trying and trying until it was my turn to be in the codes.

How do you define craftsmanship?

I think of my grandfather, and he just had a good way of building and constructing things. He'd always use tools safely, and the final result just looked good. Craftsmanship to me is in the hands of the craftsmen and when they turn out a useful product they can be proud of.

Trade Terms Introduced in This Module

Burr: A sharp, raised edge or other metal protrusion produced on the pipe end when cutting a pipe.

Bushing: A pipe fitting with both internal and external threads that can be used to connect a pipe or fitting with another pipe or fitting of a different size.

Cast iron: An alloy of iron and carbon that is cast in a mold. Cast iron fittings have a classification of either 125 or 250.

Close nipple: A nipple that is about twice the length of a standard thread that is threaded from each end with no shoulder.

Companion flange: A flange that is used to connect a fitting or flange directly to threaded pipe.

Concentric reducing coupling: A reducing coupling that maintains the same center line between the two pipes that it joins.

Coupon: The metal disc removed from a steel pipe when using a hole cutter.

Cut groove: A groove that is formed around pipe by cutting away the pipe material.

Ductile iron: A type of cast iron in which magnesium is added to the molten gray iron to reduce brittleness.

Eccentric reducing coupling: A reducing coupling that displaces the center line of the smaller of the two joining pipes to one side.

Elbow: A fitting that makes an angle between adjacent pipes. An elbow is always 90 degrees unless another angle is stated on the fitting or in the drawing specifications. Also known as an *ELL*.

Fin: A sharp, flared metal protrusion produced on a pipe end when cutting it. A fin juts out at an angle to the pipe.

Flexible sprinkler hose fittings: Flexible fittings designed to span the distance from the branch line to the sprinkler.

Gasket: A flat sheet or ring of rubber or other material used to seal a joint between metal surfaces to prevent gas or liquid from entering or escaping.

Makeup: A term used to describe the amount of pipe that screws into a threaded fitting.

Malleable iron: A metallic material that is produced by slowly reheating and cooling cast iron. Malleable iron fittings have a classification of either 150 or 300.

National Pipe Thread (NPT): The US standard for pipe threads. NPT has a $^1/_{16}$" taper per inch from back to front.

Nipple: A piece of pipe that is threaded on both ends and is less than 12" in length. Any pipe over 12" is referred to as *cut pipe*.

Pipe schedule: A pipe labeling system based on pipe wall thickness. As the schedule increases, the wall thickness also increases.

Plain-end pipe: Pipe that has no groove or threads on the end and has all imperfections removed.

Roll groove: A groove that is formed by pressing the metal around the pipe.

Standpipe: A horizontal or vertical supply main running through buildings and having hose valves on each floor.

Tee: A fitting that has one side outlet 90 degrees to the run.

Teflon® tape: Tape made of Teflon® that is wrapped around the male threads of a pipe before the pipe is screwed into a fitting. Teflon® tape serves as both a lubricant and a sealant.

Additional Resources

This module presents thorough resources for task training. The following reference material is suggested for further study.

The Handbook of Steel Pipe, Latest Edition. Washington, DC: American Iron and Steel Institute.

NFPA 13, *Standard for the Installation of Sprinkler Systems*, Latest Edition. Quincy, MA: National Fire Protection Association.

The Pipefitters Blue Book, 2002. W.V. Graves. Webster, TX: Graves Publishing Company.

Standard USAS (ASME) B1.20.1,Pipe Threads, General Purpose (Inch). New York, NY: American National Standards Institute Inc.

Figure Credits

Courtesy of RIDGID Tool Company, Figures 3, 5, 8–9, 27–32

Courtesy of Victaulic Company, Figures 7, 10–14, 15 (End-of-Run), 22–23, 51, 57–58

Anvil International & Smith-Cooper International, Figures 15 (all named images except "End-of-Run"), 16–19, 24, 36–40, 43–45, 47–50

Reliable Automatic Sprinkler Co., Inc., Figure 21

This image is provided courtesy of Southwire Company, LLC. All rights are reserved., Figures 33–34

Courtesy of ASC Engineered Solutions., Figure 42

ARGCO, Figures 52–53

Reproduced with permission of NFPA from NFPA 13, *Standard for the Installation of Sprinkler Systems*, 2019 edition. Copyright© 2018, National Fire Protection Association. For a full copy of NFPA 13, please go to www.nfpa.org, Table 1

Section Review Answer Key

SECTION 1.0.0

Answer	Section Reference	Objective
1. b	1.1.0	1a
2. a	1.2.2	1b
3. b	1.3.2	1c
4. a	1.4.1	1d

SECTION 2.0.0

Answer	Section Reference	Objective
1. c	2.1.3	2a
2. a	2.2.0	2b

SECTION 3.0.0

Answer	Section Reference	Objective
1. d	3.1.1	3a
2. c	3.2.1	3b
3. b	3.3.4	3c

SECTION 4.0.0

Answer	Section Reference	Objective
1. d	4.1.0	4a
2. b	4.2.1; *Table 3*	4b
3. c	4.3.4	4c
4. b	4.4.0	4d

SECTION 5.0.0

Answer	Section Reference	Objective
1. a	5.1.6	5a
2. b	5.2.0	5b
3. b	5.3.4	5c

SECTION 6.0.0

Answer	Section Reference	Objective
1. c	6.1.0	6a
2. a	6.2.0	6b

NCCER CURRICULA — USER UPDATE

NCCER makes every effort to keep its textbooks up-to-date and free of technical errors. We appreciate your help in this process. If you find an error, a typographical mistake, or an inaccuracy in NCCER's curricula, please fill out this form (or a photocopy), or complete the online form at **www.nccer.org/olf**. Be sure to include the exact module ID number, page number, a detailed description, and your recommended correction. Your input will be brought to the attention of the Authoring Team. Thank you for your assistance.

Instructors – If you have an idea for improving this textbook, or have found that additional materials were necessary to teach this module effectively, please let us know so that we may present your suggestions to the Authoring Team.

NCCER Product Development and Revision

13614 Progress Blvd., Alachua, FL 32615

Email: curriculum@nccer.org
Online: www.nccer.org/olf

❑ Trainee Guide ❑ Lesson Plans ❑ Exam ❑ PowerPoints Other ______________________

Craft / Level: Copyright Date:

Module ID Number / Title:

Section Number(s):

Description:

Recommended Correction:

Your Name:

Address:

Email: Phone:

CPVC Pipe and Fittings

Overview

This module covers the special type of chlorinated polyvinyl chloride (CPVC) pipe and fittings used in sprinkler systems. It also describes the procedures for cutting, beveling, cleaning, and joining CPVC pipe and fittings, and provides an overview of how to test the installed sprinkler system.

Module 18104

Trainees with successful module completions may be eligible for credentialing through the NCCER Registry. To learn more, go to **www.nccer.org** or contact us at 1.888.622.3720. Our website, **www.nccer.org**, has information on the latest product releases and training.

Your feedback is welcome. You may email your comments to **curriculum@nccer.org**, send general comments and inquiries to **info@nccer.org**, or fill in the User Update form at the back of this module.

This information is general in nature and intended for training purposes only. Actual performance of activities described in this manual requires compliance with all applicable operating, service, maintenance, and safety procedures under the direction of qualified personnel. References in this manual to patented or proprietary devices do not constitute a recommendation of their use.

18104 V4.0

18104
CPVC Pipe and Fittings

Objectives

Successful completion of this module prepares you to do the following:

1. Size and select chlorinated polyvinyl chloride (CPVC) pipe.
   a. Identify CPVC pipe characteristics.
   b. Identify the Listed applications of CPVC pipe.
   c. Describe the storage and handling requirements for CPVC pipe.
   d. Describe the protection requirements for CPVC pipe in concealed and exposed installations.
2. Cut and join CPVC pipe.
   a. Describe how to cut, deburr, bevel, and clean CPVC pipe.
   b. Describe the procedure for solvent-cementing CPVC pipe and fittings.
3. Install and test CPVC sprinkler systems.
   a. Identify support spacing requirements for CPVC pipe.
   b. Describe how to install sprinklers and test the system.
   c. Identify installation requirements for special applications.

Performance Tasks

Under supervision, you should be able to do the following:

1. Cut, deburr, bevel, and clean CPVC pipe.
2. Join CPVC pipe and fittings.
3. Install CPVC pipe using the correct support spacing.
4. Complete a manufacturer training course for CPVC.

Trade Terms

Ambient temperature
Armover
Beveling
Chlorinated polyvinyl chloride (CPVC)
CPVC compound
CPVC resin
Iron pipe size (IPS)
Light hazard
Ordinary hazard
Standard dimensional ratio (SDR)
Thermal expansion
Thermoplastic
Transverse

Industry Recognized Credentials

If you are training through an NCCER-accredited sponsor, you may be eligible for credentials from NCCER's Registry. The ID number for this module is 18104. Note that this module may have been used in other NCCER curricula and may apply to other level completions. Contact NCCER's Registry at 1.888.622.3720 or go to **www.nccer.org** for more information.

You can also show off your industry-recognized credentials online with NCCER's digital badges. Transform your knowledge, skills, and achievements into badges that you can share across social media platforms, send to your network, and add to your resume. For more information, visit **www.nccer.org**.

> **NOTE**
>
> Unless otherwise specified, references in parentheses following figure and table numbers refer to NFPA 13.

Contents

Figures and Tables

Section One

1.0.0 Sizing and Selecting Chlorinated Polyvinyl Chloride (CPVC) Pipe

Objective

Size and select chlorinated polyvinyl chloride (CPVC) pipe.

a. Identify CPVC pipe characteristics.
b. Identify the Listed applications of CPVC pipe.
c. Describe the storage and handling requirements for CPVC pipe.
d. Describe the protection requirements for CPVC pipe in concealed and exposed installations.

Trade Terms

Ambient temperature: The surrounding air temperature.

Chlorinated polyvinyl chloride (CPVC): A common thermoplastic pipe used in the sprinkler industry.

CPVC compound: An orange thermoplastic material used to produce sprinkler pipe and fittings.

CPVC resin: A white thermoplastic material that accounts for 85 percent of the weight of finished CPVC products.

Iron pipe size (IPS): The nominal pipe size of iron pipe.

Light hazard: An occupancy or portions of other occupancies where the quantity and/or combustibility of contents is low and fires with relatively low rates of heat release are expected.

Ordinary hazard: Group 1 (OH1): An area in which the combustibility of the contents is low, the quantity of the combustibles is moderate, and stockpiles of contents with low combustibility do not exceed 8'. Group 2 (OH2): Occupancies or portions of other occupancies where the quantity and combustibility of contents are moderate to high, stockpiles of contents with moderate to high combustibility do not exceed 12', and stockpiles of contents with high rates of heat release do not exceed 8'.

Standard dimensional ratio (SDR): The pipe wall thickness in proportion to its outside diameter.

Thermoplastic: Refers to plastic piping that is soft and pliable when heated, but hard when cooled.

Chlorinated polyvinyl chloride (CPVC) is a light, flexible, strong, and corrosion-resistant **thermoplastic** piping material. With its ability to withstand higher temperatures than polyvinyl chloride (PVC), CPVC can be used in sprinkler systems, hot water systems, hot acid distribution systems, and waste systems.

> **NOTE**
> CPVC manufacturers offer free installation training. Sprinkler fitters should receive manufacturer training for the system(s) in use and retrain every two years.

CPVC systems are quicker and easier to install than metal systems. Although CPVC systems have many advantages, they require careful attention to the manufacturer's installation instructions, all applicable standards, and NFPA requirements.

> **CAUTION**
> All nonmetallic piping systems, including CPVC pipe, fittings, and solvent-cement, must be Listed for use in fire sprinkler systems.

1.1.0 CPVC Characteristics

CPVC resin is produced by adding chlorine to white PVC resin. The addition of chlorine improves temperature performance and increases the material's fire and corrosion resistance. Coloring and other performance-enhancing ingredients are then added to the white CPVC resin to produce an orange **CPVC compound**. CPVC compounds come in two forms—pellets and powder. Pellets are used in the process of injection-molding fittings, while powder is extruded into pipe.

CPVC pipe for sprinkler systems is manufactured in a **standard dimensional ratio (SDR)** of 13.5 for all diameters in accordance with the American Society for Testing and Materials (ASTM) International Standard F442, *Standard Specification for Chlorinated Poly(Vinyl Chloride) (CPVC) Plastic Pipe (SDR–PR)*. SDR means the pipe wall thickness is directly proportional to the outside diameter. For SDR 13.5 pipe, the outside diameter of the pipe is 13.5 times the wall thickness. All diameters of SDR 13.5 pipe have the same pressure-bearing capability (175 pounds per square inch, or psi, at 150°F).

Did You Know?

In 1960, the B.F. Goodrich Chemical Corporation (now known as *Lubrizol*) introduced a family of specialty thermoplastic materials. Today, Lubrizol thermoplastics are used in a variety of products, including athletic shoes, CPVC sprinkler piping, and 3D printing.

SDR 13.5 CPVC pipe has a larger inside diameter than Schedule 40 steel pipe for pipe sizes up to and including 1¼". CPVC pipe also has a lower hydraulic friction loss than steel pipe. Because of this, it may be possible to use a smaller size CPVC pipe for some applications.

In addition to its inherent smoothness when new, CPVC also maintains its initial friction loss better than steel over time, as it resists both corrosion and scaling (mineral deposits) that can form on the inside of the pipe. CPVC pipe weighs approximately one-sixth that of steel and approximately one-half that of copper, which makes it much easier to lift and carry. *Table 1* provides a weight comparison of CPVC pipe versus steel pipe and copper tube.

CPVC pipe and fittings (*Figure 1*) are rated for a system design operating pressure of 175 psi at a temperature of 150°F. In accordance with the Listing requirement for a sprinkler piping system to have a safety factor of five, CPVC pipe and fittings have a minimum required quick-burst strength of 5 × 175 psi or 875 psi.

CPVC pipe and fittings are manufactured in **iron pipe size (IPS)** only. Fittings in diameters of ¾" to 1¼" are produced to Schedule 40 dimensions per ASTM F438, *Standard Specification for Socket-Type Chlorinated Poly(Vinyl Chloride) (CPVC) Plastic Pipe Fittings, Schedule 40.* In diameters of 1½", 2", 2½", and 3", the fittings are produced to Schedule 80 dimensions per ASTM F437, *Standard Specification for Threaded Chlorinated Poly(Vinyl Chloride) (CPVC) Plastic Pipe Fittings, Schedule 80.*

CAUTION

Like any piping material, CPVC is subject to degradation when exposed to incompatible products. CPVC can be damaged by incompatible thread cutting oil, pipe thread compounds, firestopping, and other construction materials (see *Appendix*). However, many of these products have special formulations that are compatible with CPVC. Refer to the CPVC pipe manufacturer for their most current list of compatible products. NFPA 13 and NFPA 13R require materials used in conjunction with nonmetallic piping to be compatible.

1.2.0 Listings for CPVC Pipe and Fittings

CPVC pipe is Listed for use in wet pipe sprinkler systems and certain types of dry and preaction systems. The following are Listings for CPVC pipe and fittings based on occupancy or use:

- **Light hazard** occupancies as defined in NFPA 13, *Standard for the Installation of Sprinkler Systems.* Examples of light hazard occupancies include churches, schools, dorms, patient rooms in hospitals, nursing homes, offices, museums, theaters, and high rises (hotels and condos over four stories in height).

Figure 1 CPVC pipe and fittings.

Table 1 Weight Comparison of CPVC Versus Metal Pipe

Nominal Pipe Size (inches)	Pounds per Foot		
	CPVC SDR 13.5	Steel Pipe Schedule 40	Copper Tube Type M
¾	0.17	–	0.33
1	0.26	1.68	0.46
1¼	0.42	2.27	0.68
1½	0.55	2.72	0.94
2	0.86	3.65	1.46
2½	1.26	5.79	2.03
3	1.87	7.59	2.68

Did You Know?

One characteristic of CPVC that makes it suitable for fire protection is its high limiting oxygen index (LOI), which is the minimum concentration of oxygen required for a material to burn. CPVC has an LOI of 60, which means it needs 60 percent oxygen in the surrounding air to sustain a flame. Since the Earth's atmosphere only contains 21 percent oxygen, CPVC will not burn and instead helps to prevent the spread of fire. Rather than consuming the pipe, fire creates a charred layer on the outside surface of CPVC, providing a thermal barrier that reduces heat conduction and enables the uninterrupted flow of water through the pipe.

- **Ordinary hazard** areas of otherwise light-hazard occupancies where the room does not exceed 400 square feet (NFPA 13, *Section 16.3.9.6*), including kitchens, laboratories, mechanical rooms, and laundry facilities.
- Residential occupancies as defined in NFPA 13D, *Standard for the Installation of Sprinkler Systems in One- and Two-Family Dwellings and Manufactured Homes.*
- Residential occupancies as defined in NFPA 13R, *Standard for the Installation of Sprinkler Systems in Low-Rise Residential Occupancies.*
- Underground water service as defined in NFPA 24, *Standard for the Installation of Private Fire Service Mains and Their Appurtenances.*
- Underground installations using pipe ranging in size from 3/4" to 3" installed in accordance with NFPA 13 and ASTM D2774, *Standard Practice for Underground Installation of Thermoplastic Pressure Piping.*
- Return air plenum installations as described in NFPA 90A, *Standard for the Installation of Air-Conditioning and Ventilating Systems,* the *International Mechanical Code (IMC), and* the *Uniform Mechanical Code (UMC).*
- System risers in accordance with NFPA 13, 13R, and 13D when installed concealed and protected by either one layer of 3/8" thick gypsum wallboard or 1/2" thick plywood (when installed exposed, additional limitations apply).

Listings vary among pipe and fitting manufacturers. CPVC sprinkler pipe and fittings from different manufacturers should not be combined unless that combination has been evaluated by their Listing. Consult the pipe and/or fitting manufacturer's installation instructions for requirements.

1.3.0 Handling and Storage of CPVC

CPVC pipe and fittings have specific handling and storage requirements, some of which are described below. Always read and follow the manufacturer's instructions to prevent damage.

1.3.1 Handling of CPVC

CPVC piping products have a lower impact strength than metal piping products. CPVC pipe fittings, whether in cartons or loose, must never be tossed or thrown to the ground. Do not drop or drag pipe, such as when unloading it from a truck, and keep the pipe in its original packaging until ready for use. Never throw heavy or sharp objects into or against CPVC pipe or fittings when unloading materials. When handling CPVC pipe, ensure that the pipe is well supported so that sagging is minimized. Very cold weather can make CPVC pipe and fittings brittle. Use extra care during handling to prevent damage.

Always inspect pipe and fittings for damage before installation. Cracks, splits, or scratches can weaken or damage the pipe and fittings. Pipe and fittings with cuts, gouges, scratches, splits, or other signs of damage from improper handling or storage must not be used. Damaged sections on lengths of pipe can easily be cut out using proper techniques for cutting CPVC pipe.

1.3.2 Storage of CPVC

Store CPVC pipe and fittings in the original packaging to prevent contamination and reduce the possibility of damage. Store CPVC pipe on a clean, flat surface that provides support for the entire length of the pipe. When palletized CPVC pipe is stored, ensure that the wooden pallet bracings are in full contact with each other. Store new CPVC pipe separately from previously used CPVC pipe. Racks used for storing CPVC pipe must have continuous or close support arms to prevent the pipe from sagging.

Did You Know?

Lubrizol offers an FBC System Compatible Product Finder app that can be used to search construction materials to see whether they are compatible with Lubrizol (e.g., BlazeMaster®) CPVC piping and fittings. It is available in versions for both Apple® and Android™ devices at **www.fbcsystemcompatible.com.**

When storing CPVC inside, keep it in a well-ventilated area, away from steam lines or other heat sources. When storing CPVC outside, it must be covered with a non-transparent material. Extended uncovered storage outdoors will result in color fading and may make CPVC brittle and prone to breakage when cutting. Check with the manufacturer's literature and company policy regarding the use of faded piping.

> **CAUTION**
>
> Take special care to avoid contamination of CPVC pipe and fittings with petroleum-based products, such as cutting or packing oils that may be present on metallic system components. Never mix CPVC fittings in storage bins with metal fittings.

1.4.0 CPVC Protection

CPVC pipe and fittings may be installed in both concealed and exposed sprinkler systems. CPVC run through metal studs should be protected against abrasion using stud inserts. Nail plates should be used to protect CPVC when run through any studs. Before penetrating fire walls and partitions, consult local building codes and the Authority Having Jurisdiction (AHJ).

CPVC systems must be designed and installed so that the piping is not exposed to excessive temperatures from specific heat-producing sources such as light fixtures, ballasts, transformers for fluorescent lights, and steam lines. Consult the manufacturer for CPVC distances from specific heat sources.

CPVC pipe and fittings are suitable for use in areas where the **ambient temperature** is between 35°F and 150°F. CPVC pipe can be installed in areas such as attic spaces where the temperatures will exceed 150°F if ventilation is provided or insulation is used around the pipe to maintain a pipe temperature of 150°F or lower. Protection from freezing must also be provided.

1.4.1 Concealed Installations

Over 90 percent of CPVC piping systems are installed concealed. The sprinklers used in these systems must be rated at 225°F or less. The sprinklers may be Listed quick response pendent or sidewall (standard or extended coverage) or Listed residential sprinklers. Underwriters Laboratories® (UL) requires that CPVC installations be protected by one of the following barriers to be considered concealed:

- $^3/_8$" gypsum wallboard
- $^1/_2$" plywood soffit
- Suspended (membrane) ceiling with lay-in panels or tiles having a weight of not less than 0.35 pounds per square foot when installed with metal support grids

For residential occupancies, as defined by NFPA 13R and NFPA 13D, the minimum protection may be $^1/_2$" plywood.

1.4.2 Exposed Installations

Listed CPVC pipe may be installed in exposed installations in residential occupancies as defined in NFPA 13D and 13R, and in light hazard occupancies as defined in NFPA 13. The following restrictions apply to standard coverage sprinklers installed in an exposed installation:

- *Upright sprinklers* — Listed 155°F maximum temperature-rated upright sprinklers may be installed in exposed locations with piping installed on a smooth, flat, horizontal, fixed ceiling. (Note that a pitch of 2 in 12 is considered a flat ceiling.) The deflectors must be installed within 4" of the ceiling, with no more than $7^1/_2$" from the ceiling to the center line of the pipe. The maximum distance from the center line of a sprinkler to the closest hanger is 3", and the maximum distance between sprinklers is 15'.
- *Pendent sprinklers* — Listed 170°F maximum temperature-rated pendent sprinklers may be installed in exposed locations with the piping mounted directly to the ceiling and deflectors installed within 8" of the ceiling. The maximum distance between sprinklers is 15'.
- *Horizontal sidewall sprinklers* — Listed 200°F maximum temperature-rated horizontal sidewall sprinklers may be installed in exposed locations with piping mounted directly to the wall. The deflectors must be installed within 12" of the ceiling and within 6" of the side walls. The maximum distance between sprinklers is 14'.

The requirements differ for extended coverage and residential sprinklers. Refer to the manufacturer's installation instructions for additional Listings and limitations.

1.0.0 Section Review

1. CPVC pipe weighs approximately _____.
   a. the same as steel pipe
   b. three-fourths that of steel pipe
   c. one-half that of steel pipe
   d. one-sixth that of steel pipe

2. CPVC sprinkler pipe and fittings from different manufacturers can be used interchangeably.
   a. True
   b. False

3. CPVC pipe should be stored _____.
   a. near a heat source to keep it pliable
   b. on a clean, flat surface that provides support for the entire length of the pipe
   c. standing on end or leaning against a wall whenever possible
   d. uncovered in direct sunlight to prevent condensation

4. CPVC pipe is suitable for use where ambient temperatures do *not* exceed _____.
   a. 150°F
   b. 200°F
   c. 250°F
   d. 300°F

Section Two

2.0.0 Cutting and Joining CPVC Pipe

Objective

Cut and join CPVC pipe.

a. Describe how to cut, deburr, bevel, and clean CPVC pipe.
b. Describe the procedure for solvent-cementing CPVC pipe and fittings.

Performance Tasks

1. Cut, deburr, bevel, and clean CPVC pipe.
2. Join CPVC pipe and fittings.

Trade Term

Beveling: Creating a slight angle on the end of a pipe to facilitate insertion into a fitting.

CPVC pipe and fittings are joined using solvent-cement. Often referred to as *interference fittings* since they have tapered sides to ensure a tight fit, CPVC fittings provide a strong, leak-proof connection when installed properly. The first step in creating a good connection is cutting and preparing the pipe.

2.1.0 Cutting and Preparing CPVC Pipe

CPVC pipe must be prepared before joining. The four steps required to prepare CPVC pipe are:

- Cutting
- Deburring
- Beveling
- Cleaning

2.1.1 Cutting CPVC Pipe

CPVC pipe can be cut using ratchet cutters, tubing cutters, fine-tooth pipe saws, and power saws. Regardless of which tool is used, it is extremely important that a square cut be made (*Figure 2*). This is critical to making a good solvent-cement connection. Keep all cutting blades sharp, and never use tools with chipped or broken blades. Damaged blades may weaken or split the pipe, while dull blades can compress rather than cut the pipe. If, after making the cut, there is any indication of damage or cracking at the pipe end, cut off at least 2" beyond any visible crack.

The following sections describe the tools used for cutting CPVC pipe.

Tubing Cutters

All tubing cutters must be rotated completely around the pipe to make a cut. Some tubing cutters have rollers that are tightened to force the pipe against the cutting wheel. Others have a cutting blade that is forced against the pipe. Tubing cutters are sized according to pipe capacity. *Figure 3* shows tubing cutters.

NOTE

The correct cutting wheel for CPVC pipe must be used.

Ratchet Cutters

Ratchet cutters (*Figure 4*) are designed for smooth, one-hand ratchet cutting action. Generally, ratchet cutters are used on pipe sizes of ¾" and 1" when temperatures are above 50°F. Avoid using ratchet cutters on material that has been stored outdoors uncovered for an extended period. This may make the pipe brittle and cause it to crack when cutting.

CPVC Pipe Saws

CPVC pipe saws have very fine teeth and are used on pipe sizes up to 3" in diameter. A miter box is recommended to ensure a square cut. *Figure 5* shows CPVC pipe saws.

Chop Saws

Electric chop saws may also be used for cutting CPVC pipe. The saw blade and speed must be matched to the pipe being cut. Proper safety precautions must be taken to avoid personal injury. *Figure 6* shows a chop saw.

2.1.2 Reaming CPVC Pipe

The rough edges of cut CPVC pipe must be smoothed prior to assembly using a reamer or deburring tool (*Figure 7*). A reamer or deburring tool is rotated on the inside of the pipe to remove burrs, which can prevent proper contact between the pipe and fitting during assembly.

2.1.3 Beveling CPVC Pipe

After deburring, CPVC pipe must be beveled on the outside edge using an outside beveling tool or inside-outside reamer (*Figure 7*).

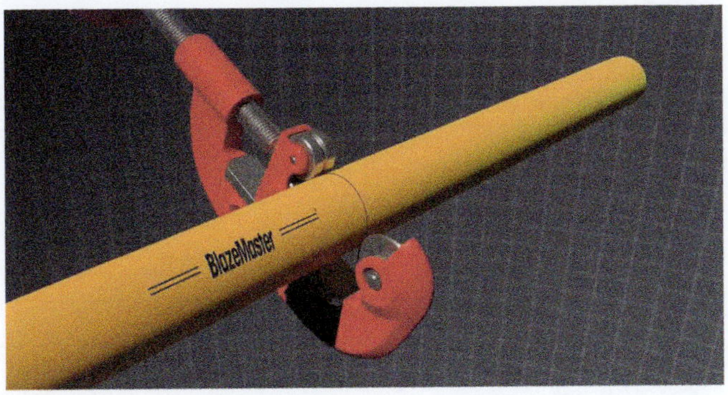

Figure 2 Cutting CPVC pipe.

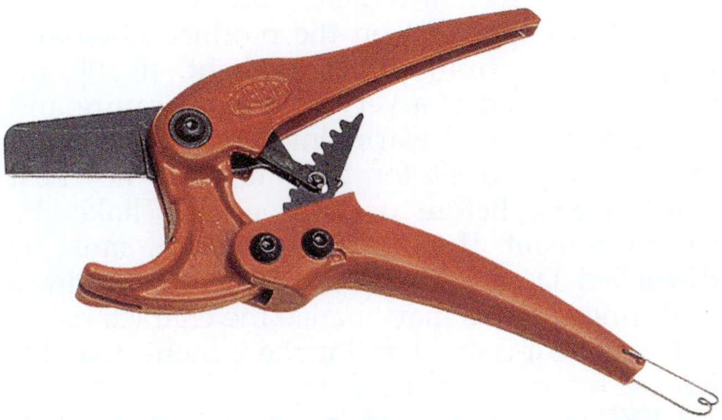

Figure 3 Tubing cutters.

Figure 4 Ratchet cutters.

Figure 5 CPVC pipe saws.

Figure 6 Chop saw.

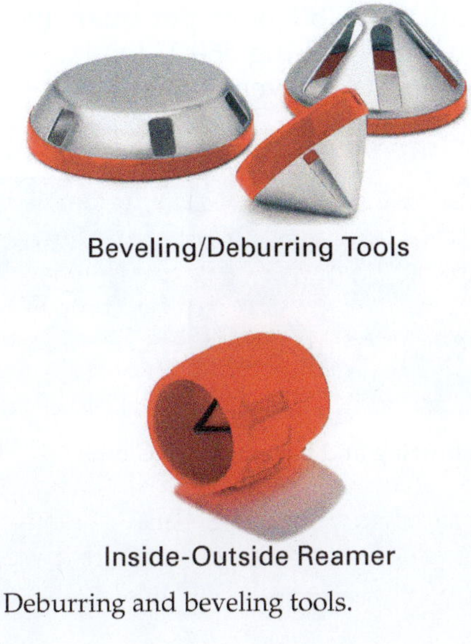

Figure 7 Deburring and beveling tools.

Beveling creates a slight angle on the end of the pipe that allows the pipe to slide into the fitting easier and prevents the pipe edge from wiping the solvent-cement from the fitting as it is pushed in. An inside-outside reamer can perform both reaming and beveling using a single tool, as shown in *Figure 8*.

2.1.4 Cleaning CPVC Pipe and Fittings

The deburred and beveled pipe must be wiped clean of all foreign material to ensure a good seal. Loose particles caused by reaming and deburring can prevent a good connection and result in leaks. Use a clean, dry shop cloth to wipe loose material and moisture from both the fitting socket and the pipe end. After cleaning, you are ready to begin the joining process.

> **CAUTION**
> Moisture or other contaminants will slow or prevent proper adhesion. Make certain both the pipe and fitting are dry and clean prior to using solvent-cement. Faulty joints caused by contamination may not show up for weeks.

2.2.0 Using Solvent-Cement

CPVC pipe and fittings are joined using a one-step solvent-cement. Always use the Listed solvent-cement recommended by the pipe manufacturer.

Plan ahead by estimating the amount of solvent-cement needed for the job. Count the number of joints to be made and divide the total by the joints/quart listed in the manufacturer's instructions for the solvent-cement in use. *Table 2* shows typical joints per quart for various fitting sizes. For example, if you need to make 90 joints using 1" fittings, you will use about half a quart of solvent-cement at 180 joints per quart (90 ÷ 180 = 0.5). If you were making 360 1" joints, you would need two quarts of solvent-cement (360 ÷ 180 = 2).

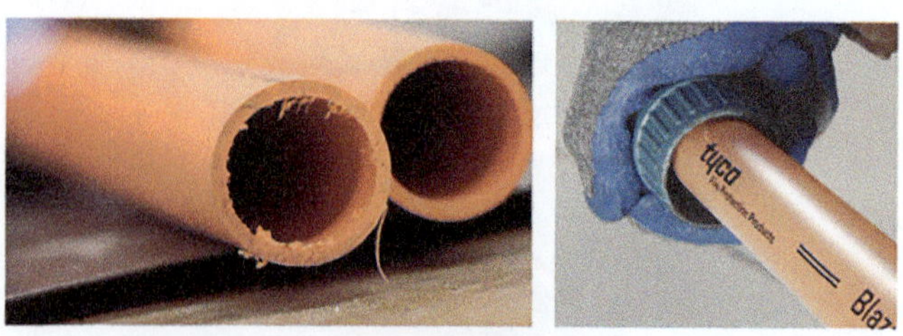

Figure 8 Deburring and beveling CPVC pipe.

Table 2 Joints per Quart of One-Step Solvent-Cement

Fitting Size (inches)	Solvent-Cement (joints/quart)
$\frac{3}{4}$	270
1	180
$1\frac{1}{4}$	130
$1\frac{1}{2}$	100
2	70
$2\frac{1}{2}$	50
3	40

Store the solvent-cement between 40°F and 110°F (or as specified on the product label) and keep it away from direct sunlight. If solvent-cement is stored at a very cold temperature and gels, bring it into a warm environment (60–90°F) and allow it to sit for 24 hours. Do not heat it artificially. Before use, vigorously shake the solvent-cement. If it remains gelled, it must be discarded. Do not attempt to dilute it with primer or thinner, as this may change the characteristics of the solvent-cement and make it ineffective. Do not use frozen solvent-cement.

CPVC solvent-cement has a maximum shelf life of two years from its date of manufacture (*Figure 9*). Always check for the date stamped on the solvent-cement can and do not use it beyond its expiration date.

2.2.1 Solvent-Cement Safety Precautions

Before using any CPVC solvent-cement, read the directions on the container label and follow all precautions listed on the product's safety data sheets (SDS) and ASTM F402, *Standard Practice for Safe Handling of Solvent-Cements, Primers, and Cleaners Used for Joining Thermoplastic Pipe and Fittings*.

Observe the following general safety precautions when using solvent-cement:

- Eliminate all ignition sources and do not smoke when using solvent-cement.
- Avoid breathing solvent-cement vapors.
- Use only with adequate ventilation. Explosion-proof general mechanical ventilation or local exhaust is recommended to maintain vapor concentrations below the recommended exposure limits.
- In confined or partially enclosed areas, use a National Institute for Occupational Safety and Health (NIOSH)-approved organic vapor cartridge respirator with a full facepiece.
- Do not eat or drink while using solvent-cement.

- Keep containers of solvent-cement tightly closed when not in use and cover them as much as possible when they are in use.
- Avoid contact with skin. Wear appropriate protective clothing and gloves.
- Wear wraparound safety goggles to protect against splashes.
- Never work overhead as this may cause solvent-cement to drip onto your skin or in your eyes.

For a complete list of safety hazards and required personal protective equipment, refer to the SDS for the solvent-cement in use.

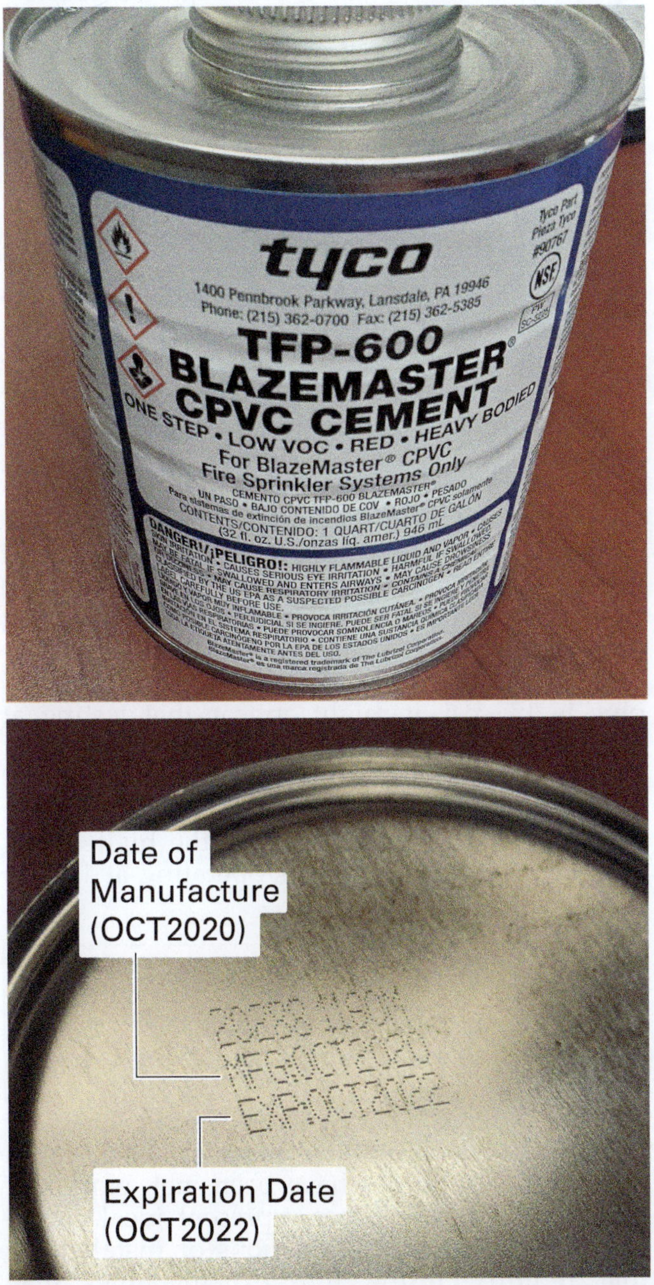

Figure 9 Solvent-cement expiration date.

2.2.2 Solvent-Cement Application and Pipe Assembly

One-step solvent-cement is designed to be used without primer. Always refer to the manufacturer's instructions before using solvent-cement. To create a good connection, joining surfaces must first be penetrated and softened (see *Figure 10*).

Before proceeding, select the correct solvent-cement dauber size for the pipe in use. Using a dauber that is too small will not spread enough solvent-cement to make a good connection, while using one that is too large may result in pooling or puddling. Refer to the manufacturer's instructions regarding dauber sizes. A typical recommendation is to use a $^{3}/_{4}$"-diameter dauber for $^{3}/_{4}$" and 1" pipe, and a $1^{1}/_{2}$"-diameter dauber for $1^{1}/_{4}$" to 3" pipe.

A typical solvent-cementing procedure is shown in *Figure 11*.

> **WARNING!**
> Before using CPVC solvent-cement, review ASTM F402 and follow all precautions found on the container label and product SDS. Provide proper ventilation and avoid breathing the vapors. Avoid skin or eye contact with the solvent-cement and keep the container closed when not in use.

Step 1 Apply a heavy, even coat of solvent-cement to the outside of the pipe end using the proper dauber size. It is important to ensure sufficient penetration of the solvent-cement into the pipe and fitting surfaces. Usually, three to five rotations around the pipe with the dauber are sufficient to achieve proper softening.

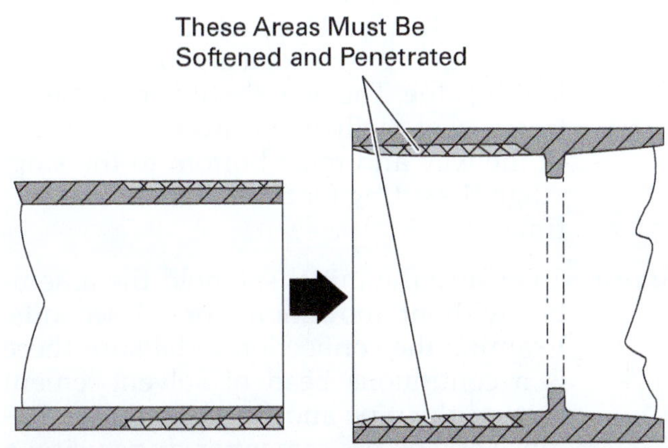

Figure 10 Area of coverage (bonding area).

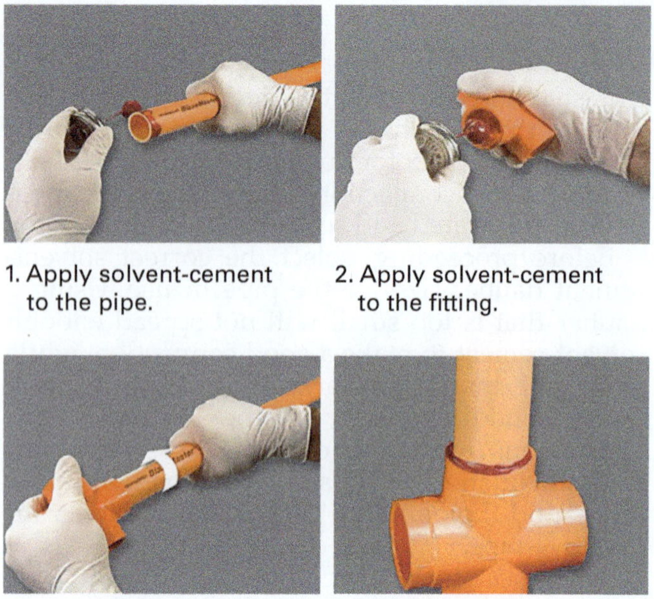

1. Apply solvent-cement to the pipe.
2. Apply solvent-cement to the fitting.
3. Insert pipe into fitting and rotate ¼ turn.
4. Hold for 30 seconds.

Figure 11 Solvent-cementing procedure.

Step 2 Without re-dipping the dauber, immediately apply a coat of solvent-cement to the fitting socket. Do not allow the solvent-cement to puddle in the fitting. A second dipping and application on the pipe end is required for pipe sizes $1\frac{1}{4}$" and above. The proper sequence is to apply solvent-cement on the pipe end, then in the fitting socket, and then again on the pipe for pipe larger than $1\frac{1}{4}$".

Step 3 Immediately after applying the solvent-cement, insert the pipe into the fitting socket and rotate the pipe one-quarter turn. Rotating the pipe eliminates tracks made from inserting the pipe into the fitting. Most of the joint strength is located close to the pipe stop, which is why it is so important to make a square cut when joining pipe. The pipe should enter the fitting socket easily one-third to two-thirds of the way and must bottom to the stop. Align the fitting for the installation at this time.

Step 4 To ensure the initial set, hold the assembly without movement for 30 seconds. Examine the connection to be sure there is a continuous bead of solvent-cement around the pipe and fitting juncture. Remove any excess cement or drips using a shop cloth.

CAUTION

If there is any break in the bead around the juncture, it may indicate that insufficient solvent-cement was applied. If this occurs, both the pipe and the fitting must be discarded, and the process repeated.

Before installing sprinklers, let the fittings cure for a minimum of 30 minutes. When installing sprinklers, the pipe drop must be held or anchored securely to avoid disturbing the pipe in previous connections.

CAUTION

Do not install sprinklers on the sprinkler adapter fittings, and then solvent-cement the sprinkler adapter fittings to the drop. This may allow the solvent-cement to run into the sprinkler orifice and plug the sprinkler.

2.2.3 Solvent-Cement Cure Times Before Pressure Testing

After assembly, CPVC connections must be allowed to cure undisturbed for a specific period prior to pressure testing. Solvent-cement cure times depend on pipe size and length, ambient temperature, relative humidity, test pressure, and the type of solvent-cement in use. Cure times are faster in dry environments than in high-humidity environments. They are also faster for smaller pipe sizes, higher ambient temperatures, and lower test pressures. *Table 3* shows minimum cure times for Listed CPVC pipe and fittings using one-step solvent-cement.

Since the design operating pressure of CPVC pipe is 175 psi, it might appear that testing at pressures above 175 psi is not allowed. This is not the case. NFPA 13 requires a test pressure of 200 psi *or* 50 psi above the system operating pressure, whichever is largest. Therefore, the highest test pressure is 225 psi. Systems operating at 150 psi or less are tested at 200 psi.

NOTE

NFPA 13D permits systems in one- and two-family dwellings and manufactured homes to be tested at line pressure (100 psi or less).

The system cannot be pressure tested until minimum cure times have been met. Note that manufacturer cure times represent ideal conditions. High-humidity conditions or installations that require cutting into existing lines may increase the cure time significantly.

Table 3 Typical Cure Times for One-Step Solvent-Cement

Test Pressure	Pipe Size (inches)	Ambient Temperature During Cure Period		
		60°F to 120°F	40°F to 59°F	0°F to 39°F
225 psi	¾	1 hour	4 hours	48 hours
	1	1½ hours	4 hours	48 hours
	1¼ and 1½	3 hours	32 hours	10 days
	2	8 hours	48 hours	*
	2½ and 3	24 hours	96 hours	*
200 psi	¾	45 minutes	1½ hours	24 hours
	1	45 minutes	1½ hours	24 hours
	1¼ and 1½	1½ hours	16 hours	120 hours
	2	6 hours	36 hours	*
	2½ and 3	8 hours	72 hours	*
100 psi	¾	15 minutes	15 minutes	30 minutes
	1	15 minutes	30 minutes	30 minutes
	1¼	15 minutes	30 minutes	2 hours

*For these sizes, the solvent-cement can be applied at temperatures below 40°F. However, the sprinkler system temperature must be raised to a temperature of 40°F or above and allowed to cure per the above recommendations prior to pressure testing.

2.0.0 Section Review

1. Generally, ratchet cutters are used on pipe sizes of ¾" and 1" when temperatures are above _____.
   a. 0°F
   b. 15°F
   c. 32°F
   d. 50°F

2. The typical cure time for 2" CPVC pipe installed at 72°F with a test pressure of 225 psi is _____.
   a. 1 hour
   b. 4 hours
   c. 8 hours
   d. 16 hours

Section Three

3.0.0 Installing and Testing CPVC Sprinkler Systems

Objective

Install and test CPVC sprinkler systems.

a. Identify support spacing requirements for CPVC pipe.
b. Describe how to install sprinklers and test the system.
c. Identify installation requirements for special applications.

Performance Tasks

3. Install CPVC pipe using the correct support spacing.
4. Complete a manufacturer training course for CPVC.

Trade Terms

Armover: A horizontal pipe that extends from a branch line to a single sprinkler.

Thermal expansion: The change in pipe length due to changes in temperature.

Transverse: The lateral shifting of pipe.

CPVC has special support spacing and testing requirements. In addition, there are specific installation requirements for special applications, including temperature extremes, repair and retrofit installations, and connecting CPVC to other types of piping systems.

3.1.0 CPVC Pipe Support Spacing

Support spacing for CPVC pipe is different than other piping materials due to the flexibility of the pipe itself. CPVC is prone to a greater degree of **thermal expansion** and must be protected against mechanical damage and excessive stress.

Most hangers and supports that are Listed for metal piping systems are also acceptable for CPVC. Residential occupancies can use plumbing-type hangers under NFPA 13D. When plumbing-type hangers are used, they must comply with local plumbing codes. Consult the manufacturer's installation instructions for specific requirements.

3.1.1 Thermal Expansion

All materials expand as the temperature increases, and contract as the temperature decreases. This causes pipe to change length as the temperature changes. The expansion and contraction of CPVC is greater than that of steel and must be considered in the design of the system. A 25°F change in temperature in CPVC piping will cause an expansion of 1" for every 100' of straight pipe. It is important to allow for expansion and contraction in CPVC sprinkler systems to prevent movement of the sprinklers.

Expansion and contraction can be accommodated using a variety of methods, including pipe offsets, changes in direction, and expansion loops. Regardless of which type of device is used, make sure the installation allows for pipe movement. Do not overtighten hangers or straps and avoid butting the pipe up against a structural member (joist or stud) at changes in direction.

Systems installed in very cold weather may require additional expansion devices to account for the large change in temperature once the building is fully constructed and heated and/or outside temperatures increase. For details, refer to the manufacturer's instructions.

3.1.2 Hanger Spacing for CPVC Pipe

Listed hangers for CPVC pipe must be free of sharp edges, correctly sized, and must not prevent pipe movement (unless Listed to do so). Pipe hangers must comply with the requirements of NFPA 13 and NFPA 13R (NFPA 13D does not require Listed hangers except as identified in other Listings).

Spacing between pipe hangers is determined by the pipe size and manufacturer. *Table 4* shows one manufacturer's recommended maximum support spacing for in-line hangers, while *Table 5* shows the support spacing for end-of-line and **armover** supports. Note that the spacing increases with larger pipe sizes and lower operating pressures. Always refer to the manufacturer's spacing requirements for the piping in use.

3.1.3 Restraints, Anchors, and Guides

In addition to pipe hangers and supports, CPVC pipe requires the use of special restraints, anchors, and guides. These are used on long runs of pipe, where there are directional changes, or when expansion joints are used.

Table 4 In-Line Hanger Spacing for CPVC Pipe

Maximum Support Spacing Distance In-Line, Sprinkler Drop Tee		
Nominal Pipe Size (inches)	**Less than 100 psi (feet)**	**More than 100 psi (feet)**
$\frac{3}{4}$	4	3
1	5	4
$1\frac{1}{4}$	6	5
$1\frac{1}{2}$ to 3	7	7

Table 5 End-of-Line and Armover Hanger Support for CPVC Pipe

Maximum Support Spacing Distance End of Line, Sprinkler Drop Elbow		
Nominal Pipe Size (inches)	**Less than 100 psi (inches)**	**More than 100 psi (inches)**
$\frac{3}{4}$	9	6
1	12	9
$1\frac{1}{4}$	16	12
$1\frac{1}{2}$ to 3	24	12

Anchors direct the movement of the pipe within a defined reference frame. An anchor prevents both vertical movement and **transverse** (lateral) movement at the anchoring point. Guides prevent transverse movement of the pipe but allow for expansion and contraction. Restraints prevent only vertical movement of the pipe.

Materials that are commonly used for restraints, anchors, and guides include special fittings, plastic poly strapping, and U-bolt clamps. There are also strap hangers specifically engineered for CPVC pipe. These are combination hanger-restraint devices with flared edges to prevent pipe abrasion. They are made of galvanized sheet metal and meet the requirements of NFPA 13, 13R, and 13D. *Figure 12* shows CPVC pipe straps.

When a sprinkler discharges, a significant reactive force is exerted on the pipe, especially at system pressures greater than 100 psi. This reactive force will cause the pipe to lift vertically if it is not properly secured, especially if the sprinkler drop is from a small diameter pipe.

When a sprinkler drop is from a $\frac{3}{4}$", 1", or $1\frac{1}{4}$" pipe, the closest hanger must brace the pipe against the vertical lift. A number of techniques can be used to brace the pipe, such as using a standard band hanger to position the threaded support rod to $\frac{1}{16}$" above the pipe or using a split ring or wraparound hanger.

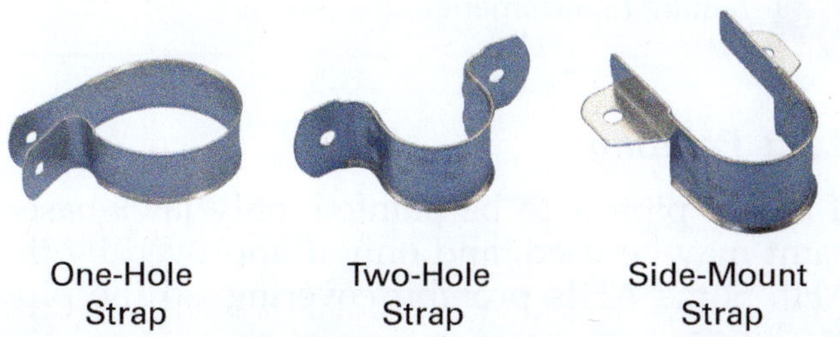

Figure 12 CPVC pipe straps.

When piping is suspended from a deck, hangers are required to suspend the pipe and provide vertical lift restraint. One support can serve as both. Drop locations between supports are acceptable in any location if support spacing requirements are followed.

When the piping is supported by wood joists or trusses, the structure provides the support, especially when the joists are close together (*Figure 13*). The only requirement with this type of construction is to provide vertical restraint. As with a decking installation, drop location supports are acceptable in any location if support spacing requirements are followed.

Make sure that all pipe hangers are sized for the same pipe size or one size larger. When using a two-hole strap near a fitting or when suspending pipe from a ceiling, UL allows for upsizing the hanger by one size over the pipe diameter to relieve stress on the pipe.

CAUTION

Do not use undersized pipe hangers. Undersized hangers will create stress on the pipe and may result in pipe deflection or damage.

NOTE

The requirements for seismic bracing are similar to those for steel, and can be found in NFPA 13, ***Chapter 18*** and ***Appendix A.18.3***.

3.2.0 Installing Sprinklers and Testing the System

Sprinklers may be installed only after all the fittings, including the sprinkler adapters, are solvent-cemented to the piping and have been allowed to cure for a minimum of 30 minutes or as recommended by the manufacturer.

3.2.1 Installing Sprinklers

Prior to sprinkler installation, sprinkler fittings must be visually inspected and probed with a wooden dowel to ensure that the pipe and threads are clear of any excess solvent-cement.

Figure 13 Pipe supports.

To assemble the joint on a standard (non-gasketed) adapter, use only Teflon® tape or chemically compatible pipe thread compound (also referred to as *thread sealant* or *pipe dope*). Never use both as the excess stress caused by the additional material will result in a leak or cracked fitting. When using tape, apply two or three wraps in the direction of the threads to prevent it from unraveling when the sprinkler is installed. The wrap is clockwise for right-hand threads. Tighten per the manufacturer's instructions.

When installing sprinklers in gasketed adapters, tighten threaded joints one to two turns beyond finger tight. Note that a hold-back device approved for use with CPVC pipe and fittings must be used when torquing the body of the sprinkler adapter. Avoid over-torquing.

Refer to the manufacturer's installation instructions for the sprinklers and sprinkler adapters in use. After installing all sprinklers, inspect the work for proper alignment and hanger placement prior to performing hydrostatic tests.

3.2.2 Hydrostatic Testing

Once the installation has cured for the specified amount of time, hydrostatic (water pressure) testing must be performed to ensure that the connections are leak-free.

To perform hydrostatic testing, fill the system slowly with water and bleed air from the farthest and highest sprinklers before the test pressure is applied. Pressure test in accordance with the applicable NFPA standards and the manufacturer's installation instructions. If a leak is found, the associated fitting(s) must be cut out and replaced.

WARNING!

Some CPVC manufacturers allow a low-pressure air test and low-pressure air may also be used in dry or preaction systems. Strict adherence to the manufacturer's recommended air pressures is required to prevent serious injuries or explosion when air is introduced into a piping system. Where low-pressure air is allowed, it must be supplied by an oil-free compressor. Oil-laden air will compromise the system and require replacement of the piping.

3.2.3 Painting

If CPVC pipe is to be painted, only latex-based paint may be used, and only if approved by the AHJ. Some AHJs prohibit covering up the pipe markings.

> **WARNING!** Do not use oil-based paint on CPVC. Oil or solvent-based paints may be chemically incompatible with CPVC pipe and fittings, which could cause damage. Check with the pipe manufacturer for a list of compatible coatings.

3.2.4 Installing CPVC Signage

Before starting the installation, install signage indicating the presence of a CPVC sprinkler system and cautioning other construction personnel of the common materials that can degrade or damage CPVC systems. The signage should be placed in a heavily trafficked area on the job site. A typical CPVC placard is shown in *Figure 14*. These can be found on the CPVC manufacturer websites for printing and also form part of the contract submittal.

3.3.0 Special Applications

Special application considerations for CPVC pipe include installation in hot or cold weather, repair and modification of existing systems, and connecting CPVC pipe to other types of piping systems.

3.3.1 Joining CPVC in Cold Weather

CPVC pipe and fittings are Listed for installation at temperatures down to 0°F. However, cold temperatures make CPVC pipe and fittings brittle, so extra care must be taken to avoid damage during installation. In addition, solvents penetrate and soften the pipe and fitting surfaces more slowly in cold weather, which means that fittings must be held in place longer for the initial set. Colder temperatures also result in longer cure times before pressure testing.

> **CAUTION** Installation temperatures below 0°F invalidate the Listing of the CPVC and solvent-cement, as surfaces below this temperature cannot soften enough to achieve a good connection.

The following precautions should be observed when solvent-cementing during cold weather:

- Note that cold pipe will expand after installation. Refer to the pipe manufacturer's information for installation at very cold temperatures.
- Ensure that the pipe, fittings, and valves are at the same temperature prior to solvent-cementing.
- Prefabricate as much of the system as possible in a heated work area.
- Store the solvent-cement in a warm area when it is not in use and make sure it remains fluid. Frozen solvent-cement must be discarded.
- Take special care to remove moisture, including ice and snow.
- Allow the manufacturer's recommended cure period for the ambient temperature before the system is pressurized.

3.3.2 Joining CPVC in Hot Weather

CPVC solvent-cements contain volatile solvents. Higher temperatures and wind accelerate evaporation. Pipe stored in direct sunlight may have surface temperatures of 20°F to 30°F above air temperatures. Solvents attack these hot surfaces more deeply and it is very important to avoid puddling the solvent-cement inside the fitting socket. Always ensure that the excess solvent-cement is wiped from the outside of the joint. Follow the manufacturer's installation instructions and observe the following precautions:

- Note that heated pipe will contract after installation. Refer to the pipe manufacturer's information for installation at very warm temperatures.
- Store solvent-cements in a cool, shaded area prior to use.
- If possible, store pipe and fittings, or at least the ends to be joined, in a shady area before solvent-cementing.
- Make sure that both surfaces to be joined are still wet with solvent-cement when putting them together. With larger sizes of pipe, the assistance of a second fitter may be required to complete the application successfully.

3.3.3 Repair and Modification to Existing CPVC Systems

Cutting into active systems requires extra caution due to the risk of water damage. Follow the manufacturer's cut-in procedures and consult with your company's service department before doing this type of work.

> **CAUTION** One of the most frequent types of insurance claims by sprinkler contractors is failure of cut-ins and modifications to existing CPVC systems. These claims can be especially large if the failure occurs in the upper floors of a high-rise building.

The following is a general cut-in procedure for modifications or repairs to existing systems.

Step 1 Review the manufacturer's solvent-cementing and cut-in procedures to ensure that proper joining techniques are used. Always use the correct tools for the task.

Step 2 To obtain faster cure times before filling the system with water, begin by making the cut-in on the smallest diameter pipe, as close as possible to the area to be repaired. Completely drain all existing lines prior to applying solvent-cement, as moisture increases cure times and results in weaker joints.

Step 3 Measure and cut pipe to the proper length, making sure that it is square. Then deburr and bevel it to ensure a proper fit. Before applying the solvent-cement, clean any moisture and dirt out of fitting sockets and pipe ends using a clean, dry shop cloth. Always use a fresh, unexpired can of solvent-cement for cut-in connections.

Step 4 Follow the manufacturer's instructions for assembling the cut-in. Remember that a quarter-turn is required when inserting pipe into a fitting. You may have to assemble some components using socket unions, flanges, or grooved coupling adapters to allow the quarter-turn. Carefully inspect each connection to ensure there is a complete, even bead of solvent-cement around the fitting.

Step 5 Allow the joints to cure completely before filling the system with water. Note that cure times for cut-ins are longer than for new installations. Refer to the manufacturer's guidelines for cut-in cure times as determined by pipe size and ambient temperature.

Step 6 Slowly fill the system with water and bleed air from the furthest and highest sprinklers. Perform pressure testing in an isolated area to avoid water damage if the new connections fail. Do not exceed 50 psi over the system operating pressure. If any leaks are detected, cut out and replace the affected fittings.

WARNING!

Some CPVC manufacturers allow a low-pressure air test and low-pressure air may also be used in dry or preaction systems. Strict adherence to the manufacturer's recommended air pressures is required to prevent serious injuries or explosion when air is introduced into a piping system. Where low-pressure air is allowed, it must be supplied by an oil-free compressor. Oil-laden air will compromise the system and require replacement of the piping.

Specialty Sprinkler Adapters

Some specialty sprinkler adapters have plastic threads and use a gasket to achieve a watertight connection. These fittings install faster because they do not require the use of either Teflon® tape or pipe thread compound. The brass-free design also eliminates corrosion in hard-water environments and meets state and federal low-lead requirements for sprinkler systems.

Figure Credit: "Used with permission of Johnson Controls"

NOTICE

This building contains a CPVC fire sprinkler system. This CPVC fire sprinkler system is a Life Safety Assembly and must be treated carefully. Please read the following before any activity which could contact this system:

CPVC piping components may be damaged by certain substances and construction practices.

- DO NOT stack, support, hang equipment, or hang flexible wire/cable, especially communications cable, or other material on the fire sprinkler system.
- ONLY system compatible materials including, but not limited to solvent cements, caulks, sealants, cutting oils and thread pastes as noted by the CPVC fire sprinkler piping system manufacturer's installation instructions should be used in contact with this system.
- DO NOT expose CPVC products to incompatible substances, such as cutting oils, non-water based paints, packing oils, traditional pipe thread paste and dope, fungicides, termiticides, insecticides, detergents, building caulks, adhesive tape, solder flux, flexible wire/cable (with special consideration for communications cabling), and non-approved spray foam insulation materials.
- DO NOT expose CPVC products to edible oils, solvents, or glycol-based anti-freeze fluids.
- DO NOT expose CPVC products to open flame, solder, and soldering flux.
- DO NOT drop, distort, or impact CPVC products or allow objects to be dropped on them.
- DO NOT handle CPVC products with gloves contaminated with oils (hydrocarbons) or other incompatible materials.

Failure to follow this notice may cause cracks or fractures to develop in CPVC products resulting in property damage due to leaks or flooding. The presence of any visible cracks may require partial or full system replacement. For additional information contact the general contractor or the fire sprinkler system installer.

Figure 14 CPVC placard.

3.3.4 Connecting CPVC to Other Materials

CPVC can be connected to other materials using male threaded adapters, female threaded adapters, flanges, or grooved coupling adapters. *Figure 15* shows CPVC threaded and grooved adapters. Grooved adapters must be used in conjunction with flexible couplings.

Teflon® tape is the recommended thread compound and should be used with all threaded connections. The fitting manufacturer must be consulted before pipe thread compound is used. If using a compatible pipe thread compound, cover the male threads only and exercise caution to avoid clogging the opening with excess sealant.

Never use both thread tape and pipe thread compound on the same connection. When used together, the stress created by the excess product can result in a leak or cracked fitting.

Before making any connections, make certain that any thread cutting oil or other residue has been completely removed from both the inside and outside of the metal pipe, fittings, and flanges.

CAUTION

Always make sure that all products in contact with CPVC are chemically compatible. Consult the CPVC manufacturer for guidance. See the *Appendix* for additional information.

NOTE

In addition to the installations described here, some CPVC pipe is Listed for use in other special locations, including combustible concealed spaces, attic spaces, basements, and other areas. Always follow both the Listing requirements and the shop drawings when installing CPVC in special locations.

Figure 15 CPVC threaded and grooved adapters.

3.0.0 Section Review

1. The *maximum* in-line support spacing for 2" CPVC pipe at an operating pressure of more than 100 psi is _____.
   a. 5'
   b. 6'
   c. 7'
   d. 8'

2. During hydrostatic testing, the system is filled with _____.
   a. nitrogen
   b. compressed gas
   c. atmospheric air
   d. water

3. The Listing of the CPVC and solvent-cement is invalidated at installation temperatures below _____.
   a. 32°F
   b. 15°F
   c. 10°F
   d. 0°F

Review Questions

1. All diameters of SDR 13.5 pipe have the same pressure-bearing capability of _____.
   a. 150 psi
   b. 175 psi
   c. 200 psi
   d. 225 psi

2. CPVC pipe and fittings have a required *minimum* quick-burst strength of 5 times the rated pressure or _____.
   a. 525 psi
   b. 700 psi
   c. 875 psi
   d. 1,050 psi

3. Which of the following is *true* regarding CPVC pipe?
   a. CPVC has a higher impact strength than steel pipe.
   b. CPVC can become brittle at very cold temperatures.
   c. CPVC is unlikely to be damaged by rough handling.
   d. If a length of pipe has damage on one end, the entire pipe must be thrown away.

4. If a pipe end shows any cracking after cutting, _____.
   a. disregard any damage that will be hidden within the fitting
   b. brush solvent-cement over the damaged area
   c. lightly sand the damaged area
   d. cut off at least 2" beyond any visible crack

5. Because the solvent-cement joining process is a chemical reaction, _____.
   a. foreign material is unlikely to affect the final joint
   b. cleaning is recommended but not required
   c. poor joints will be immediately apparent
   d. thorough cleaning is necessary to prevent contamination

6. The number of joints that can be made on 1" pipe using a quart of one-step solvent-cement is _____.
   a. 60
   b. 120
   c. 180
   d. 270

7. The *minimum* test pressure for a sprinkler system with a system operating pressure of 175 psi is _____.
   a. 150 psi
   b. 175 psi
   c. 200 psi
   d. 225 psi

8. The maximum in-line support spacing for 1½" CPVC pipe operating at less than 100 psi is _____.
   a. 5'
   b. 6'
   c. 7'
   d. 8'

9. Which of the following prevents *both* transverse (lateral) movement and vertical movement of a pipe?
   a. Restraints
   b. Guides
   c. Anchors
   d. Saddles

10. Which of the following is *correct* regarding the painting of CPVC?
    a. If CPVC pipe is to be painted, a latex-based paint should be used.
    b. If CPVC pipe is to be painted, an oil-based paint should be used.
    c. CPVC may be painted using any type of acrylic or oil-based paint.
    d. NFPA 13 prohibits painting of CPVC under any circumstances.

Trade Terms Quiz

Fill in the blank with the correct term learned from your study of this module.

1. _______________ is a white thermoplastic material that accounts for 85 percent of the weight of finished CPVC products.
2. A common thermoplastic pipe used in the sprinkler industry is _______________.
3. When pipe has shifted laterally, it has done a(n) _______________ shift.
4. The nominal pipe size of iron pipe is referred to as _______________.
5. _______________ is the name for the surrounding air temperature.
6. _______________ refers to plastic piping that is soft and pliable when heated, but hard when cooled.
7. _______________ is an orange thermoplastic material used to produce sprinkler pipe and fittings.
8. Angling the end of a pipe to facilitate insertion into a fitting is called _______________.
9. _______________ is the pipe wall thickness in proportion to its outside diameter.
10. An occupancy or portions of other occupancies where the quantity and/or combustibility of contents is low and fires with relatively low rates of heat release are expected is a(n) _______________ occupancy.
11. A horizontal pipe that extends from the branch line to a single sprinkler is a(n) _______________.
12. An occupancy that is divided into two groups depending on the quantity and type of hazard is a(n) _______________occupancy.
13. The change in pipe length due to changes in temperature is known as _______________.

Trade Terms

Ambient temperature
Armover
Beveling
Chlorinated polyvinyl chloride (CPVC)
CPVC compound
CPVC resin
Iron pipe size (IPS)
Light hazard
Ordinary hazard
Standard dimensional ratio (SDR)
Thermal expansion
Thermoplastic
Transverse

Cornerstone of Craftsmanship

Mac McGaffagan

Fire Sprinkler Service Division Manager
Piper Fire Protection

How did you choose a career in the industry?

I was born and raised on a barrier island 25 miles west of Tampa. Growing up, I watched developments come along and after seeing all the construction, I decided to become a property manager. I remember walking through the buildings and looking at the rusty pipes thinking "someone could make a lot of money there." Before I knew it, I was recruited by Piper Fire Protection, and I've been there ever since.

Who inspired you to enter the industry?

I was inspired by the need and ready for a change in my life. I'd taken a year off to check building a house off my bucket list and I thought to myself, "What am I going to do next?" Before I knew it, my house needed a fire sprinkler inspection, and I thought: "Well, that could be a cool thing to do."

What types of training have you been through?

I've been a NICET Level III inspector since 2006, and I've been trained in foam, pump, valve, and pre-action schools. I also have extensive nitrogen training.

How important is education and training in construction?

Extremely, extremely important. The world has changed considerably since I started, and there's no way you can do it without education. If you don't have some education, you won't stay up with the times.

How important are NCCER credentials to your career?

Those credentials gave me the ability to teach. Last year, I taught at a technical college, and now I'm teaching out of the office and break room at my company.

How has training/construction impacted your life?

It's given me an opportunity to provide young people with a direction and allow them to make a decent amount of money in a reasonable amount of time. If you follow the program and NCCER/AFSA books, you can make great money. My son grew up helping me in the industry and later went on to win an ABC Gold Medal and become a successful craft professional in the industry.

What kinds of work have you done in your career?

By trade, I'm a NICET Level III inspector. I've been with the same company my entire career. Six months after I started on my first big construction, I became an inspector and was able to sell a lot of accounts. Those were the early years that put Piper on the map.

Tell us about your present job.

I oversee inspections and service for a very busy, large-scale department. And then I'm teaching at the same time. The trade gave me something, and I want to give back to the trade. I truly believe in what we do.

What do you enjoy most about your job?

I still like going to work every day. It's so multi-faceted. Some days I deal with staff and HR, and other days I'm in customer service and sales. I love the challenges.

What factors have contributed most to your success?

There's definitely the ability to get training if you want it. If someone wants to succeed and gain knowledge, they can do it. My success came from always being honest with customers. It keeps them coming back and really builds your reputation.

Would you suggest construction as a career to others? Why?

Absolutely. It offers so many opportunities, and you can even move to different fields.

What advice would you give to those new to the field?

Study, learn, and never stop.

Interesting career-related fact or accomplishment?

When I started, the company had 100 accounts. We now have over 15,000. Honesty and hard work get you there. Five out of seven of my original staff still work for me, and it's really cool to see those relationships develop.

How do you define craftsmanship?

An acute attention to detail, understanding, and execution.

Appendix

Chemical Compatibility

CPVC domestic, water, sprinkler, and industrial piping systems have been used successfully for more than 45 years in new construction, re-pipe, and repair. CPVC products are ideally suited for these applications due to their outstanding corrosion resistance. Occasionally, however, CPVC can be damaged by contact with chemicals found in some construction products (and site preparations). Reasonable care needs to be taken to ensure that any products in contact with CPVC systems are chemically compatible. Chemical compatibility with CPVC should be confirmed with the manufacturer of the product in contact with CPVC piping systems. If chemical compatibility with CPVC is in question, it is recommended to isolate the suspect product from contact with CPVC pipe or fittings.

For BlazeMaster® products, a list of compatible and incompatible products is available in a searchable form via their FBC System Compatible Product Finder app available for both Apple® and Android™ devices at **www.fbcsystemcompatible.com**. It is important to note that this program applies only to BlazeMaster® products, not other brands of CPVC. Contact the CPVC manufacturer to obtain the latest compatibility information, as changes are made as more tests are conducted. Always be mindful of these concerns while on the jobsite. If you suspect product contact that could compromise the integrity of CPVC systems, notify your supervisor at once.

Scan this QR code with the camera app on your smartphone to access **lubrizol.com**, which has the latest information on system compatibility as well as installation videos for BlazeMaster® products.

Trade Terms Introduced in This Module

Ambient temperature: The surrounding air temperature.

Armover: A horizontal pipe that extends from a branch line to a single sprinkler.

Beveling: Creating a slight angle on the end of a pipe to facilitate insertion into a fitting.

Chlorinated polyvinyl chloride (CPVC): A common thermoplastic pipe used in the sprinkler industry.

CPVC compound: An orange thermoplastic material used to produce sprinkler pipe and fittings.

CPVC resin: A white thermoplastic material that accounts for 85 percent of the weight of finished CPVC products.

Iron pipe size (IPS): The nominal pipe size of iron pipe.

Light hazard: An occupancy or portions of other occupancies where the quantity and/or combustibility of contents is low and fires with relatively low rates of heat release are expected.

Ordinary hazard: Group 1 (OH1): An area in which the combustibility of the contents is low, the quantity of the combustibles is moderate, and stockpiles of contents with low combustibility do not exceed 8'. Group 2 (OH2): Occupancies or portions of other occupancies where the quantity and combustibility of contents are moderate to high, stockpiles of contents with moderate to high combustibility do not exceed 12', and stockpiles of contents with high rates of heat release do not exceed 8'.

Standard dimensional ratio (SDR): The pipe wall thickness in proportion to its outside diameter.

Thermal expansion: The change in pipe length due to changes in temperature.

Thermoplastic: Refers to plastic piping that is soft and pliable when heated, but hard when cooled.

Transverse: The lateral shifting of pipe.

Additional Resources

This module presents thorough resources for task training. The following reference material is suggested for further study.

BlazeMaster® Online Training Program, **www.blazemastertraining.com**.

NFPA 13, *Standard for the Installation of Sprinkler Systems*, Latest Edition. Quincy, MA: National Fire Protection Association.

NFPA 13D, *Standard for the Installation of Sprinkler Systems in One- and Two-Family Dwellings and Manufactured Homes*, Latest Edition. Quincy, MA: National Fire Protection Association.

NFPA 13R, *Standard for the Installation of Sprinkler Systems in Low-Rise Residential Occupancies*, Latest Edition. Quincy, MA: National Fire Protection Association.

UL 203, *Standard for Pipe Hanger Equipment for Fire Protection Service*, Latest Edition. Northbrook, IL: Underwriters Laboratories.

Figure Credits

Courtesy of Blazemaster® | Lubrizol Advanced Materials, Inc., Module Opener, Figures 1, 2, 8, 11, Appendix

Courtesy of Reed Manufacturing, Figures 3–5

Image property of Stanley Black & Decker. Used with permission., Figure 6

Courtesy of RIDGID Tool Company, Figure 7

Mac McGaffagan, Figures 9, 15

Used with permission of Johnson Controls, Figures 12, 13

Courtesy of Victaulic Company, Figure 14

Section Review Answer Key

SECTION 1.0.0

Answer	Section Reference	Objective
1. d	1.1.0	1a
2. b	1.2.0	1b
3. b	1.3.2	1c
4. a	1.4.0	1d

SECTION 2.0.0

Answer	Section Reference	Objective
1. d	2.1.1	2a
2. c	2.2.3; *Table 3*	2b

SECTION 3.0.0

Answer	Section Reference	Objective
1. c	3.1.2; *Table 4*	3a
2. d	3.2.2	3b
3. d	3.3.1	3c

NCCER CURRICULA — USER UPDATE

NCCER makes every effort to keep its textbooks up-to-date and free of technical errors. We appreciate your help in this process. If you find an error, a typographical mistake, or an inaccuracy in NCCER's curricula, please fill out this form (or a photocopy), or complete the online form at **www.nccer.org/olf**. Be sure to include the exact module ID number, page number, a detailed description, and your recommended correction. Your input will be brought to the attention of the Authoring Team. Thank you for your assistance.

Instructors – If you have an idea for improving this textbook, or have found that additional materials were necessary to teach this module effectively, please let us know so that we may present your suggestions to the Authoring Team.

NCCER Product Development and Revision
13614 Progress Blvd., Alachua, FL 32615

Email: curriculum@nccer.org
Online: www.nccer.org/olf

❑ Trainee Guide ❑ Lesson Plans ❑ Exam ❑ PowerPoints Other ____________________

Craft / Level: Copyright Date:

Module ID Number / Title:

Section Number(s):

Description:

Recommended Correction:

Your Name:

Address:

Email: Phone:

Copper Tubing and Fittings

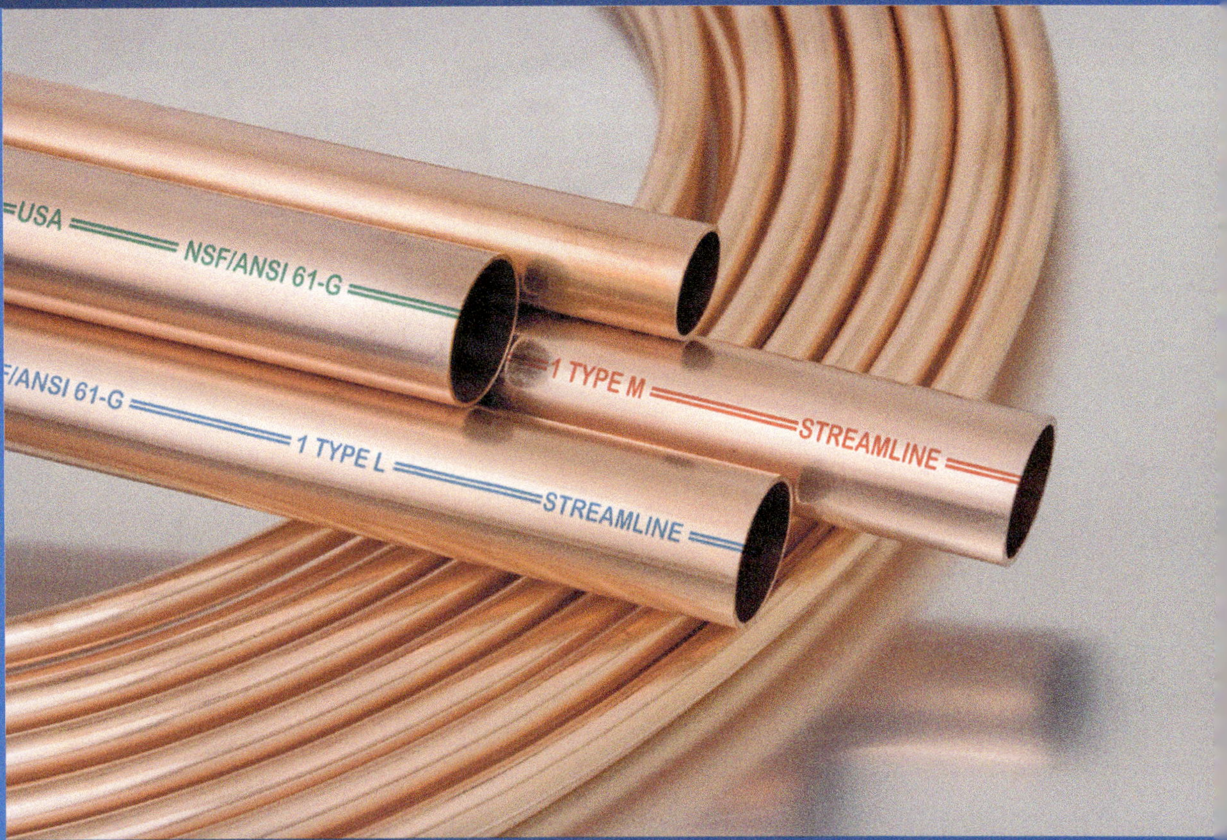

Overview

This module describes the installation of copper tubing and fittings, including support spacing requirements. It explains how to measure, cut, ream, clean, and solder copper tubing. The brazing process is described, as are brazing metals, fluxes, and heating equipment. It also covers mechanical connections.

Module 18105

Trainees with successful module completions may be eligible for credentialing through the NCCER Registry. To learn more, go to **www.nccer.org** or contact us at 1.888.622.3720. Our website, **www.nccer.org**, has information on the latest product releases and training.

Your feedback is welcome. You may email your comments to **curriculum@nccer.org**, send general comments and inquiries to **info@nccer.org**, or fill in the User Update form at the back of this module.

This information is general in nature and intended for training purposes only. Actual performance of activities described in this manual requires compliance with all applicable operating, service, maintenance, and safety procedures under the direction of qualified personnel. References in this manual to patented or proprietary devices do not constitute a recommendation of their use.

18105 V4.0

18105

Copper Tubing and Fittings

Objectives

Successful completion of this module prepares you to do the following:

1. Size and select copper tubing and fittings.
   a. Identify copper tubing types and sizes.
   b. Identify the fittings used with copper tubing.
   c. Explain how to cut and bend copper tubing.
   d. Identify support spacing requirements for copper tubing.
2. Describe the process for soldering copper tubing.
   a. Identify the characteristics of solder metal and fluxes.
   b. List the steps in the soldering process.
3. Describe the process for brazing copper tubing.
   a. Identify the safety precautions associated with brazing.
   b. Identify the characteristics of brazing metal and fluxes.
   c. List the steps in the brazing process.
4. Identify mechanical connection methods for copper tubing.
   a. Explain how to install a grooved fitting.
   b. Describe how to make a press-fit connection.

Performance Tasks

Under supervision, you should be able to do the following:

1. Identify various fittings used with copper tubing.
2. Cut, ream, clean, and bend copper tubing in preparation for joining.
3. Make a soldered joint.
4. Join copper using mechanical connectors.

Trade Terms

Alloy
Borates
Brazing
Capillary action
Dielectric
Dimple
Flux
Galvanic
Liquidus
Oxidation
Oxyfuel equipment
Potable
Solder
Soldering
Solidus
Sweat fittings
Sweat joint
Wick
Wrought fittings

Industry Recognized Credentials

If you are training through an NCCER-accredited sponsor, you may be eligible for credentials from NCCER's Registry. The ID number for this module is 18105. Note that this module may have been used in other NCCER curricula and may apply to other level completions. Contact NCCER's Registry at 1.888.622.3720 or go to **www.nccer.org** for more information.

You can also show off your industry-recognized credentials online with NCCER's digital badges. Transform your knowledge, skills, and achievements into badges that you can share across social media platforms, send to your network, and add to your resume. For more information, visit **www.nccer.org**.

NOTE

Unless otherwise specified, references in parentheses following figure and table numbers refer to NFPA 13.

Contents

Figures and Tables

Section One

1.0.0 Copper Tubing and Fittings

Objective

Size and select copper tubing and fittings.

a. Identify copper tubing types and sizes.
b. Identify the fittings used with copper tubing.
c. Explain how to cut and bend copper tubing.
d. Identify support spacing requirements for copper tubing.

Performance Tasks

1. Identify various fittings used with copper tubing.
2. Cut, ream, clean, and bend copper tubing in preparation for joining.

Trade Terms

Alloy: Any substance made up of two or more metals.

Brazing: A method of fusing metals with a nonferrous filler metal using heat above 840°F, but below the melting point of the base metals being joined.

Capillary action: The spontaneous flow of a liquid into a narrow space between two solid surfaces.

Dielectric: A nonconductor of electricity.

Dimple: A slight surface depression, or the act of creating one.

Flux: A chemical substance that aids the flow of solder or brazing metal by removing and preventing the formation of oxides on the pieces to be joined.

Galvanic: Refers to direct-current electricity, especially when produced chemically.

Potable: Water that is fit for human consumption.

Soldering: A method of joining metals with a nonferrous filler metal using heat below 840°F and below the melting point of the base metals being joined.

Sweat fittings: Thin-wall copper fittings suitable for making soldered connections.

Wrought fittings: Heavy-duty copper fittings suitable for making brazed connections.

Copper tubing is used less often than steel pipe, but it does provide certain advantages. Copper resists corrosion much better than steel pipe and is used in **potable** water supplies. Copper also has better flow characteristics than steel. Because of this, copper tubing can be smaller than steel pipe in a given situation. For example, if an application required a minimum 1" diameter steel pipe, a ¾" diameter copper tube could be used instead.

Copper also has a few disadvantages. It is expensive and less resistant to the internal buildup of scaling (water-hardness products) than plastic pipe. Copper also has a high thermal conductivity and tends to freeze more readily than steel.

The use of copper tubing in sprinkler systems requires different joining and support methods than those used for steel or plastic pipe.

1.1.0 Copper Tubing Types

Copper tubing is available in five different types: Type K, Type L, Type M, ACR (air conditioning and refrigeration), and DWV (drain, waste, and vent). Only Type K, Type L, and Type M copper tubing are allowed for use in sprinkler systems per the *Standard for the Installation of Sprinkler Systems* (NFPA 13), *Standard for the Installation of Sprinkler Systems in One- and Two-Family Dwellings and Manufactured Homes* (NFPA 13D), and *Standard for the Installation of Sprinkler Systems in Low-Rise Residential Occupancies* (NFPA 13R).

Type K and Type L are very ductile (pliable) and come in straight lengths, as well as annealed or soft rolls. Type M is hard-drawn copper and is available in straight sections only. Hard-drawn copper tube can be bent without kinking using tools designed for this purpose.

Did You Know?

According to the United States Geological Survey (USGC), the copper industry recovered an estimated 160,000 tons of copper from post-consumer scrap in 2019. Sources of scrap copper include discarded plumbing pipes, appliances, copper roofing, wiring, and various other building materials and equipment. Many recycling stations pay for scrap copper.

The cost of copper tubing is directly related to wall thickness and the thinnest wall permitted is normally used. Type K tubing has the thickest wall and Type M tubing has the thinnest wall. The actual internal diameter of copper tubing is very close to the nominal tube size. Copper specifications are often referred to as *CTS*, meaning copper tube size, as differentiated from *IPS*, originally standing for iron pipe size. *Table 1* shows the differences in wall thickness.

1.2.0 Fittings

A variety of fittings are used with copper tubing, including **sweat fittings**, cast bronze and wrought copper fittings, and **dielectric** fittings. (Grooved and compression fittings may also be used and are discussed later in this module.) Copper tubing can also be connected without the use of fittings using a special tool known as a *T-drill*.

1.2.1 Sweat Fittings

The walls of copper tubing are too thin to be threaded, so they are commonly joined using a soldered connection and a special type of fitting known as a sweat fitting. *Figure 1* shows common sweat fittings.

1.2.2 Cast Bronze and Wrought Copper Fittings

Cast bronze and wrought (sometimes referred to as *wrot*) copper fittings (*Figure 2*) are commonly used when making brazed connections in copper. **Brazing** operates at a higher temperature than **soldering** and requires the use of heavier fittings.

Cast bronze fittings for copper tubing are specified in ASTM B16.18, *Cast Copper Alloy Solder-Joint Pressure Fittings*. Cast fittings require the use of **flux** when brazing.

When brazing **wrought fittings**, the use of flux is optional. Wrought fittings for copper tubing are specified in *Wrought Copper and Copper Alloy Solder-Joint Pressure Fittings* (ASME B16.22).

Did You Know?

To solder copper fittings, you must heat the fitting until the soldering paste starts to melt. This melting paste looks like beads of sweat, which led to the name *sweat fittings*.

Table 1 Differences in Wall Thickness

Nominal Tube Size	Outside Diameter	Type K		Type L		Type M	
		Inside Diameter	Wall Thickness	Inside Diameter	Wall Thickness	Inside Diameter	Wall Thickness
¾	0.875	0.745	0.065	0.785	0.045	0.811	0.032
1	1.125	0.995	0.065	1.025	0.050	1.055	0.035
1¼	1.375	1.245	0.065	1.265	0.055	1.291	0.042
1½	1.625	1.481	0.072	1.505	0.060	1.527	0.049
2	2.125	1.959	0.083	1.985	0.070	2.009	0.058
2½	2.625	2.435	0.095	2.465	0.080	2.495	0.065
3	3.125	2.907	0.109	2.945	0.090	2.981	0.072
4	4.125	3.857	0.134	3.905	0.110	3.935	0.095

Note: All dimensions in inches.

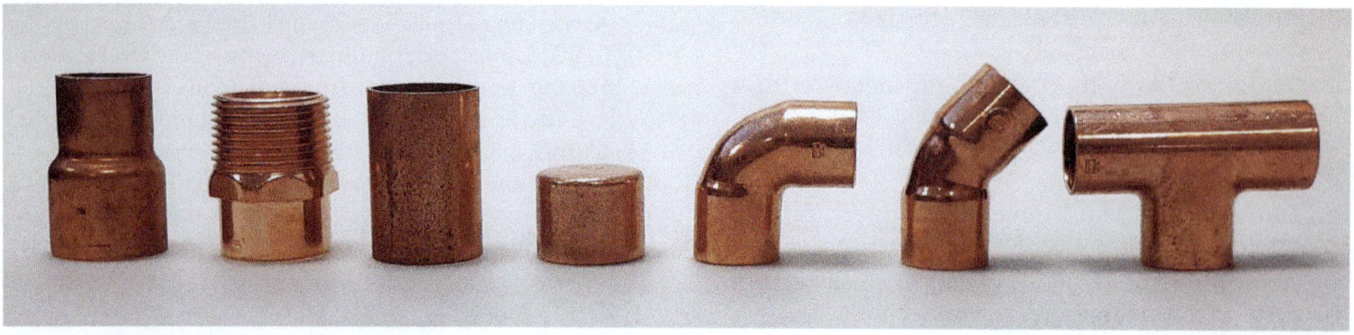

Figure 1 Sweat fittings.

Cast Bronze Wrought Copper

Figure 2 Cast bronze and wrought copper fittings.

1.2.3 Dielectric Fittings

When a connection is made between dissimilar metals such as steel pipe and copper tubing, a **galvanic** cell can be formed. This cell is similar to a battery in that it can generate electric current. When water is present in the tubing, it provides an electrolyte that allows the current to flow between the two different metals. The galvanic action produced can cause corrosion of the steel pipe and fittings and early failure of the system. To prevent this, a dielectric fitting is used in systems with dissimilar metal connections (e.g., steel pipe and copper tubing).

> **NOTE**
> Dielectric unions are not needed for connections between CPVC and copper.

In a dielectric fitting, an insulating washer or other nonconductive material is used to prevent electric current from passing between the dissimilar metals. *Figure 3* shows a typical dielectric fitting.

Brass fittings can also be used as insulating fittings. Because brass is an **alloy** of copper and zinc, the same nonreactive material used to coat galvanized steel, brass fittings protect against corrosion in a similar way.

1.2.4 Branching without Fittings

A T-drill can be used to create branches without fittings. A T-drill drills a hole in the tube on the run. Once the T-drill has penetrated the tube, collaring pins are extended, and the head extracts itself. This leaves an extruded outlet on the main. *Figure 4* shows the use of a T-drill.

Figure 3 Dielectric fitting.

> **Did You Know?**
>
> Copper is one of the most plentiful metals. It has been in use for thousands of years. In fact, a piece of copper pipe used by the Egyptians more than 5,000 years ago is still in good condition. When the first Europeans arrived in the New World, they found the Native Americans using copper for jewelry and decoration. Much of this copper was from the region around Lake Superior, which separates northern Michigan and Canada.
>
> One of the first copper mines in North America was established in 1664 in Massachusetts. During the mid-1800s, new deposits of copper were found in Michigan. Later, when miners went west in search of gold, they uncovered some of the richest veins of copper in the United States.

After using a T-drill to cut the hole, a tube end notcher (*Figure 5*) is used to notch the outlet tube to avoid obstruction of the waterway. At the same time, it forms a **dimple** in the tube so that it can only penetrate the collar enough to provide the required overlap. This process produces a Listed assembly when soldered or brazed. The use of the notcher is extremely important because it controls the depth that the branch penetrates into the main. This method of branching is economical because it eliminates a fitting and two brazed or soldered joints. The outlet point can be readily tailored to the specifics of the installation.

Figure 4 T-drill.

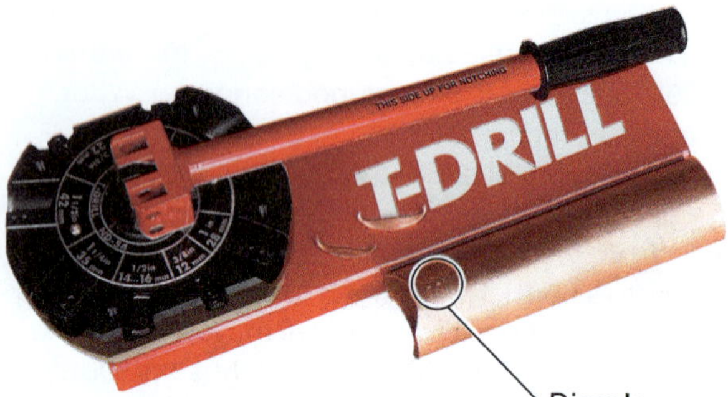

Figure 5 Tube end notcher and notched tube end.

1.3.0 Cutting and Bending Copper Tube

Copper tube is soft and easily cut and bent by hand. However, care must be taken to ensure straight cuts and smooth bends.

1.3.1 Cutting Copper Tube

Copper tube must be cut square to the run of the tube. Thin-wall tube from ⅛" to 3" in diameter can be cut using various tube-cutting tools, such as those shown in *Figure 6*.

When using manual tube cutters, clamp the tool onto the tube and rotate it, tightening the adjusting screw as you proceed (*Figure 7*).

Power saws can also be used to cut tube. The chop saw, which uses an abrasive blade, has a built-in vise to hold the tube in place while cutting it (*Figure 8*).

WARNING!

Because of their high speed and extremely sharp blades, power saws are inherently dangerous. Do not attempt to operate a power saw unless you have been trained to use it.

Hacksaws can also be used for cutting copper tubing, but it is more difficult to get a square cut this way. Cutting with a hacksaw also tends to produce a lot of burrs. The hacksaw blade must have 24 or 32 teeth per inch. A coarser blade will tear and strip the thin copper tube wall.

Chop Saw Blades

The abrasive cutting blades for chop saws have maximum rated speeds according to the abrasive blades they will accept. Choose a blade rated for the speed of the saw. For example, the typical maximum safe speed for a 12" blade is 5,000 revolutions per minute (rpm).

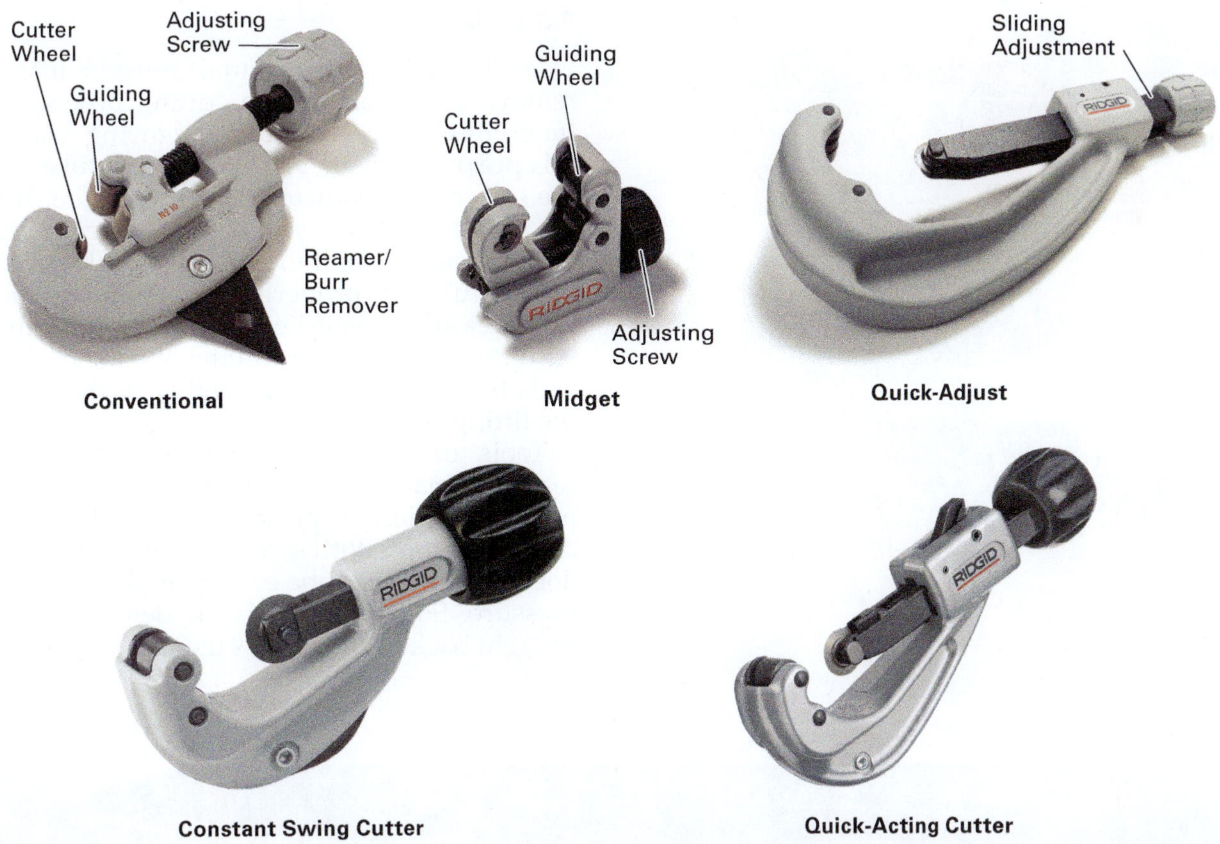

Figure 6 Tube cutters.

Figure 7 Using a handheld tube cutter.

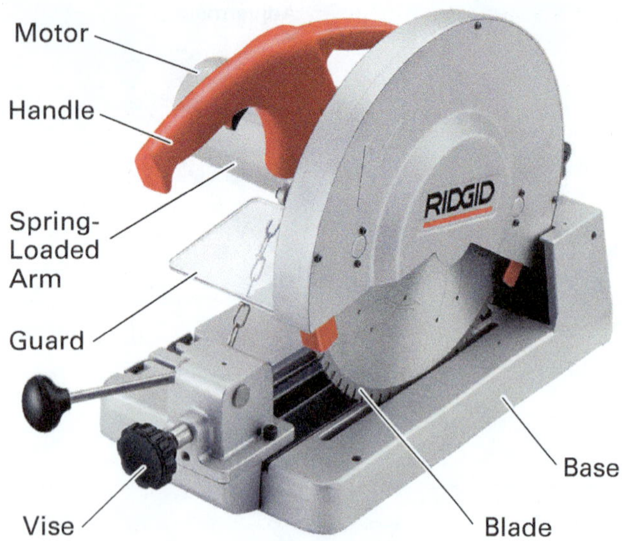

Figure 8 Power saw.

1.3.2 Reaming

Once the tube has been cut, it must be reamed to remove burrs left by the cutting process. Some tube cutters have built-in deburring blades for this purpose (*Figure 9*). Ream all cut tube ends to the full inside diameter (ID) of the tube. If burrs are not removed, corrosion may occur due to local turbulence and increased local flow velocity in the tube. A properly reamed piece of tube provides an undisturbed surface for smooth flow. Also remove any burrs on the outside of the tube ends to ensure proper assembly of the tube into the fitting cup.

Tools used to ream tube ends include the reaming blade on the tube cutter, half-round or round files, a pocketknife, and a suitable deburring tool. With soft tube, care must be taken not to deform the tube end by applying too much pressure. Soft-temper tube, if deformed, can be brought back to roundness using a sizing tool.

Figure 9 Reaming copper tube.

1.3.3 Bending Copper Tube

Bending reduces the number of joints and fittings in a plumbing system. This can reduce leaks, as well as installation time. When you are using soft tubing, it is best to bend rather than cut it.

Small-diameter tube is easily bent by hand. Take care not to flatten it with a bend that is too sharp. Tube-bending springs can be used on the outside of the tubing to prevent it from collapsing while bending (*Figure 10*). Tube-bending equipment can also be used to make accurate and reliable bends in both soft and hard tubing. This equipment has various sizes of forming attachments for use in bending tubing with diameters up to $^{7}/_{8}$" at any angle up to 180 degrees (*Figure 11*).

Per NFPA 13, ***Section 16.3.8.2***, the minimum bend radius of copper tubing must be no less than six times the tube diameter for tube sizes 2" and smaller, and five times the tube diameter for tube sizes $2^{1}/_{2}$" and larger. For example, the minimum bend radius for 1" copper is 6 × 1" = 6".

1.3.4 Cleaning

Copper tubing and fittings must be properly cleaned before soldering. Cleaning removes oxides and surface soil to ensure proper flow of filler metal into the joint. These contaminants interfere with **capillary action**, reducing the strength of the joint and contributing to joint failure.

After cutting and reaming, lightly abrade (clean) the tube ends using sand cloth or nylon abrasive pads for a distance slightly more than the depth of the fitting cup. Clean the fitting cups using abrasive cloth, abrasive pads, or a properly sized fitting brush.

Take care when cleaning copper as it is a relatively soft metal. If too much material is removed, a loose fit may occur, resulting in a poor joint. Chemical cleaning may be used if the tube ends and fittings are thoroughly rinsed after cleaning according to the manufacturer's instructions. Do not touch the cleaned surface with bare hands or oily gloves. Skin oils, lubricating oils, and grease will impair the adherence of the filler metal.

1.4.0 Support Spacing for Copper Tube

Support spacing for copper tube is different than for other piping materials because of the differences in the tube itself. Copper tube also requires the use of special precautions to prevent corrosion between the copper tube and steel hangers or other ferrous metal supports.

1.4.1 Spacing Limitations for Copper Tube

Strength requirements are based on maximum spacing between hangers for copper tube. NFPA 13 recognizes that the mechanical strength of copper tube is less than that of steel pipe. This means copper tube hangers must be spaced closer together than they would be on steel pipe. *Table 2* shows the maximum allowable hanger spacing for copper tube from NFPA 13.

Figure 10 Tube-bending spring.

Two other differences need to be recognized between copper tube and steel pipe support requirements. For maximum operating pressures over 100 psi, the maximum unsupported distance between an end sprinkler (pendent position or drop nipple) and the last hanger on the branch line is limited to 12" or less for steel pipe, or 6" or less for copper tube.

In an armover situation with maximum pressures less than 100 psi, the maximum unsupported armover to a sprinkler, sprinkler drop, or sprig-up must be 24" or less for steel pipe, or 12" or less for copper tube. With maximum pressures exceeding 100 psi, the maximum unsupported armover for steel pipe is 12" and 6" for copper tube.

If these configurations present a problem, the branch line can always be extended to pick up the next structural member.

1.4.2 Protecting Copper Tubing from Hangers in High-Humidity Environments

As mentioned previously, any time two dissimilar metals such as copper and steel come in contact and moisture is present, a galvanic cell may be formed and cause corrosion. The installation of copper in high-humidity environments requires the use of a protective barrier between the tubing and the steel hangers to prevent galvanic corrosion over time. One way to accomplish this is to use galvanized (zinc-coated) hanger material. Another method is to insulate the connection so there is no direct contact between the tubing and the hanger. Listed insulating tape or wraps can be used to provide this protection.

NOTE

NFPA 13 requires that any hanger material attaching directly to pipe or tubing be Listed by a recognized testing laboratory.

Figure 11 Tube-bending equipment.

Table 2 Hanger Spacing for Copper Tube [Data from NFPA 13, *Table 17.4.2.1(a)*]

Copper Tube Size (inches)	Maximum Hanger Spacing (feet)
¾–1	8
1¼–1½	10
2–3	12
3½–8	15

Protecting the National Zoo Amazonia Exhibit

The Smithsonian National Zoo in Washington, DC uses copper tubing for the fire sprinkler system in its glass-enclosed Amazonia Exhibit. Plastic separators were used between the copper tube and steel hangers to prevent the galvanic corrosion that occurs between dissimilar metals in high-humidity environments. In addition to complying with NFPA 13 requirements, this system was designed to meet the special requirements found in NFPA 150, *Fire and Life Safety in Animal Housing Facilities Code.* This code describes life and safety requirements for both humans and animals in all types of animal housing facilities, including zoos.

Figure Credit: Strickland Fire Protection, Inc.

1.0.0 Section Review

1. The copper tubing with the *thinnest* wall is _____.
   a. Type K
   b. Type L
   c. Type M
2. A connection between CPVC and copper requires a dielectric fitting.
   a. True
   b. False
3. Touching freshly cleaned copper tubing with bare hands can result in _____.
   a. impaired adherence of the filler metal
   b. purple staining on the copper
   c. severe pitting over time
   d. immediate corrosion
4. The *maximum* hanger spacing for 2" copper tubing is _____.
   a. 8'
   b. 10'
   c. 12'
   d. 15'

Section Two

2.0.0 Soldering Copper Tubing

Objective

Describe the process for soldering copper tubing.

a. Identify the characteristics of solder metal and fluxes.
b. List the steps in the soldering process.

Performance Task

3. Make a soldered joint.

Trade Terms

Liquidus: The state of solder when it has melted by heating and is flowing readily.

Oxidation: The process by which the oxygen in the air combines with metal to produce tarnish and rust.

Solder: An alloy that is melted and used to join metallic surfaces.

Solidus: The solid or stable phase of solder.

Sweat joint: A soft solder joint, or a joint soldered below 840°F.

Wick: To draw solder into a joint.

Soldering is a simple and economical method of creating a joint in copper tubing. Soldering is the process of joining (bonding) copper tube and fittings using a dissimilar metal or an alloy of two or more dissimilar metals that has a melting temperature (**liquidus** temperature) of 840°F or lower.

Soldering is distinguished from brazing and welding by the temperature at which joining takes place. To be a soldering operation, the liquidus temperature of the **solder** must be 840°F or below. A soldered joint is also referred to as a **sweat joint**.

A variety of tools and materials are used in the soldering process, including tools for cutting and cleaning the tube and fittings, a heat source, and solder metal. See *Figure 12*.

Figure 12 Tools for soldering copper tubing to a fitting.

2.1.0 Solder Metal and Fluxes

Solder is an alloy of metals selected to produce specific melting, bonding, and physical characteristics. Solders approved for sprinkler systems are also used for potable (drinking) water connections because they contain no lead. A characteristic of many alloys used for solder is that they are formulated to have a temperature range between the **solidus** (or *solid state*) and the liquidus (or *liquid state*). This temperature range is usually called the *working range* or the *pasty range*. The true liquidus state is reached when the solder begins to **wick** or be drawn into the joint. At that point, the joining area has reached the proper temperature and should not be heated further.

In soldering, molecules of the solder are attached to those in the material being soldered. The molecular bond can take place only if the surface of the material is clean. This requires mechanical and chemical cleaning of the joint pieces. The use of a flux cleans and prevents **oxidation** while the material is being heated to the liquidus temperature of the solder. When the liquidus temperature is reached, the solder metal is wicked into the joint by capillary action, producing a pressure-tight seal when it cools (*Figure 13*). The capillary space between the tube and fitting is approximately 0.004". This spacing is critical because it determines whether there is a proper flow of the filler metal into the gap, ensuring a strong joint. Maintaining the temperature of the two pieces of metal at or slightly above the liquidus point of the solder long enough for the bonding to take place is sometimes referred to as *sweating*.

Under NFPA 13, NFPA 13D, and NFPA 13R, any solder meeting *Standard Specification for Solder Metal* (ASTM B32), having a melting point above 400°F, and containing no more than 0.2 percent lead (Pb) would be accepted. The use of solder in fire protection is restricted to use in wet pipe systems. It is typically a tin-antimony alloy.

CAUTION

When making soldered connections in copper/CPVC combination systems, make sure that the flux used is compatible with CPVC.

Solder can be elastic in nature and should not be depended upon for mechanical strength. The main purpose of the solder is to provide a seal against leakage. The principal mechanical strength of a soldered joint depends on the tight fit and overlap between the tube and fitting. The integrity of the solder joint is a function of the character of the solder itself, how carefully the joint parts have been cleaned, the clearance between the two parts being joined, and the amount of overlap between the two pieces being joined. The tube must be fully inserted into the fitting and the clearance between the tube and fitting must be minimal. Generally, the tighter the fit, the better the joint. Due to the softness of solder, cold flow or *creep* can cause the solder to move over time. This can destroy the seal under some load conditions.

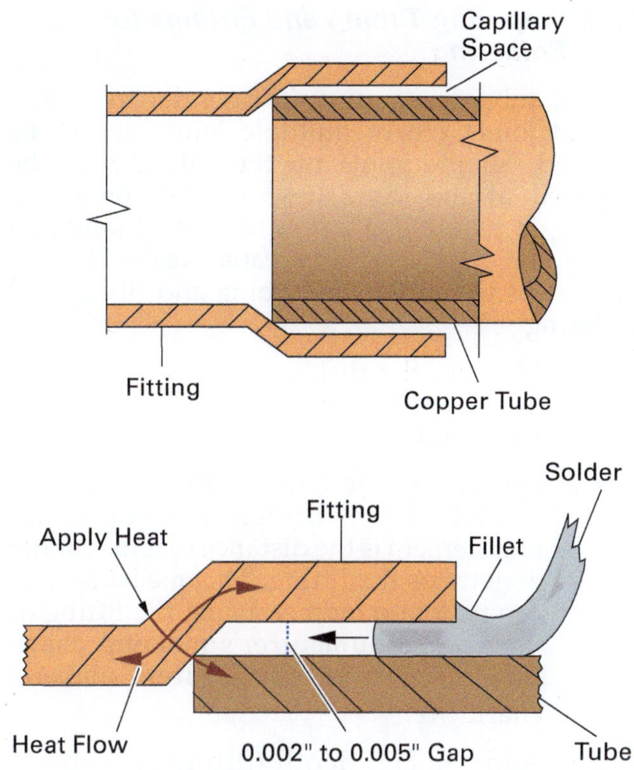

Figure 13 Capillary action.

2.2.0 Soldering Process

The soldering process involves cutting and preparing the tube and then making the soldered joint. Although the soldering operation is inherently simple, the deletion or misapplication of a single step in the process may mean the difference between a good joint and a failure. Patience and practice are required for good results. Solder joints, where permitted, shall be fabricated in accordance with the methods and procedures listed in ASTM B828, *Standard Practice for Making Capillary Joints by Soldering of Copper and Copper Alloy Tube and Fittings*.

2.2.1 *Preparing Tubing and Fittings for Soldering*

Careful tube preparation is essential to ensure a good joint. Where multiple joints are to be soldered, all the joints on the fitting must be prepared at the same time so that only one heating is required. Heating a second joint can remelt the first joint and cause leaks. Follow these steps to prepare the tubing and fittings for soldering:

Step 1 Measure the distance between the faces of the two fittings, using the face-to-face method (*Figure 14*).

Step 2 Determine the cup depth engagement of each of the fittings. The cup depth engagement is the distance that the tubing penetrates the fitting. This measurement can be found by measuring the fitting or using a manufacturer's makeup chart. *Table 3* shows a manufacturer's makeup chart.

Step 3 Add the cup depth (fitting socket) engagement of both fittings to the measurement found in *Step 2* in order to find the length of tubing needed.

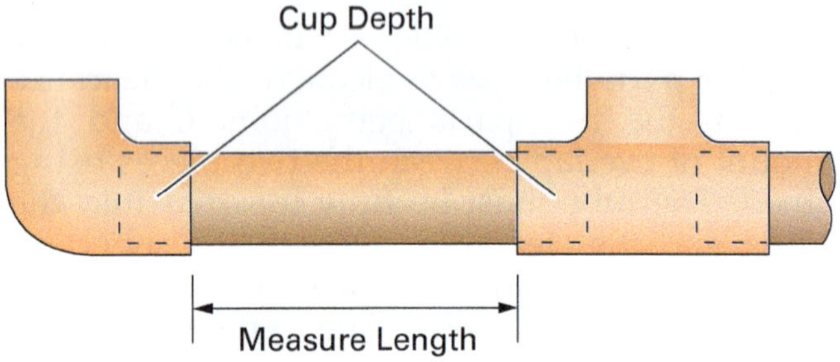

Figure 14 Face-to-face measuring method.

Step 4 Cut the copper tubing to the correct length, making sure that the cut is square to the tube and the tube is not deformed while being cut. Accuracy is important to joint quality. If the tube is too short, it will not reach all the way into the cup of the fitting and a proper joint cannot be made. If the tube segment is too long, system strain may be introduced, which could affect service life.

Step 5 Ream the inside and outside of both ends of the copper tube using a deburring tool, file, or reamer.

Step 6 Test the fit of the two items to be soldered. The fitting should slide over the tube, but the fit should be tight. Loose fits produce weak joints.

Step 7 Clean the tubing and the fitting to a bright finish using No. 00 steel wool, an abrasive pad, an emery cloth, and/or a tubing brush, as shown in *Figure 15* (*A*) and *Figure 15* (*B*). Examine the fitting and tube to ensure they are clean, and no significant flaws exist in the finish, as shown in *Figure 15* (*C*).

CAUTION

Care must be taken when cleaning the tubing and fittings to remove all oxides on the copper without removing a large amount of metal. Do not touch or brush away filings from the tube or fitting with your fingers because the metal slivers can cause injury and the oil on your fingers will contaminate the freshly cleaned metal.

Table 3 Manufacturer's Makeup Chart for Sweat Fittings

Tubing Size	Depth of Cup
1⁄2	1⁄2
5⁄8	5⁄8
3⁄4	3⁄4
1	29⁄32
1 1⁄4	31⁄32
1 1⁄2	1 3⁄32
2	1 11⁄32
2 1⁄2	1 15⁄32
3	1 21⁄32
4	2 5⁄32

Note: All dimensions in inches.

Figure 15 Preparing the tube and applying flux.

Step 8 Apply flux to the outside of the copper tubing and the inside of the copper fitting socket immediately after cleaning them. See *Figure 15* (*D*) and *Figure 15* (*E*).

WARNING!

Some types of flux produce harmful fumes and/or contain acid. Always make sure the work area is properly ventilated. Wear appropriate personal protective equipment, including safety glasses or goggles. Avoid eye and skin contact with flux. Wash flux off immediately if contact is made.

CAUTION

Brazing flux and soldering flux are not the same. Do not allow these fluxes to become mixed or interchanged. Carelessness can ruin quality work.

Step 9 Insert the tube into the fitting socket with a slight twisting motion until the tube touches the inside shoulder of the fitting.

Step 10 Wipe away any excess flux from the joint, as shown in *Figure 15* (*F*).

Step 11 Check the tube and fitting for proper alignment and support before soldering.

2.2.2 Soldering Joints

The most common method of heating a copper joint on a construction site is with a propane torch. Air/fuel torches using acetylene, butane, or other gases are also used.

WARNING!

Always solder in a well-ventilated area. Fumes from the flux can irritate your eyes, nose, throat, and lungs.

The use of torches and welding equipment is a major source of construction fires. Always follow these guidelines when soldering:

- Ensure that a fire extinguisher is close by when using a torch.
- Shield the structure and devices such as valves from the flame and excessive heat.
- Keep the heat and the flame directed only towards the piece to be heated.
- Use wet rags wrapped around the tube block to confine the heat to the joint.

The temperature produced by the heating source and the heat capacity of the source are critical. Torches produce temperatures well in excess of what is needed for soldering. The torch tip needs to be sufficiently large so that the volume of heat generated offsets the heat loss down the tube and into the atmosphere. Move the torch flame around to produce uniform heating and to avoid extreme hot spots that result in premature vaporization or burning of the flux.

Heating

Obtaining a proper solder joint requires the filler metal to be heated to the liquidus temperature. Melting of the solder alone is not adequate. Check the temperature of the tube as it is being heated using stick or wire solder. Touch the solder to the metal at a point away from the flame to determine when the metal is hot enough to melt the solder. Melting solder with the flame does not help. The bonding (sweat joint) will not occur until the metal itself is hot enough to melt the solder. Begin heating with the flame perpendicular to the tube. The copper tube conducts the initial heat into the fitting cup for even distribution of heat in the joint area. The extent of this preheating depends on the size of the joint.

Preheating of the assembly should include the entire circumference of the tube in order to bring the entire assembly up to a suitable preheat condition. However, for joints in the horizontal position, avoid directly preheating the top of the joint to prevent the soldering flux from burning. The natural tendency for heat to rise will ensure adequate preheat of the top of the assembly. Experience will indicate the amount of heat and the time needed.

Next, move the flame onto the fitting cup. Sweep the flame alternately between the fitting cup and the tube a distance equal to the depth of the fitting cup. Again, while preheating the circumference of the assembly as described above, with the torch at the base of the fitting cup, touch the solder to the joint. If the solder does not melt, remove it, and continue heating.

CAUTION

Do not overheat the joint or direct the flame into the face of the fitting cup. Overheating could burn the flux, which will destroy its effectiveness, and the solder will not enter the joint properly.

When the solder melts, apply heat to the base of the cup to aid capillary action in drawing the molten solder into the cup toward the heat source.

Applying Solder

Solder joints depend on capillary action in which free-flowing molten solder is wicked or drawn into the narrow clearance between the fitting and the tube. The amount of solder consumed depends on the size of the tube. Molten solder metal is not dependent on gravity and is drawn into the joint by capillary action regardless of whether the solder flow is upward, downward, or horizontal. Capillary action is most effective when the space between the surfaces to be joined is between 0.002" and 0.005".

The following is an overview of the soldering process. Because soldering requires relatively low heat, heating equipment that mixes acetylene, butane, or propane directly with air is all that is needed. Only one tank of gas and a torch are required for soldering.

Follow these steps to solder a joint:

WARNING!

Soldering involves the use of an open flame and highly flammable gases. Always read and follow the torch manufacturer's safety instructions and operating procedures. Wear appropriate personal protective equipment and ensure that a fire extinguisher is easily accessible. Never operate any equipment unless you have been fully trained in its safe use.

Step 1 Apply heat blocks or heat shields to the tube and between the tube and structure as appropriate. *Figure 16* shows a typical heat shield.

Step 2 Set up the heating equipment according to the manufacturer's instructions.

Figure 16 Heat shield.

Step 3 Light the heating equipment according to the type of equipment being used and the manufacturer's instructions.

Step 4 Preheat the tubing first, and then move the flame onto the fitting.

Step 5 Move the flame to the back of the fitting cup.

Step 6 For joints in the horizontal position, start applying the solder metal slightly off-center at the bottom of the joint (*Figure 17*). Touch the end of the solder to the area between the fitting and the tube. When the solder begins to melt from the heat of the tube and fitting, the solder will be drawn into the joint by capillary action. The solder can be fed upward or downward into the joint while keeping the torch at the base of the fitting and slightly ahead of the point of application of the solder. If the solder does not melt on contact with the joint, remove the solder and heat the joint further. Do not melt the solder with the flame.

Figure 17 Applying solder to the heated joint.

Thermal Shields

In addition to physical heat shields, thermal shields are also available as both a liquid spray and a gel. These substances evaporate after use and can be used to protect hard-to-reach surfaces and combustible elements of different shapes. The spray shown here is non-staining and can be used on both soft and rigid surfaces during soldering or brazing.

Figure Credit: The Mill-Rose Company

Step 7 Continue to feed the solder into the joint until a ring of solder appears around the joint, indicating that the joint is filled. The now-solidified solder at the bottom of the joint has created an effective dam that will prevent the solder from running out of the joint as the sides and top of the joint are being filled.

Step 8 Return to the point of beginning, overlapping slightly, and proceed up the uncompleted side to the top, again overlapping slightly. While soldering, small drops may appear behind the point of solder application, indicating that the joint is full to that point and will take no more solder. Throughout this process, you are using all three physical states of the solder: solid, paste, and liquid. For joints in the vertical position, make a similar sequence of overlapping passes, starting wherever is convenient.

Step 9 Allow the joint to cool without moving it.

Step 10 Wipe the joint clean with a cloth after the joint has cooled (*Figure 18*).

Did You Know?

Many jobsites require the use of a hot work permit for any type of work producing a spark, flame, or heat. A hot work permit provides a checklist of fire safety procedures and must be posted on the jobsite for the duration of the hot work activity. The submittal procedure for hot work can be complex and time consuming, which is one reason why many fitters prefer to use mechanical connections whenever possible.

NOTE

Follow the manufacturer's instructions when joining copper tube to solder cup valves. The valve should be in a partially open position before applying heat, and the heat should be applied primarily to the tube. Commercially available heat-sink materials can also be used to protect temperature-sensitive components during the joining operation.

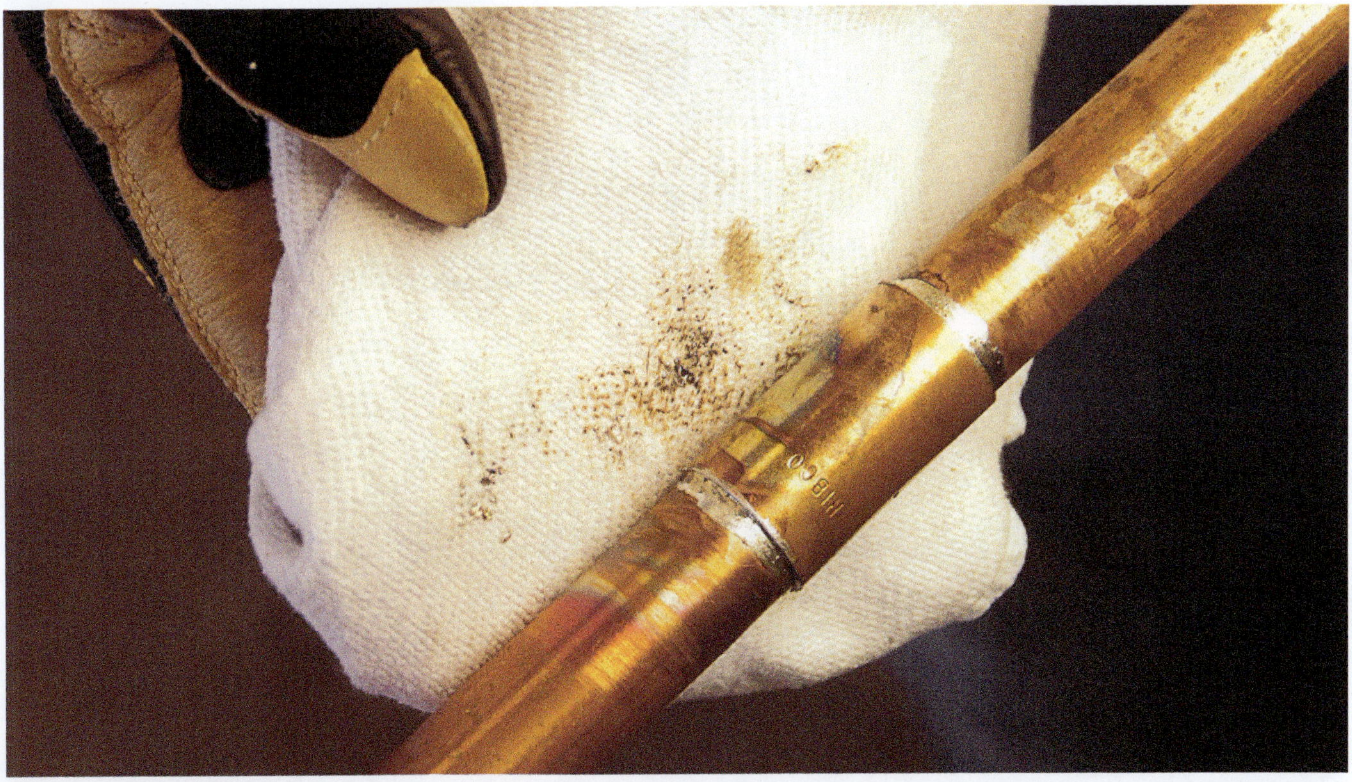

Figure 18 Cleaning the soldered joint.

2.0.0 Section Review

1. The capillary space between the tube and fitting is approximately _____.
   a. 0.004"
   b. 0.04"
   c. 0.4"
   d. 4"

2. The proper application of soldering flux involves _____.
   a. dipping the tubing end into the flux container
   b. applying the flux to the tube end only
   c. applying the flux to both the tube end and fitting
   d. applying the flux to the fitting cup only

SECTION THREE

3.0.0 BRAZING COPPER TUBING

Objective

Describe the process for brazing copper tubing.

a. Identify the safety precautions associated with brazing.
b. Identify the characteristics of brazing metal and fluxes.
c. List the steps in the brazing process.

Trade Terms

Borates: Boric acid salts used as a flux.

Oxyfuel equipment: A type of brazing equipment that uses oxygen and fuel gas mixed in a nozzle and directed at the work surface.

Brazing is similar to soldering, but at a higher temperature. Soldering takes place at temperatures at or below 840°F. Brazing takes place at temperatures above 840°F but below the melting point of the base metal. As with soldering, brazing is the joining of two pieces of metal using a nonferrous, dissimilar metal that has a lower melting temperature than the pieces being joined.

Brazing produces mechanically strong, pressure-resistant joints. The strength of a brazing joint results from the ability of the filler metal to penetrate the base metal. Penetration can only occur if the base metals are properly cleaned, the proper flux and filler metals are selected, and the clearance gap between the outside of the tubing and the inside of the fitting is 0.002" to 0.005".

Brazing is the required process for joining copper tubing in dry pipe and preaction systems. It is also the approved method for installing an outlet or branch when creating a tee on a run using a T-drill or other similar method.

3.1.0 Brazing Safety Precautions

Brazing is done using a combination of oxygen and fuel gas. Although the hazards are similar to those encountered in soldering, they are compounded through the use of two gases, each of which can be hazardous if handled improperly. In addition, when fuel gases are directly combined with oxygen instead of air, the temperature of the flame rises significantly. The **oxyfuel equipment** used to braze copper produces flame temperatures up to 5,300°F. This is much hotter than the flame used when soldering and can cause serious personal injury or material damage if not handled carefully. The following safety guidelines apply to brazing:

- Read and follow all warnings and instructions on cylinder labels.
- Never use equipment for which you have not been trained.
- Wear clothing made of non-flammable fabric, such as cotton. Avoid synthetic fabrics such as nylon, which can melt and adhere to the skin.
- Wear a long-sleeve shirt to avoid burns from sparks or molten metal. Pants should not have cuffs.
- Wear eye and face protection. Brazing should be performed using safety glasses as well as a face shield, one of which should be tinted. For brazing, a #3 or #4 tint is recommended. Refer to *OSHA Standard 1926.102(a)(8)*.
- Wear flame-resistant gloves and high-top work boots.
- Do not ignite the torch while oxygen is flowing to it.
- When turning the torch off, always turn off the oxygen before turning off the fuel gas. Turning off the fuel gas while oxygen is still flowing could result in a flashback and explosion.
- Store fuel and oxygen cylinders in well-ventilated areas. Do not store near any flame or source of heat or in a room where people live or sleep.
- Keep fuel and oxygen cylinders out of reach of children.
- Avoid prolonged exposure of the cylinders to sunlight. This can cause them to heat up and explode.
- Always detach the torch from the cylinders when not in use.
- Do not breathe gas, fumes, vapor or spray from the fuel or oxygen cylinders. Always braze in a well-ventilated area and wear respiratory equipment, if necessary.
- Avoid contact of cylinder contents with eyes or skin. In case of accidental contact, rinse immediately with water for 15 minutes and then seek medical advice.
- When cylinders are empty, dispose of them properly. It is illegal and dangerous to refill these cylinders.
- Do not puncture or incinerate cylinders.
- Make sure gas cylinders are securely turned off when not in use.
- Keep an appropriate fire extinguisher close by while brazing in case material near the work area ignites. When working close to flammable material such as wood, use a fire blanket or other flame/heat blocking material to protect it.

- Oxygen can ignite even when there is no obvious flame or spark to set it off, especially when it comes in contact with oil or grease. Never handle oxygen cylinders with oily hands or greasy gloves. Keep grease away from the cylinders and do not use oil or grease on cylinder attachments or valves.

3.2.0 Brazing Metals and Fluxes

Filler metals used to join copper tubing are copper alloys that contain phosphorus (the BCuP series). The brazing alloys used to braze copper tube and fittings typically melt between 1,200°F and 1,500°F. When joining copper tubing, only these filler metals can be used. *Table 4* shows brazing filler materials for copper tubing.

Brazing fluxes can be applied using the same methods as soldering fluxes. Brazing fluxes operate at much higher temperatures than soldering fluxes, so care must be taken never to mix a soldering flux with a brazing flux. For best results, use the flux recommended by the manufacturer of the brazing filler metals.

Brazing fluxes are chemically active at brazing temperatures and are usually **borates**. Borates are a class of white, water-soluble crystalline salts, the familiar Borax being one of them. Brazing fluxes are available in paste, powder, or liquid form. The usual method of flux application is to brush it on the inside of the fitting cup and the tube end with a brush in the same manner as soldering flux is applied. Where flux is required, the residues may be removed after the brazing operation by cleaning with hot water and brushing with a stainless-steel brush.

NOTE

In accordance with NFPA 13D, dry pipe and preaction systems must be brazed with BCuP-3 or BCuP-4 brazing filler metal; 50 percent tin/50 percent lead solder is no longer allowed.

3.3.0 Brazing Process

The brazing process involves cutting and obtaining the correct brazing equipment, preparing the tube and fittings, and then brazing the joints.

3.3.1 Brazing Equipment

The brazing heating procedure differs from soldering in that different equipment is required to raise the temperature of the metals to be joined above 840°F. Because of the higher temperatures needed, oxygen/liquid propane or oxygen/propylene (MAP/Pro™) brazing equipment is used. This combination is often referred to as *oxyfuel* because the flame is produced by burning a fuel gas mixed with pure oxygen. *Figure 19* shows typical oxyfuel equipment. MAP/Pro™ by itself can also be used to braze smaller-diameter tubing (*Figure 20*).

A variety of brazing rods are available. Copper brazing is typically performed using copper-phosphorus brazing rods.

Always use a friction striker (*Figure 21*), also known as a *spark-lighter,* to ignite the cutting torch. A friction striker works by rubbing a piece of flint on a steel surface to create sparks. When the torch is lit, the brazing flame should have a bright blue inner cone of about ½" long at the tip.

WARNING!

Do not use a match or a gas-filled lighter to light a torch. This could result in severe burns and/or cause the lighter to explode.

Table 4 Brazing Filler Materials for Copper Tubing

	Percent of Principal Element						
AWS Classification	**Silver**	**Phosphorus**	**Zinc**	**Cadmium**	**Lead**	**Tin**	**Copper**
BCuP-3	4.75–5.25	5.75–6.25	–	–	–	–	Balance
BCuP-4	5.75–6.25	7–7.50	–	–	–	–	Balance

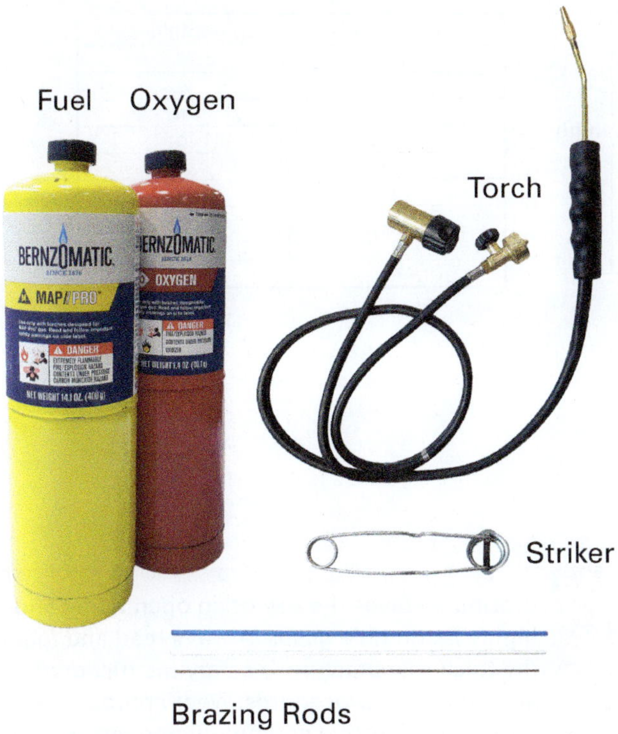

Figure 19 Oxyfuel brazing equipment.

Figure 20 MAP/Pro^{TM} torch kit.

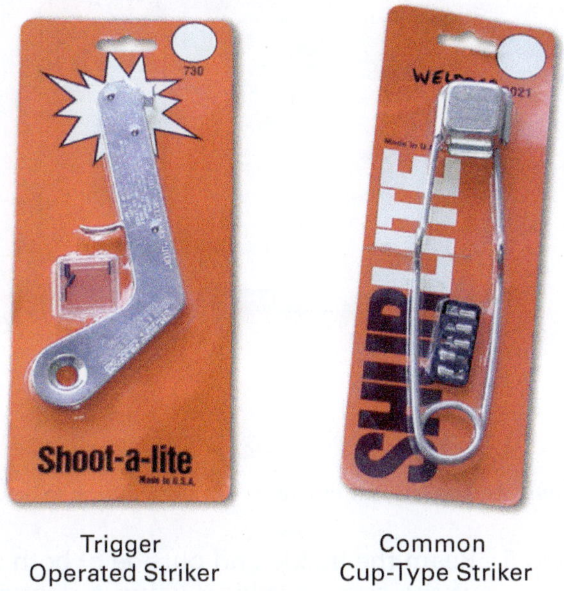

Figure 21 Typical friction strikers.

3.3.2 Preparing Tubing and Fittings for Brazing

To prepare tubing and fittings for brazing, follow the same procedures for preparing tubing and fittings for soldering. It is critical that proper cleaning techniques be used in order to produce a solid, leak-proof joint. Follow these steps to prepare the tubing and fittings for brazing:

Step 1 Measure the distance between the faces of the two fittings.

Step 2 Determine the cup depth engagement of each of the fittings. The cup depth required for brazing is much shallower than that for soldering. Brazing alloy penetration that is three times the thickness of the tube is all that is needed to provide a stronger joint than the tube. This is known as the *3-T Rule*. Short-cup brazing fittings are not readily available and most brazed joints will be made with solder-depth fittings. The cup depth can be found by measuring the fitting or using a manufacturer's makeup chart. *Figure 22* shows a typical manufacturer's brazing fitting makeup chart.

Step 3 Add the cup depth engagement of both fittings to the measurement found in *Step 1* in order to find the length of tubing needed.

Step 4 Cut the copper tubing to the correct length, making sure the cut is square.

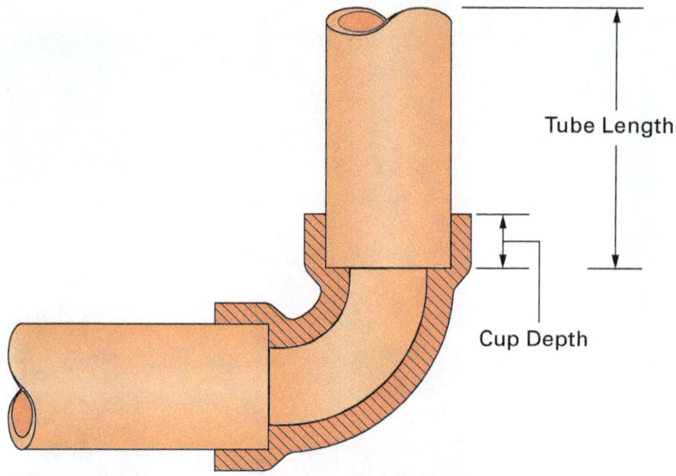

Tube Size (in)	Cup Depth (in)
1/2	3/8
3/4	13/32
1	7/16
1 1/4	1/2
1 1/2	5/8
2	21/32
2 1/2	25/32
3	53/64
4	29/32

Figure 22 Typical manufacturer's brazing fitting makeup chart.

Step 5 Ream the inside and outside of both ends of the copper tubing using a deburring tool, file, or reamer.

Step 6 Clean the tubing and the fitting using an emery cloth or a special copper-cleaning tool.

Step 7 Manually check the fit of the two items to be joined. The fitting should slide over the tube, but the fit should be tight. Loose fits produce weak joints.

CAUTION

Care must be taken when cleaning the tubing and fittings to remove all oxides on the copper without removing a large amount of metal. Do not touch or brush away filings from the tube or fitting with your fingers because the metal slivers can cause injury and the oil on your fingers will contaminate the freshly cleaned metal.

Step 8 Apply flux to the copper tubing and to the inside of the copper fitting socket immediately after cleaning them.

Step 9 Insert the tube into the fitting socket with a slight twisting motion until the tube touches the inside shoulder of the fitting.

Step 10 Wipe away any excess flux from the joint.

Step 11 Check the tube and fitting for proper alignment and support before brazing.

3.3.3 Brazing Joints

Except for the temperatures involved, brazing is very much like soldering. Capillary action is required to produce a satisfactory joint. Follow these steps to braze a joint:

WARNING!

Brazing involves the use of an open flame and highly flammable gases. Always read and follow the torch manufacturer's safety instructions and operating procedures. Wear appropriate personal protective equipment and ensure that a fire extinguisher is easily accessible. Never operate any equipment unless you have been fully trained in its safe use.

Step 1 Set up the brazing equipment per the manufacturer's instructions.

Step 2 Ensure that the tubing and fitting are properly aligned and supported. Restrain the two parts so they remain in the proper place while brazing.

Step 3 Apply heat blocks/shields to the tube and between the tube and structure as appropriate.

Step 4 Apply heat to the tubing about 1" from the joint and, if brazing flux is used, observe the flux. The flux will bubble and turn white and then melt into a clear liquid.

Step 5 Switch the flame to the fitting at the base of the cup. Heat uniformly, sweeping the flame alternately from the fitting to the tube until the flux stops bubbling. Avoid excessive heating of cast fittings, due to the possibility of cracking. When the flux appears liquid and transparent, start sweeping the flame back and forth along the axis of the joint to maintain heat on the parts tobe joined, especially toward the base of the cup of the fitting. It is important to keep the flame in motion and not remain on any one point long enough to damage the tube or fitting.

> **NOTE**
> A mild preheating of the entire fitting is recommended for larger sizes, and the use of a second torch to retain a uniform preheating of the entire fitting assembly may be necessary in the largest diameters.

Step 6 Touch the filler metal rod to the joint to test whether the filler metal will melt. Keep the flame away from the filler metal itself. The temperature of the tube and fitting at the joint should be high enough to melt the filler metal. Do not use the flame to melt the filler metal.

> **NOTE**
> If the filler metal melts upon contact with the joint, the brazing temperature has been met and you should proceed to *Step 8*. If the filler metal does not melt upon contact, continue to heat and test the joint until the filler metal melts.

Step 7 Hold the filler metal rod to the joint and allow the filler metal to enter into the joint while holding the torch slightly ahead of the filler metal and directing most of the heat to the shoulder of the fitting (*Figure 23*).

Step 8 Apply filler metal around the entire joint and continue applying heat until wicking takes place. Wicking is the capillary action that draws the filler metal into the joint between the tube and the fitting.

> **NOTE**
> If the filler metal fails to flow or balls up, it indicates oxidation on the metal surfaces or insufficient heat on the parts to be joined. If the tube or fitting starts to oxidize during heating, there is too little flux. If the filler metal does not enter the joint and instead flows over the outside of either member of the joint, it indicates that one member is overheated or the other is underheated.

Step 9 Continue to fill the joint with the filler metal until the filler metal bead is at least three times the thickness of the tubing. This will not fill the joint, but a joint with this amount of penetration will be stronger than the tubing. For larger joints, small sections of the joint can be heated and brazed. Be sure to overlap the previously brazed section while continuing to work around the fitting. *Figure 24* illustrates how to work in overlapping sectors.

Step 10 Shut off the torch and allow the joint to cool naturally without disturbing it.

Step 11 Wash the joint with warm water and use a stainless-steel brush to clean excess, dried, or hardened brazing flux from the joint. If the flux is too hard to be removed with water, chip the excess flux off using a small chisel and light peen hammer.

Step 12 Test all completed assemblies for joint integrity.

Figure 23 Filler metal entering the joint.

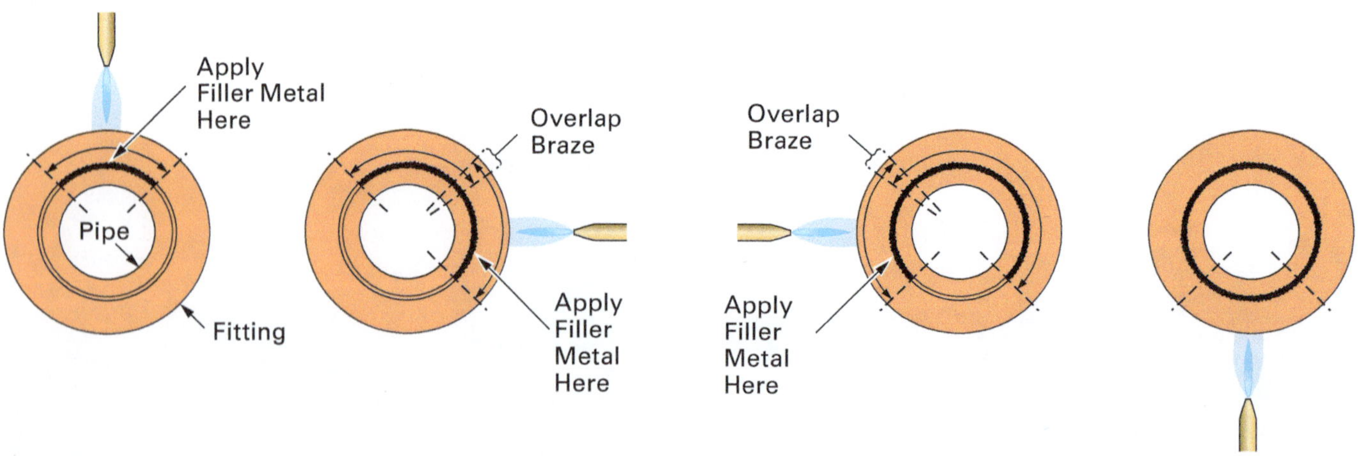

Figure 24 Working in overlapping sectors.

3.0.0 Section Review

1. Oxyfuel equipment produces flame temperatures up to _____.
   a. 1,200°F
   b. 2,600°F
   c. 4,200°F
   d. 5,300°F

2. Soldering fluxes can be used in place of brazing fluxes if necessary.
   a. True
   b. False

3. Which of the following is *true* with regard to brazing?
   a. Focus all of the heat on the fitting and avoid moving the torch.
   b. Cast fittings should never be heated.
   c. Keep the flame away from the filler metal itself.
   d. Fill the joint until the filler metal bead is the same thickness as the tubing.

Section Four

4.0.0 Mechanical Connections for Copper Tubing

Objective

Identify mechanical connection methods for copper tubing.

a. Explain how to install a grooved fitting.
b. Describe how to make a press-fit connection.

Performance Task

4. Join copper using mechanical connections.

Copper tube may also be joined using mechanical connections. Two types of mechanical connections for copper include grooved fittings/couplings and compression fittings.

4.1.0 Grooved Couplings and Fittings

Grooved fittings and couplings specifically Listed for copper tube may be used to make mechanical connections. These are Listed for an operating pressure of 175 psi for joining 2" through 8" copper tubing.

Grooving copper tube requires a special copper roll groover. It follows the conventional methods for roll grooving of steel pipe. The tube is cut to length and reamed, then roll-grooved to the dimensions indicated by the fitting manufacturer. The gasket seating area must be free of indentations, projections, or roll marks to provide a leak-proof seal. *Figure 25* shows some of the grooved fittings available for copper tube.

4.2.0 Compression Fittings

Compression fittings, also known as *press fittings*, use a special tool to crimp a gasketed fitting into place around a plain-end pipe or tubing. *Figure 26* shows common compression fittings. Only those fittings specifically Listed for copper tube may be used to make mechanical connections. These fittings are Listed for an operating pressure of 175 psi for joining $^3/_4$" through 4" copper tubing.

A general procedure for installing compression fittings is described here. Always follow the manufacturer's instructions for the fittings in use.

Step 1 Examine the copper tubing and fittings for cracks or defects. Do not use tubing or fittings with defects.

Step 2 Cut copper tubing using a wheeled tubing cutter or other approved cutting tool. Cut the tubing square to permit proper joining with the fittings.

Step 3 Ream the tubing using an appropriate device.

Step 4 Check the manufacturer's insertion depth chart and mark this distance from the end of the tubing. Insert the tubing into the fitting to the marked depth.

Step 5 Press the joints using the tool(s) approved by the manufacturer. *Figure 27* shows a battery-powered compression tool.

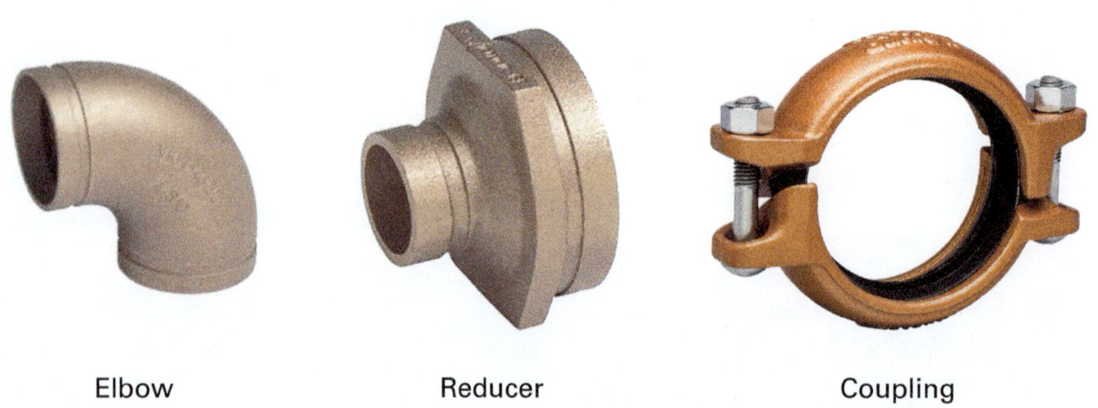

Figure 25 Grooved fittings and coupling.

Figure 26 Compression fittings.

Figure 27 Installing a compression fitting.

Using Compression Fittings to Prevent Heat Damage

The National Science Foundation's Daniel K. Inouye Solar Telescope in Maui, Hawaii, is the world's largest solar telescope. It contains more than seven miles of piping, much of it installed alongside multimillion-dollar optics and other highly sensitive electronic equipment. This necessitated the use of fittings that did not require heat or produce fumes. Copper compression fittings were selected to complete the installation without damaging the electronics.

Figure Credit: Photo courtesy of NSO/AURA/NSF

4.0.0 Section Review

1. Grooved fittings and couplings for copper tubing are Listed for up to _____.
   a. 175 psi
   b. 200 psi
   c. 250 psi
   d. 300 psi
2. Listed compression fittings for copper tubing are rated for an operating pressure of 250 psi.
   a. True
   b. False

Review Questions

1. Which of these types of copper tube are permitted for use in sprinkler systems by NFPA 13, NFPA 13D, and NFPA 13R?
   a. DWV, ACR, and K
   b. K, L, and ACR
   c. K, L, and M
   d. ACR, M, and L

2. Galvanic action, which can lead to corrosion of copper tube, is caused by _____.
   a. overheating the tubing
   b. leaving the tubing out in the rain
   c. joining dissimilar metals
   d. using the wrong flux

3. Tube cutters can be used to cut copper tubing up to _____.
   a. 1" in diameter
   b. 2" in diameter
   c. 3" in diameter
   d. 6" in diameter

4. Copper tube ends *must* be cleaned because oxides and surface soil can _____.
   a. interfere with capillary action
   b. change the structure of the copper
   c. prevent the solder from solidifying
   d. result in staining

5. The *maximum* hanger spacing for 1" copper tubing is _____.
   a. 8'
   b. 10'
   c. 12'
   d. 15'

6. The cup depth for a 1" copper sweat fitting is _____.
   a. $^{7}/_{16}$"
   b. $^{1}/_{2}$"
   c. $^{2}/_{3}$"
   d. $^{29}/_{32}$"

7. The cup depth engagement is the _____.
   a. distance the tubing penetrates the fitting
   b. outside diameter of the tubing
   c. inside diameter of the fitting
   d. distance between fittings

8. When soldering a joint, _____.
   a. only the tubing is heated
   b. only the fitting is heated
   c. heat is applied directly to the tip of the solder
   d. heat is first applied to the tubing and then to the fitting

9. Clothing worn when performing soldering or brazing should be made from _____.
   a. nylon
   b. cotton
   c. polyester
   d. acrylic

10. The filler metal used for brazing usually melts at a temperature of _____.
    a. 375°F to 500°F
    b. 750°F to 850°F
    c. 850°F to 1,200°F
    d. 1,200°F to 1,500°F

11. Copper brazing is usually performed using _____.
    a. copper-phosphorus brazing rods
    b. nickel-silver brazing rods
    c. titanium brazing rods
    d. aluminum brazing rods

12. The color of the inner cone in a brazing flame is _____.
    a. bright blue
    b. white
    c. orange
    d. yellow

13. The 3-T Rule refers to _____.
    a. cup depth engagement
    b. fitting radius
    c. flame length
    d. brazing alloy penetration
14. Grooved couplings and fittings are Listed for use on copper tubing from 2" to _____.
    a. 4" in diameter
    b. 6" in diameter
    c. 8" in diameter
    d. 10" in diameter
15. The fitting shown in *Figure RQ01* is a(n) _____.
    a. reducing female adapter
    b. eccentric female connector
    c. male adapter
    d. hose connector

Figure RQ01

Trade Terms Quiz

Fill in the blank with the correct term learned from your study of this module.

1. A(n) ______________ is a soft solder joint, or a joint soldered below 840°F.
2. ______________ refers to water that is fit for human consumption.
3. A descriptive term for solder that is flowing readily and that has been melted by heating is ______________.
4. A(n) ______________ is a nonconductor of electricity.
5. Any substance made up of two or more metals is called a(n) ______________.
6. ______________ is an alloy that is melted and used to join metallic surfaces.
7. The process by which the oxygen in the air combines with metal to produce tarnish and rust is ______________.
8. ______________ refers to direct-current electricity, especially when produced chemically.
9. ______________ is a method of fusing metals with a nonferrous filler metal using heat above 840°F, but below the melting point of the base metals being joined.
10. ______________ is a method of joining metals with a nonferrous filler metal using heat below 840°F and below the melting point of the base metals being joined.
11. A type of brazing equipment that uses oxygen and fuel gas mixed in a nozzle and directed at the work surface is ______________.
12. ______________ is a chemical substance that aids the flow of solder by removing and preventing the formation of oxides on the pieces to be joined by soldering.
13. Boric acid salts used as a flux are called ______________.
14. Heavy-duty copper fittings that are suitable for making brazed connections are ______________.
15. Thin-wall copper fittings that are suitable for making soldered connections are called ______________.
16. A slight surface depression is called a(n) ______________.
17. To ______________ is to draw solder into a joint.
18. ______________ is the spontaneous flow of a liquid into a narrow space.
19. A descriptive term for solder in the solid or stable phase is ______________.

Trade Terms

Alloy
Borates
Brazing
Capillary action
Dielectric
Dimple
Flux
Galvanic
Liquidus
Oxidation
Oxyfuel equipment
Potable
Solder
Soldering
Solidus
Sweat fittings
Sweat joint
Wick
Wrought fittings

Cornerstone of Craftsmanship

Robert (Bob) Dunn
Service & Inspections Training Support
Fire & Life Safety America

How did you choose a career in the industry?

In 1981, I was working as a groom for a horse stable in Virginia making $125.00 a week. I wanted a career that was a craft and rewarding with the potential for growth. I answered an ad for entry-level sprinkler fitters in Maryland.

Who inspired you to enter the industry?

After interviewing for the sprinkler apprentice position, I was inspired by the trade itself and getting to the end product. The sprinkler industry was booming at that time and I was eager to learn all aspects of how a fire sprinkler system works and the importance of its purpose.

What types of training have you been through?

I completed a registered apprenticeship program in 1985 through Sprinkler Fitters Local 669. I also wanted to learn sprinkler system design, so I attended classes in the evenings at a local college. I also received Naval Facilities Engineering Command (NAVFAC) and Army Corps of Engineers (COE) certifications for quality control, including project management, OSHA-40, and NICET training as well.

How important is education and training in construction?

I believe it's imperative to educate yourself regardless of what your career path is. Knowledge is power. You will be fabricating and installing life safety systems to protect people's lives and property. In addition, safety is first and foremost when working in the construction industry.

How important are NCCER credentials to your career?

The construction industry is evolving, and an important part is having an accredited, recognized training program to develop a qualified workforce. NCCER does this with career development.

How has training/construction impacted your life?

I would not be where I am today without the training I have received from my apprenticeship and the training I am still receiving to this day. The importance of working safely (including proper handling of materials and tool safety), communication skills, and installation best practices are relevant and in my "toolbox" today.

What kinds of work have you done in your career?

I started as a sprinkler apprentice, sprinkler fitter, foreman, superintendent, project manager, designer, sales, service manager, district manager, and owned my own business for 12 years. I also enjoyed being a sprinkler service technician.

Tell us about your present job.

My current role is providing training and support for my company, including the NCCER/AFSA apprenticeship program, as well as inspection, testing, and maintenance (ITM) of fire sprinkler systems, fire alarm, suppression systems, and other areas related to our business.

What do you enjoy most about your job?

What I enjoy most is to work with individuals who have minimal or no industry experience and watch them evolve through their hard work and determination to achieve their goals and become an asset to their organization. Their success is my success.

What factors have contributed most to your success?

Some factors that have allowed me to excel are having an opportunity, training, teamwork, accountability, credibility, and integrity. Set your goal and go after it.

Would you suggest construction as a career to others? Why?

Absolutely, it is a career with so much opportunity to grow within. I always say it "takes a village" to have a successful project, from the installers, sales, management, partners, owners, etc. Anyone who decides to enter the construction industry has this opportunity to progress as far as they are willing to work for it—the sky's the limit.

What advice would you give to those new to the field?

The big two are a positive attitude and effort; they have total control of these. Always maintain your integrity, honesty, and a positive work ethic. Achieve as much knowledge as you can through training opportunities as that will be your springboard to a successful career.

Interesting career-related fact or accomplishment?

One that resonates with me today is a fire sprinkler system installation early in my career. I was the foreman and responsible for the project. Approximately three months after project completion and owner occupancy, a fire started in an area being used for storage. The fire sprinkler system activated as designed and contained/extinguished the fire with minimal property damage and no injuries. I didn't know about this until the owner sent a letter to our company thanking us for providing this life safety system that was needed and performed as expected. This has happened a few times in my career and the takeaway is I was part of something that made a difference. I never forgot that, and you won't either.

How do you define craftsmanship?

Craftsmanship is an art that exhibits skills (using tools or not) but provides a quality product or end result. All I can say is whatever you decide to do, be good at it. When I was installing sprinkler systems, I wanted the general contractor, owner, or inspector to look at the installation and say, "This is a quality installation." This is a reflection of you and your team. Remember, quality before quantity—it's your name on it.

Cornerstone of Craftsmanship

Steven Rasch

Regional Sprinkler Manager
Johnson Controls Fire Protection

How did you choose a career in the industry?

I started working in the sprinkler business as a summer job before going to medical school (Mom's dream). And then I just never left and continued to grow with the business.

Who inspired you to enter the industry?

My best friend's uncle, Roger Wilson. He offered us summer jobs to work in the fabrication shop in 1981. During that period, I really liked the direction of the business, and he inspired me to stay and saw potential in me.

What types of training have you been through?

I've been through special hazards training like foam, deluge, and preaction systems; integrity room testing; management and leadership; fire pumps; all types of sprinkler systems; and of course, NCCER.

How important is education and training in construction?

It's very important, and the lack of adequate training is one of the largest issues in the trade today. We must first attract the younger generation to the trade, and in order to do that, they need to know and be able to see what is in it for them.

How important are NCCER credentials to your career?

It helped me become certified as an instructor once we went away from the Union 669 in 1994. Then, we started our own in-house training with the Wheels of Learning through ABC.

How has training/construction impacted your life?

I hate to admit this, but training has only been whatever I made it into on my own here. Without training and mentorship from the guys I grew up with in the construction business, there is no way I would be in the leadership position that I am now.

What kinds of work have you done in your career?

I've done fabrication, installation, inspection, and management in the fire sprinkler industry. I've worked in refineries, power plants, high-rise towers, and more. One of the most fun jobs was inspector for Wormald Fire Systems where we worked on every system known to the industry.

Tell us about your present job.

I'm currently the Regional Sprinkler Manager and oversee the East Region sprinkler business (60 US offices) for SimplexGrinnell and now Johnson Controls. I focus on many things, including financials, improving and enforcing established processes, training new management and sales reps, recruiting, reviewing projects, project management, and more.

What do you enjoy most about your job?

I enjoy traveling to different offices, interacting with our district personnel, and training our new recruits. I still enjoy the "light bulb moment" with our trainees.

What factors have contributed most to your success?

My past mentors and the many guys who have taught me along the way. I can always go back to no more than five over the course of the last few years who really shaped me to who I am today.

Would you suggest construction as a career to others? Why?

Construction is always going to be there in some way. Whether it will be as robust as recent years, we don't know; however, it is what makes our economy go as it drives every other type of supply needed out there. This field offers good job security, especially if you are good at what you do.

What advice would you give to those new to the field?

Above all, come in each and every day eager to learn. Strive to be the best at what you are doing regardless of the position you are in at the time. This is how you make yourself ready for the next endeavor. Another piece of advice? Don't always ask to be paid to learn something new. Training and education aren't cheap, so if there is something that needs to be done outside of normal business hours, do it as it will pay you in the long run.

Interesting career-related fact or accomplishment?

Nothing earth shattering here. I've been in the business for almost 40 years. I'm NICET-certified, Water-Based Sprinkler Level III, Fire Alarm Level II and Special Hazards Level II.

How do you define craftsmanship?

The end product of something that was well-built with well-utilized skills and techniques. The craftsman has paid attention to detail and requirements within a given trade or craft. I refer to this as their signature.

Trade Terms Introduced in This Module

Alloy: Any substance made up of two or more metals.

Borates: Boric acid salts used as a flux.

Brazing: A method of fusing metals with a nonferrous filler metal using heat above 840°F, but below the melting point of the base metals being joined.

Capillary action: The spontaneous flow of a liquid into a narrow space between two solid surfaces.

Dielectric: A nonconductor of electricity.

Dimple: A slight surface depression, or the act of creating one.

Flux: A chemical substance that aids the flow of solder or brazing metal by removing and preventing the formation of oxides on the pieces to be joined.

Galvanic: Refers to direct-current electricity, especially when produced chemically.

Liquidus: The state of solder when it has melted by heating and is flowing readily.

Oxidation: The process by which the oxygen in the air combines with metal to produce tarnish and rust.

Oxyfuel equipment: A type of brazing equipment that uses oxygen and fuel gas mixed in a nozzle and directed at the work surface.

Potable: Water that is fit for human consumption.

Solder: An alloy that is melted and used to join metallic surfaces.

Soldering: A method of joining metals with a nonferrous filler metal using heat below 840°F and below the melting point of the base metals being joined.

Solidus: The solid or stable phase of solder.

Sweat fittings: Thin-wall copper fittings suitable for making soldered connections.

Sweat joint: A soft solder joint, or a joint soldered below 840°F.

Wick: To draw solder into a joint.

Wrought fittings: Heavy-duty copper fittings suitable for making brazed connections.

Additional Resources

This module presents thorough resources for task training. The following reference material is suggested for further study.

ASME B16.18, *Cast Copper Alloy Solder-Joint Pressure Fittings*, Latest Edition. New York, NY: American Society of Mechanical Engineers.

ASTM B32, *Standard Specification for Solder Metal*, Latest Edition. New York, NY: American Society of Mechanical Engineers.

ASTM B75/B75M, *Standard Specification for Seamless Copper Tube*, Latest Edition. West Conshohocken, PA: American Society for Testing and Materials International.

ASTM B828, *Standard Practice for Making Capillary Joints by Soldering of Copper and Copper Alloy Tube and Fittings*, Latest Edition. West Conshohocken, PA: American Society for Testing and Materials International.

NFPA 13, *Standard for the Installation of Sprinkler Systems*, Latest Edition. Quincy, MA: National Fire Protection Association.

NFPA 13D, *Standard for the Installation of Sprinkler Systems in One- and Two-Family Dwellings and Manufactured Homes*, Latest Edition. Quincy, MA: National Fire Protection Association.

NFPA 13R, *Standard for the Installation of Sprinkler Systems in Low-Rise Residential Occupancies*, Latest Edition. Quincy, MA: National Fire Protection Association.

Figure Credits

Mueller Streamline Co., Module Opener

NIBCO INC., Figures 2, 26, RQ01

Courtesy of www.wattswater.com, Figure 3

Courtesy of T-DRILL, Figures 4–5

Courtesy of RIDGID Tool Company, Figures 6, 8, 11, 12 (Tube Cutter, Fitting Brush), 27

The Mill-Rose Company, Figure 16

Courtesy of Victaulic Company, Figure 25

Reproduced with permission of NFPA from NFPA 13, *Standard for the Installation of Sprinkler Systems*, 2019 edition. Copyright© 2018, National Fire Protection Association. For a full copy of NFPA 13, please go to www.nfpa.org, Table 2

Section Review Answer Key

Section 1.0.0

Answer	Section Reference	Objective
1. c	1.1.0	1a
2. b	1.2.3	1b
3. a	1.3.4	1c
4. c	1.4.1; *Table 2*	1d

Section 2.0.0

Answer	Section Reference	Objective
1. a	2.1.0	2a
2. c	2.2.1	2b

Section 3.0.0

Answer	Section Reference	Objective
1. d	3.1.0	3a
2. b	3.2.0	3b
3. c	3.3.3	3c

Section 4.0.0

Answer	Section Reference	Objective
1. a	4.1.0	4a
2. b	4.2.0	4b

NCCER CURRICULA — USER UPDATE

NCCER makes every effort to keep its textbooks up-to-date and free of technical errors. We appreciate your help in this process. If you find an error, a typographical mistake, or an inaccuracy in NCCER's curricula, please fill out this form (or a photocopy), or complete the online form at **www.nccer.org/olf**. Be sure to include the exact module ID number, page number, a detailed description, and your recommended correction. Your input will be brought to the attention of the Authoring Team. Thank you for your assistance.

Instructors – If you have an idea for improving this textbook, or have found that additional materials were necessary to teach this module effectively, please let us know so that we may present your suggestions to the Authoring Team.

NCCER Product Development and Revision
13614 Progress Blvd., Alachua, FL 32615

Email: curriculum@nccer.org
Online: www.nccer.org/olf

❑ Trainee Guide ❑ Lesson Plans ❑ Exam ❑ PowerPoints Other ______________

Craft / Level: Copyright Date:

Module ID Number / Title:

Section Number(s):

Description:

Recommended Correction:

Your Name:

Address:

Email: Phone:

Underground Pipe

Overview

This module describes the installation of underground piping systems. It covers ductile iron and PVC piping and components, including risers, backflow preventer assemblies, hydrants, and restraints. It also covers trench safety requirements, backfilling, and testing requirements.

Module 18106

Trainees with successful module completions may be eligible for credentialing through the NCCER Registry. To learn more, go to **www.nccer.org** or contact us at 1.888.622.3720. Our website, **www.nccer.org**, has information on the latest product releases and training.

Your feedback is welcome. You may email your comments to **curriculum@nccer.org**, send general comments and inquiries to **info@nccer.org**, or fill in the User Update form at the back of this module.

This information is general in nature and intended for training purposes only. Actual performance of activities described in this manual requires compliance with all applicable operating, service, maintenance, and safety procedures under the direction of qualified personnel. References in this manual to patented or proprietary devices do not constitute a recommendation of their use.

18106 V4.0

18106

Underground Pipe

Objectives

Successful completion of this module prepares you to do the following:

1. Describe underground piping systems and their components.
   a. Identify the types of piping used in underground fire service.
   b. Identify risers, backflow preventer assemblies, hydrants, and associated equipment.
2. Identify trench safety requirements.
   a. Describe how trenches are prepared and protected.
   b. Identify safety guidelines for working in or near trenches.
3. Describe the installation of underground piping systems.
   a. Explain how bedding is placed.
   b. Describe the methods for cutting and joining underground pipe.
   c. Explain how to install thrust blocks and restraints.
   d. Identify methods of corrosion protection for pipe and fittings.
   e. Explain how to install tracer wire.
   f. Describe how a trench is backfilled.
   g. Identify the testing, inspection, and flushing requirements for underground piping systems.

Performance Task

Under supervision, you should be able to do the following:

1. Complete a Contractor's Material and Test Certificate for Underground Piping.

Trade Terms

Back siphonage
Backflow
Bedding
Benching
Competent person
Cross braces
Cross-connection
Deadman
Disturbed soil
Ditch breakers
Double-check valve backflow preventer assembly (DCV)
Dual-check valve backflow preventer
Excavation
Haunching
Manufactured air gap
Proctor curve
Reduced-pressure zone (RPZ) backflow preventer assembly
Shielding
Shoring
Sloping
Subsidence
Test cocks
Trench
Trench box
Uprights
Wales

Industry Recognized Credentials

If you are training through an NCCER-accredited sponsor, you may be eligible for credentials from NCCER's Registry. The ID number for this module is 18106. Note that this module may have been used in other NCCER curricula and may apply to other level completions. Contact NCCER's Registry at 1.888.622.3720 or go to **www.nccer.org** for more information.

You can also show off your industry-recognized credentials online with NCCER's digital badges. Transform your knowledge, skills, and achievements into badges that you can share across social media platforms, send to your network, and add to your resume. For more information, visit **www.nccer.org**.

NOTE

Unless otherwise specified, references in parentheses following figure and table numbers refer to NFPA 13.

Contents

Figures

Tables

SECTION ONE

1.0.0 UNDERGROUND PIPING SYSTEMS AND COMPONENTS

Objective

Describe underground piping systems and their components.

a. Identify the types of piping used in underground fire service.
b. Identify risers, backflow preventer assemblies, hydrants, and associated equipment.

Trade Terms

Back siphonage: A form of backflow caused by lower-than-normal upstream pressure in the potable water line.

Backflow: An undesirable condition that results when non-potable liquids enter the potable water supply by reverse flow through a cross-connection.

Cross-connection: A direct link between the potable water supply and water of unknown or questionable quality.

Double-check valve backflow preventer assembly (DCV): A backflow preventer assembly that prevents back pressure and back siphonage through the use of two spring-loaded check valves and includes test cocks for testing individual pressure zones.

Dual-check valve backflow preventer: A simple backflow preventer that uses two spring-loaded check valves to prevent back siphonage and back pressure but does not include test cocks.

Manufactured air gap: An air gap installed on a reduced-pressure zone (RPZ) backflow preventer assembly to prevent back pressure and back siphonage.

Reduced-pressure zone (RPZ) backflow preventer assembly: A backflow preventer assembly that uses two spring-loaded check valves plus a hydraulic, spring-loaded pressure differential relief valve to prevent back pressure and back siphonage.

Test cocks: Valves in a backflow preventer assembly that permit the testing of individual pressure zones.

Sprinkler fitters install underground piping to serve private fire hydrants and sprinkler supply lines. The minimum requirements for pipe and fittings used for underground pipe in fire protection are specified in NFPA 24, *Standard for the Installation of Private Fire Service Mains and Their Appurtenances*. Private fire service mains begin at the backflow preventer assembly. Everything up to that point falls under the jurisdiction of the local Water Authority.

1.1.0 Underground Piping

NFPA 24 requires pipe for fire service mains to be Listed for fire protection service and to comply with American Water Works Association (AWWA) standards, where applicable. The type and class of pipe for a particular installation must consider fire resistance, maximum working pressure, laying conditions, soil conditions, corrosion, and external loads. External loads include earth loads, buildings, and traffic or vehicle loads. Pipe used in private fire service must be designed to withstand a working pressure of not less than 150 psi.

The two most common types of piping used for underground fire service mains are ductile iron and polyvinyl chloride (PVC).

1.1.1 Ductile Iron Pipe

Ductile iron is a type of cast iron containing magnesium to enhance its strength and elasticity. It is less brittle than cast iron and has a similar shock resistance to mild carbon steel. It has a higher impact and burst strength than PVC pipe and can be welded or brazed.

Ductile iron pipe is available in a wide variety of pressure and thickness classes. The ductile iron pipe selected for use in fire service may vary depending on location. A typical example is Thickness Class 52 with a pressure rating of 350 psi (*Figure 1*).

Ductile iron pipe is available with either a flanged or a bell-and-spigot end. It is coated with an asphaltic coating per AWWA C151, *Ductile-Iron Pipe, Centrifugally Cast*. The use of ductile iron pipe in very corrosive soils may require the addition of a polyethylene wrap and/or a zinc coating.

Figure 1 Ductile iron pipe.

NFPA 13 and NFPA 24 allow a cement lining for cast iron and ductile iron pipe. This lining reduces the buildup of scale and rust on the inside of the pipe. The cement lining is cast when the pipe is manufactured. Exercise care to avoid damage to the lining. The lining will chip out with hammer blows or crack if the pipe is dropped. Any point of damage can result in the buildup of foreign material over time, reducing the flow characteristics of the pipe.

1.1.2 Polyvinyl Chloride (PVC) Pipe

The PVC pipe used in underground fire service must meet the performance requirements of AWWA C900, *Standard for Polyvinyl Chloride (PVC) Pressure Pipe, 4" through 12"*.

C900 PVC pipe is increasingly popular for use in private fire service mains (*Figure 2*). It is available with a gasketed bell-and-spigot end and colored blue to indicate use in potable water service. It is lightweight and easy to load, transport, and handle. The ease of installation also makes it cost-effective.

A PVC pipe interior will remain smooth over years of service with no loss in carrying capacity. Because it is nonmetallic, it does not lose strength due to external corrosion. PVC has a pressure rating of 150 psi and is available in sizes that are compatible with cast and ductile iron fittings.

Some disadvantages of PVC are that it is not as strong as ductile iron and is likely to become brittle in cold temperatures. It must be covered when stored outdoors to avoid damage due to ultraviolet (UV) exposure. Refer to the manufacturer's instructions for installation and compatibility information.

1.2.0 Other System Components

Underground piping system components include in-building risers, backflow preventer assemblies, hydrants, hose stations, and related equipment.

Figure 2 C900 PVC pipe.

1.2.1 In-Building Risers

The in-building riser (*Figure 3*) is a Listed fitting used to connect the main fire supply to the building's overhead fire protection system. It consists of a single extended 90-degree fitting of fabricated stainless-steel tubing with a maximum working pressure of 175 psi. The fitting passes under the foundation without joints and extends up through the floor. It has a grooved or flanged connection on the outlet (building) side and a cast iron pipe size (CIPS) coupler on the underground (inlet) side for connection to the underground water supply. It includes installation tabs or eyes to lock in push-type joints. In lieu of an in-building riser, a ductile iron stub-up can also be used.

1.2.2 Backflow Preventer Assemblies

Piping systems deliver fresh water and take away wastewater. Normally, these two water sources are separated. However, sometimes problems or malfunctions can force water to flow backward through a system. When this happens, wastewater and other liquids can be siphoned into the fresh water supply. This can cause contamination, sickness, and even death.

The reverse flow of non-potable liquids into the potable water supply is called **backflow**. Backflow can contaminate a fixture, a building, or an entire community. There are two types of backflow: back pressure and **back siphonage**. Back pressure is a higher-than-normal downstream pressure in the potable water system. Back siphonage is a lower-than-normal upstream pressure.

Backflow cannot happen unless there is a direct link between the potable water supply and another source. This condition is called a **cross-connection**. Several conditions can cause backflow:

- Cuts or breaks in the water main
- Failure of a pump
- Injection of air into the system
- Accidental connection to a high-pressure source

Backflow preventer assemblies protect against cross-connections by providing a gap or barrier to keep backflow from entering the water supply. There are many types of backflow preventer assemblies, each designed to work under different conditions. Some protect against both back pressure and back siphonage, while others can handle only one type of backflow.

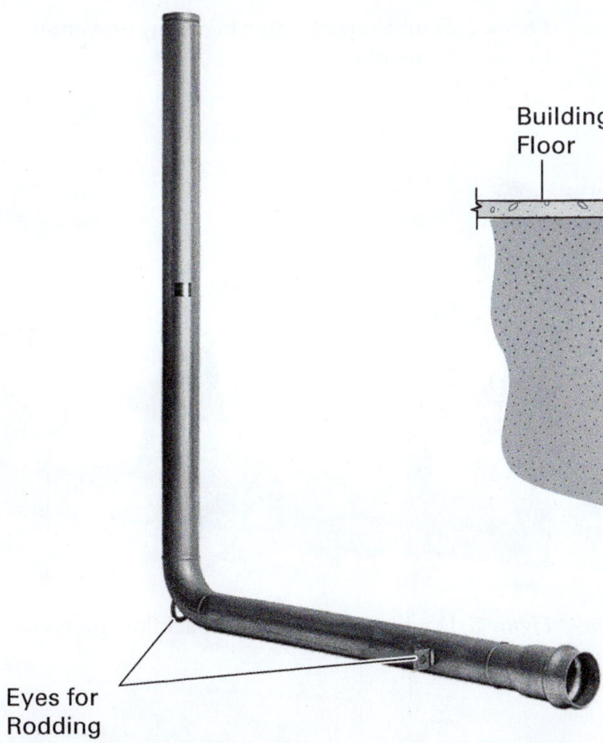

Figure 3 In-building riser.

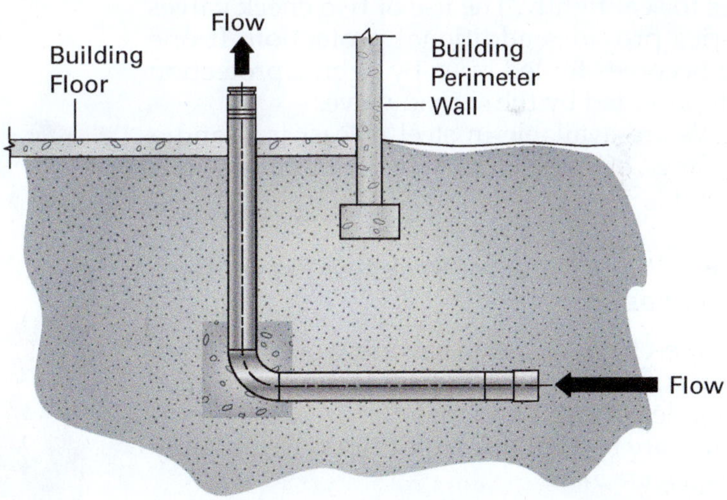

Backflow preventer assemblies are available with flanged or grooved connections and either butterfly or outside screw and yoke (OS&Y) valves. Before installing a backflow preventer assembly, always ensure that it is appropriate for the installation. Install backflow preventer assemblies at least 12" above grade and provide support to prevent sagging. Select a protected location to help prevent damage and vandalism.

NFPA 13 requires a forward-flow test on a yearly basis on all backflow preventer assemblies. The minimum flow rate shall be the system demand, including the hose stream demand where applicable. Common types of backflow preventer assemblies used in fire protection systems are described in the following sections.

Double-Check Valve Backflow Preventer Assemblies (DCVs)

A **double-check valve backflow preventer assembly (DCV)** can be used to protect against back siphonage, back pressure, or both (see *Figure 4*). This type of backflow preventer assembly has two spring-loaded check valves and includes several **test cocks**. DCVs are designed for installation where both back pressure and back siphonage are threats. This assembly has two spring-loaded check valves mounted side by side or in series. During normal operation, both check valves are open; however, backflow causes the valves to seal tightly. The use of two check valves in series provides additional protection. If one valve becomes fouled open by debris, protection is still provided by the second valve.

DCVs are available in steel or cast iron and a wide range of sizes. DCVs have a shutoff valve at each end.

Double-Check Detector Backflow Preventer Assemblies

A double-check detector backflow preventer assembly is a DCV with a bypass meter that can be used to detect leaks and unauthorized use of water. They are used in fire protection system supply mains (*Figure 5*).

Did You Know?

When your local Water Authority issues a Boil Water Order, it is in response to an event that could allow contaminants to enter the water distribution system, such as a large water main break, a loss of system pressure, or a natural disaster.

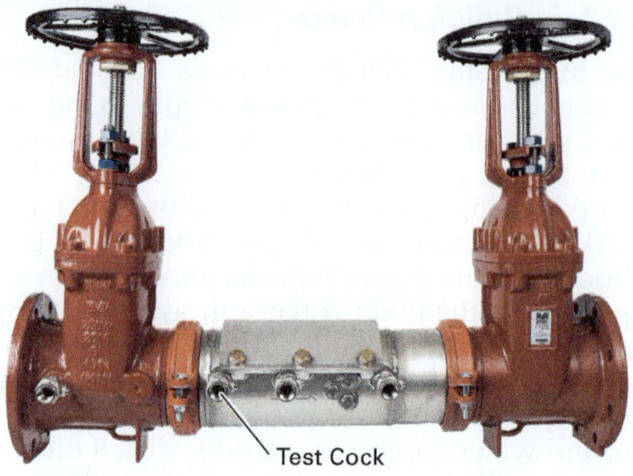

DCV with OS&Y Valves and Flanged Connections

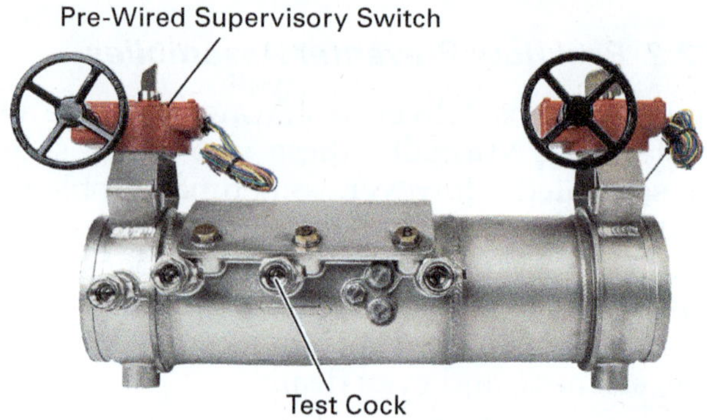

DCV with Integral Butterfly Valves and Grooved Connections

Figure 4 Double-check valve backflow preventer assemblies.

Figure 5 Double-check detector backflow preventer assembly.

Reduced-Pressure Zone (RPZ) Backflow Preventer Assemblies

A **reduced-pressure zone (RPZ) backflow preventer assembly** protects the water supply from contamination by dangerous liquids. RPZ backflow preventer assemblies are used in systems containing foaming agents, antifreeze, or other additives. These devices include two spring-loaded check valves plus a hydraulic, spring-loaded pressure differential relief valve (*Figure 6*). The relief valve is located underneath the first check valve. The device creates a low pressure between the two check valves.

If the pressure differential between the inlet and the space between the check valves drops below a specified level, the relief valve opens. This allows water to drain out of the space between the check valves. The relief valve will also operate if one or both check valves fail. This offers added protection against backflow. RPZ backflow preventer assemblies also have shutoff valves at each end and test cocks that allow testing of the individual pressure zones within the device.

A **manufactured air gap** should be installed on all RPZ installations (*Figure 7*). It acts as a drain receptor to prevent siphonage. Constant water dripping from an RPZ backflow preventer assembly indicates that the valve disc is not seating properly. Flush the unit, and the problem should correct itself. If it does not, refer to the manufacturer's instructions for maintenance procedures.

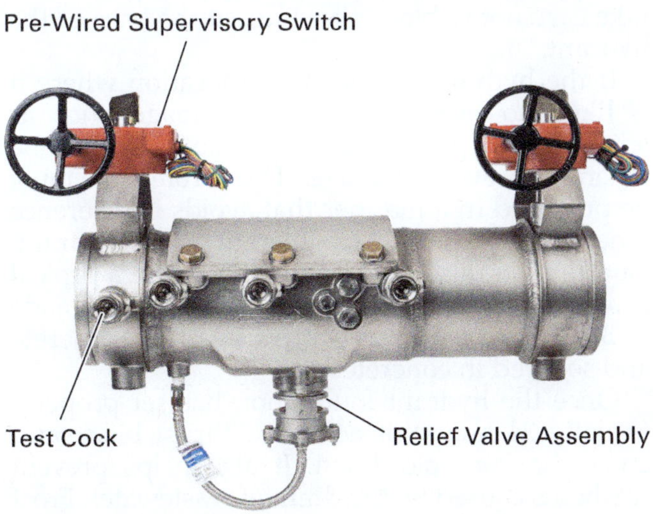

Figure 6 Reduced-pressure zone (RPZ) backflow preventer assembly.

Dual-Check Valve Backflow Preventer

A **dual-check valve backflow preventer** also protects against back pressure and back siphonage (*Figure 8*). These devices are usually installed immediately downstream of the water meter in residential installations and are available in sizes up to 2". Unlike DCVs, they are not fitted with shutoff valves or test cocks. This makes it difficult to remove, test, or replace them. For this reason, some localities do not allow the use of dual-check valve backflow preventers. A residential DCV such as the one shown in *Figure 9* is another option. It includes isolation valves, test cocks, and a removable cover for servicing.

Figure 7 Manufactured air gap.

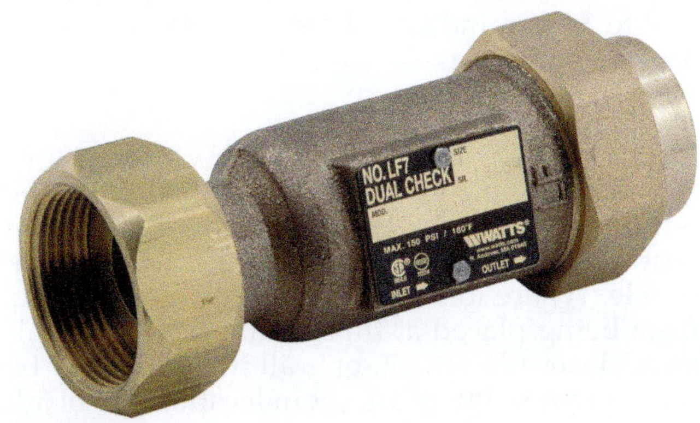

Figure 8 Dual-check valve backflow preventer.

Figure 9 Residential DCV.

1.2.3 Fire Hydrants

Fire hydrants are important in the overall fire protection system. The Fire Department must have quick access to a hydrant even if a building is sprinklered.

The first arriving firefighter must make sure the sprinkler system is working and has sufficient water pressure. The pumper is connected between a fire hydrant and the sprinkler system to raise pressure and improve water flow to the sprinkler system. This is especially important if the sprinkler system has not controlled or extinguished the fire by the time the Fire Department arrives. Firefighters can provide additional pressure to the sprinkler system by pumping into the Fire Department Connection (FDC).

Hydrant Spacing

The Fire Department is usually the Authority Having Jurisdiction (AHJ) over fire hydrants. It determines hydrant locations based on applicable fire codes and ordinances. However, insurance authorities may have more demanding requirements for hydrant placement than the local Fire Department. Spacing can vary from as much as 300' to 800' in industrial areas, to 500' or less in densely built areas.

Separation from Structures

Hydrants should be placed at least 40' from the building being protected. This prevents radiant heat or falling debris from making a hydrant unusable. Where local conditions prevent hydrants from being placed at this distance, NFPA 24 allows closer placement, or wall hydrants may be used, provided they are set in locations by blank (window and doorless) walls where the possibility of injury by falling debris is small. The requirement for placing hydrants in a location where people are not likely to be driven away by smoke or heat still applies.

Outlets

Standard fire hydrants ordinarily have a minimum of two $2\frac{1}{2}$" connections and a $4\frac{1}{2}$" pumper connection; however, in places such as California and Hawaii, wet barrel street (municipal) hydrants with only one $2\frac{1}{2}$" outlet are allowed.

NFPA specifies that national standard external threads must be used in outlets. There are localities with long-standing infrastructures that use street hydrants with threads other than the national standard thread pattern. These hydrants require the use of adapters if standard hoses are used. Check the hose thread requirements before installing hydrants.

The center of the hose connection can be no less than 18" above the final grade level. The connection to the main is specified by NFPA 24 to be at least 6" in diameter.

Hydrant Valving, Blocking, and Drainage

When hydrant runouts (piping between the main and the hydrant) are equipped with a valve, the valve can be located at the hydrant, at the main, or anywhere between the main and the hydrant. The valve is usually a gate valve in a roadway box. Regardless of the location, the hydrant must be separately supported and thrust-blocked by a concrete slab or flat stones. The hydrant must be braced in place while the pipe connections are being made and held there until the concrete foundation has been poured and has set. The hydrant must never be supported by the pipe. When pouring the base, take care not to block the drain at the base of the hydrant.

If the hydrant is placed in a location where it is likely to suffer mechanical damage, such as adjacent to a parking area, precautions must be taken to prevent damage. This protection must be provided in a manner that avoids interference when connecting to or operating the hydrant. Guard posts, also known as *bollards*, are a typical protective measure (*Figure 10*). Bollards are made of 3" to 4" pipe and are both filled with concrete and secured in concrete.

Once the hydrant foundation has set properly (usually 24 hours), it acts as a thrust block and an anchor or hold-down. It also helps prevent washouts caused by the drain of wastewater. Frost heave is common in freezing climates, and it must be prevented if at all possible. Freezing of the soil causes it to expand, producing a heaving action. The result is a shifting or raising of the fire hydrant, which creates shearing stress on the pipe. Leaking gaskets and broken pipes can result.

Figure 10 Hydrant protected by bollards.

Specifying Hydrants

When hydrants are specified, the following must be included:

- Outlet number, size, and type of threads
- Valve size
- Inlet size
- Barrel type
- Operating nut

There are two basic types of barrels in fire hydrants: wet barrel and dry barrel. Wet barrel hydrants are used in areas where the temperatures remain above freezing. Dry barrel (frostproof) hydrants are more common and are used where there is a possibility of freezing temperatures.

Dry Barrel Hydrants

The base valve on dry barrel hydrants is located below the frost line, keeping water out of the barrel under normal circumstances. Water enters the barrel only when needed and a drain valve (weep hole) at the bottom is open when the main valve is closed. This prevents water from standing in the barrel and minimizes the danger of freezing. Dry barrel hydrants must be inspected semi-annually, preferably in the early spring and the fall. The local AHJ specifies inspection schedules by adapting the requirements found in NFPA 25, *Standard for the Inspection, Testing, and Maintenance of Water-Based Fire Protection Systems*. *Figure 11* shows a typical dry barrel fire hydrant.

> **NOTE**
> Dry barrel hydrants must be open all the way or closed all the way. If not closed all the way, water will constantly drain from the weep holes.

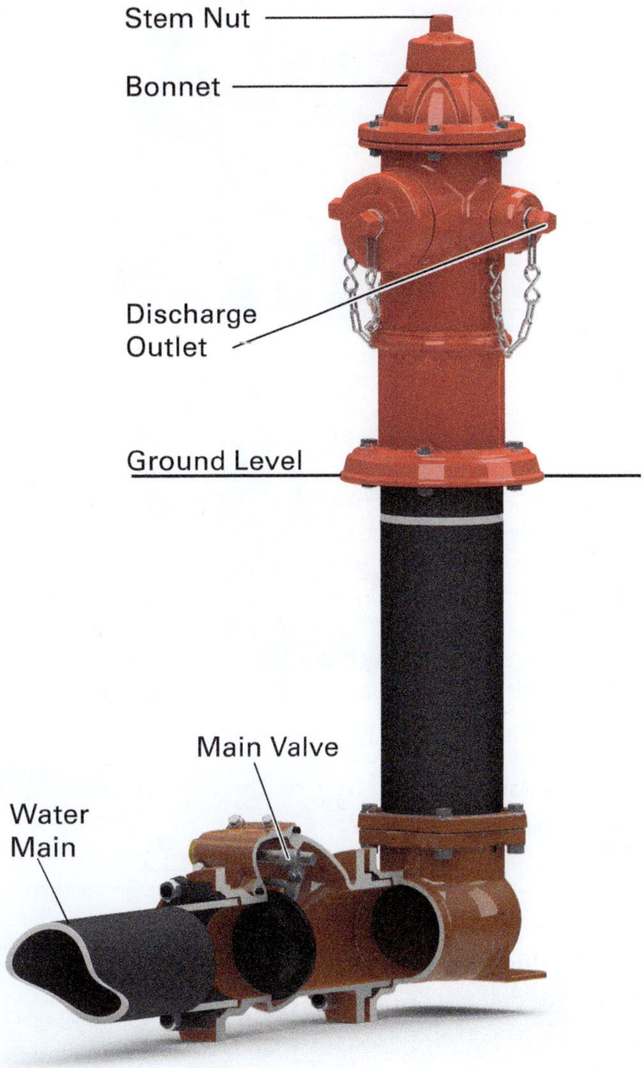

Figure 11 Dry barrel fire hydrant.

Wet Barrel Hydrants

Wet barrel hydrants are sometimes called *California hydrants*. They usually have a compression valve at each outlet and another valve in the bonnet. If constructed with a valve in the bonnet, that valve controls the flow of water to all outlets. Wet barrel hydrants must be inspected annually to ensure proper functioning per NFPA 25. Wet barrel hydrants must never be installed where there is danger of freezing temperatures. *Figure 12* shows a typical wet barrel fire hydrant.

Figure 12 Wet barrel fire hydrant.

Yard Valves

Yard valves typically have fewer outlets than a street hydrant. Yard hydrants are located on a yard loop. *Figure 13* shows a typical yard installation employing a loop. The loop has an FDC for pressurizing by a pumper from a street hydrant.

Other sources of pressurization, such as elevated tanks and surface tanks with fire pumps, may also be present.

Yard valves are valves installed outside a building to control underground water supplies. These valves allow for opening and closing water lines as needed for maintenance and testing.

Most yard valves are iron gate valves designed to be used with indicator posts. As shown in *Figure 14*, the indicator post has a handle arrangement that allows locking the valve either open or shut. A target on the side shows the status of the valve.

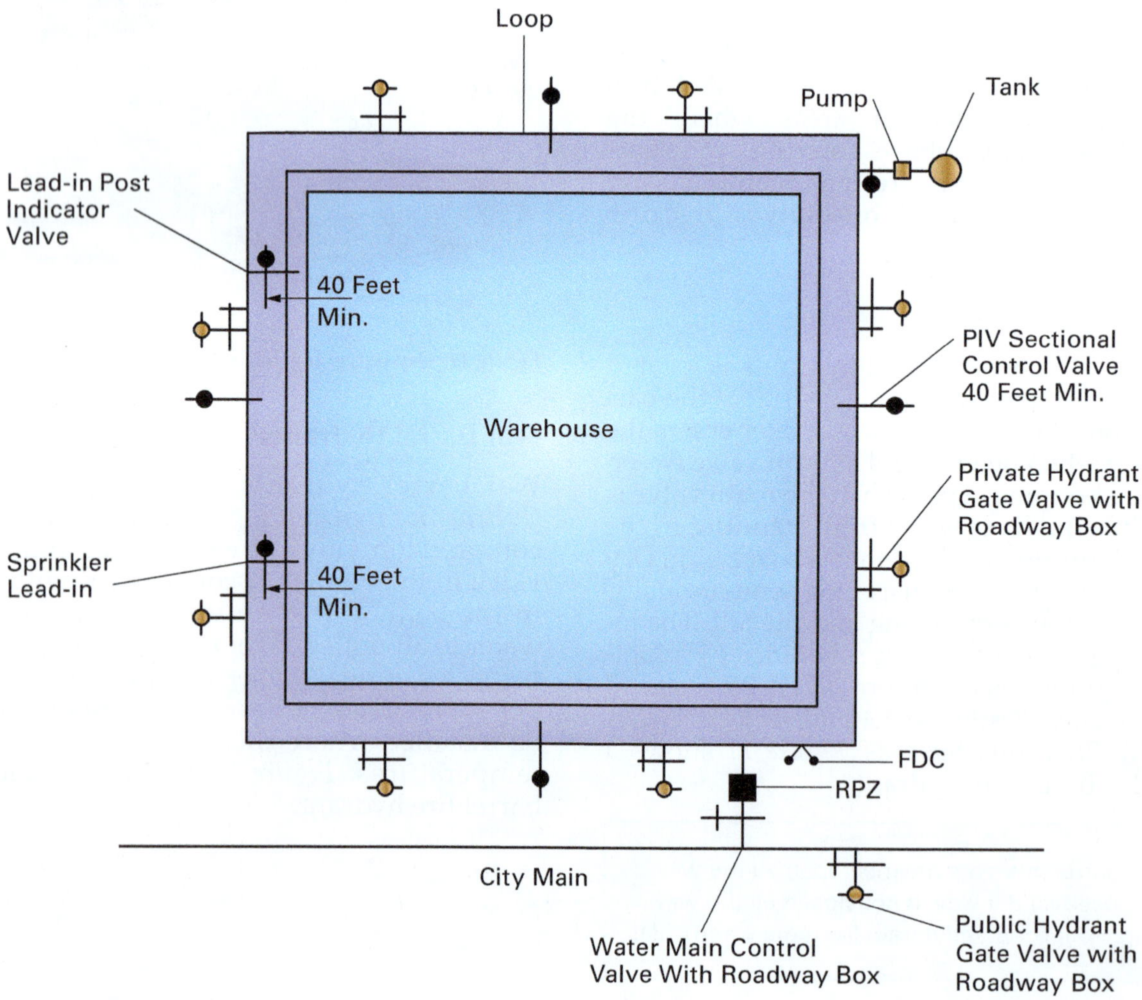

Figure 13 Typical yard loop in plan view.

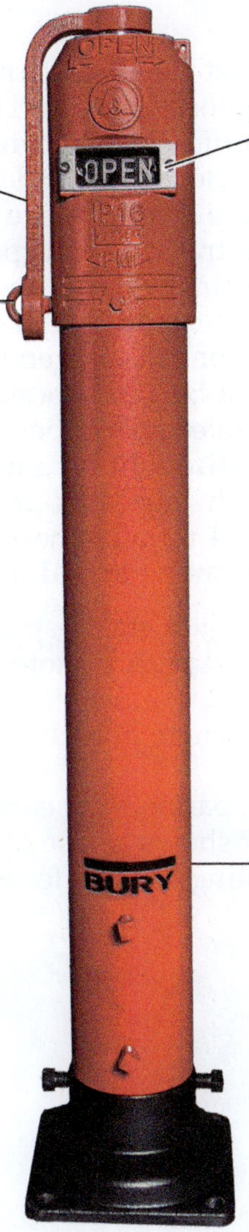

Figure 14 Indicator post.

Post indicator valves (PIVs) must be set so the top of the post is 32" to 40" above the final grade. They must be protected from mechanical damage and tamperproof. NFPA 24 requires monitoring of PIVs. Some PIVs are equipped with supervisory switches linked to a proprietary or central station fire alarm systems, while others are connected to audible alarms. If a PIV is within a fenced enclosure under the control of the owner, it may be installed without a monitor but must be inspected weekly.

NFPA 24 requires a Listed indicating valve in all water sources to a sprinkler system except FDCs. By exception, a non-indicating valve, such as a gate valve with a roadway box complete with T-wrench, are accepted by some AHJ's. *Figure 15* shows a T-wrench.

Figure 15 T-wrench.

Hydrant Color Codes

Hydrant bonnets or *caps* are color coded to indicate how much flow is available in gallons per minute (gpm) at a residual pressure of 20 psi. These flow rates are color coded per NFPA 291, *Recommended Practice for Fire Flow Testing and Marking of Hydrants*. The recommended color codes are as follows:

- *Red* — Less than 500 gpm
- *Orange* — 500 to 999 gpm
- *Green* — 1,000 to 1,499 gpm
- *Blue* — Over 1,500 gpm

The color of a hydrant barrel may also serve as an identifier. A yellow barrel indicates a public hydrant, while a red barrel indicates a private hydrant. A hydrant that is painted black or bagged indicates that it is out of service. Many communities have their own color codes.

WARNING!

Never open a municipal hydrant or municipal backflow preventer assembly without approval from your supervisor and the local Water Authority. Failure to secure permission can result in fines and possibly arrest.

1.2.4 Nozzles, Hoses, and Hose Houses

Hose houses are used to store firefighting equipment in the vicinity of a hydrant. They can also be built with the hydrant inside. This type of hose house is sometimes referred to as a *hydrant house*.

Nozzles and Hoses

Fire hose nozzles must be equipped with a shutoff valve. Shutoff valves can be a part of the nozzle itself or a separate attachment. They must be kept in good operating order and replaced if they do not function properly. NFPA 25 has established guidelines for the care, use, and maintenance of nozzles and associated accessories.

Most fire hoses have a woven jacket and are rubber-lined. The woven jacket is the outer cover of the hose and can be a single jacket or a double jacket. Most jackets are woven polyester or other synthetic material. The rubber lining adds to the life of the hose because it does not absorb water and facilitates quick drying.

The old standard for fire hose was linen hose, but it deteriorated rapidly when exposed to moisture. Unlined linen hose is not permitted in new installations or as replacement hose in existing installations. Periodic inspection and hydrostatic testing are required for proper hose maintenance; however, unlined linen hose is rarely hydrostatically tested because of the effect of moisture on the material.

Hose Houses

Hose houses must be located near a hydrant. *Figure 16* shows a hose house and hydrant.

NFPA 24 has guidelines and requirements for the construction, location, and equipment for hose houses. These hydrants are intended to be used by properly trained plant personnel, a Fire Department, or a fire brigade. They must contain an adequate amount of hose and equipment. The quantity and type of hose and equipment is determined by the number and location of hydrants relative to the protected property, extent of the fire hazard, and firefighting capabilities of the equipment users. In all cases, the AHJ must be consulted. NFPA 24 requires hose houses to have the following equipment in addition to the hose:

- Two approved adjustable-spray solid-stream nozzles equipped with shutoffs for each size of hose provided.
- One hydrant wrench, in addition to the wrench on the hydrant.
- Four coupling spanners for each size hose provided. *Figure 17* shows a hose coupling spanner.
- Two hose coupling gaskets for each size hose.

Figure 16 Hose house and yard hydrant.

In hose houses that contain more than two sizes of hose and nozzles, reducers or gated wyes must also be included in the hose house equipment. *Figure 18* shows a gated wye.

Optional equipment in a hose house includes the following:

- One fire axe with brackets
- One crowbar with brackets
- Two hose and ladder straps
- Two battery-operated hand lights
- Thread adapters

In some localities, the local FDCs do not conform to NFPA standards. In these cases, hydrants and hoses must be special ordered to meet local standards, and the fire code and/or the fire marshal must be consulted.

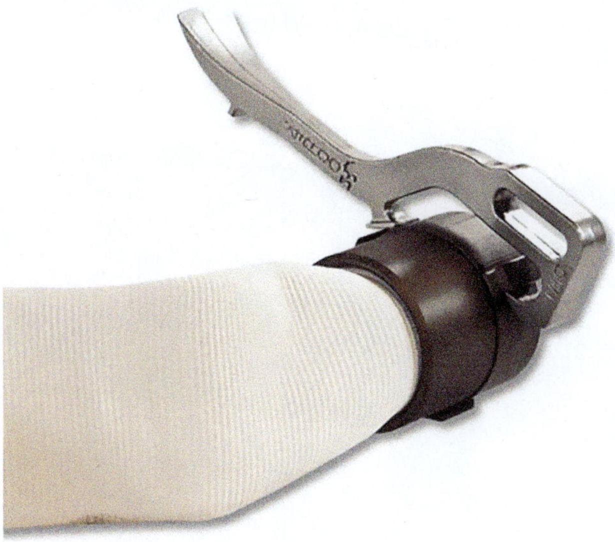

Figure 17 Coupling spanner.

Figure 18 Gated wye.

Hydrant Houses

Hydrant houses are frequently specified for use in rural and industrial areas and in wide-open spaces. NFPA 24 requires that the hydrant be as close to the front of the hydrant house as possible and still allow enough room behind the doors for the hose and other firefighting equipment. If the hydrant house requires locking, special locks with a brittle shackle must be used to facilitate entry in an emergency. A latch behind a glass plate can also be used. The glass can be broken in case of fire. *Figure 19* shows a hydrant house.

A hydrant house must be equipped with a minimum of 100' of lined hose; however, it is preferable to have a 150' hose. To determine the amount of hose needed to protect a property, the hose should be long enough to reach all sides at the full height of the building while maintaining a safe distance when fighting a fire. It is common practice to have the hose connected to one of the hydrant outlets at all times. One or more other outlets on the hydrant remain available for additional hoses to be connected if necessary.

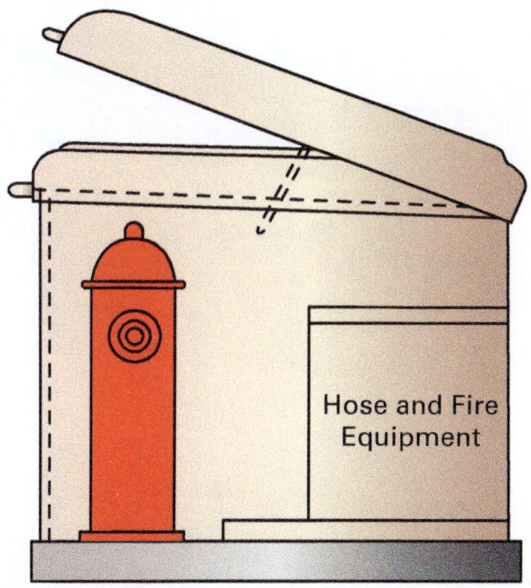

Figure 19 Hydrant house.

1.0.0 Section Review

1. A common type of plastic piping used in underground fire service mains is _____.
   a. ABS-20
   b. C900 PVC
   c. PB-1
   d. PE-RT
2. The bonnet or cap color of a hydrant with a flow rate of over 1,500 gpm at a residual pressure of 20 psi is _____.
   a. red
   b. orange
   c. green
   d. blue

Section Two

2.0.0 Trench Safety Requirements

Objective

Identify trench safety requirements.

a. Describe how trenches are prepared and protected.
b. Identify safety guidelines for working in or near trenches.

Trade Terms

Benching: A method of protecting workers from cave-ins by excavating the sides of a trench to form one or a series of horizontal levels or steps, usually with vertical or near-vertical surfaces between levels.

Competent person: A person who is capable of identifying existing and predictable hazards in the area or working conditions that are unsanitary, hazardous, or dangerous to employees, and who has the authority to take prompt corrective measures to fix the problem.

Cross braces: The horizontal members of a shoring system installed perpendicular to the sides of the excavation, the ends of which bear against either uprights or wales.

Disturbed soil: Soil that has been previously backfilled or removed from the earth and replaced by any means.

Excavation: Any manmade cut, cavity, trench, or depression in an earth surface, formed through earth removal.

Shielding: A structure that is able to withstand the forces imposed on it by a cave-in and thereby protect employees within the structure. Shielding can be permanent structures or designed to be portable and moved along as work progresses.

Shoring: A structure such as a metal hydraulic, mechanical, or timber system that supports the sides of an excavation and is designed to prevent cave-ins.

Sloping: A method of protecting employees working in excavations from cave-ins by cutting the excavation walls to a specific angle at which the soil will rest without sliding.

Subsidence: A depression in the earth that is caused by unbalanced stresses in the soil surrounding an excavation.

Trench: A narrow excavation (in relation to its length) made below the surface of the ground. In general, the depth is greater than the width, but the width of a trench (measured at the bottom) is not greater than 15' (4.6 m).

Trench box: A pre-manufactured steel box that provides a shielding system by allowing workers to safely lay pipe within the box while it is inside an excavation.

Uprights: The vertical members of a trench shoring system placed in contact with the earth and usually positioned so that individual members do not contact each other.

Wales: Horizontal members of a shoring system placed parallel to the excavation face whose sides bear against the vertical members of the shoring system or the earth.

During an **excavation** for underground piping, earth is removed from the ground, creating a **trench** (*Figure 20*). The soil removed from a trench is called *spoil*. The depth of a trench depends on the location, load, and soil conditions. Water mains must be installed at least 6" below the lowest recorded frost depth for the location, which in northern climates can be 7' or more.

2.1.0 Preparing Trenches

Three main classes of equipment are used in trenching: backhoes, excavators, and trenching machines. Underground installations are based on straight runs of pipe and specific angles where bends are necessary. It is important that the trench be straight, not only for the initial installation, but for future excavations when repairs are necessary. A string line is used to create a straight trench. A grade line is used to set the trench depth.

WARNING!

Trenching equipment may only be operated by qualified individuals trained in its safe use.

WARNING!

Existing underground systems may not be restrained and can result in dangerous blowouts when disturbed. Exercise extreme caution when excavating for new systems or working on existing systems.

Figure 20 Excavation.

WARNING!

Federal law requires that the designated authority be contacted at least 72 hours before the start of the excavation in order to locate underground utility lines or cables. Failure to do so can result in your employer being liable for any injury and/or property damage that results. In addition, some localities may require a permit prior to excavating.

During excavation, you may encounter unidentified hazardous materials, such as asbestos piping or old sewer or septic systems. If this occurs, stop work and notify your supervisor immediately.

Trenching Machine

The trenching machine shown here is specifically designed for trench excavation.

Figure Credit: "HT300 Track Trencher" by Ditch Witch is licensed under CC BY 2.0

2.1.1 Checking for Underground Utilities

Before beginning to dig, you must determine if there are any utilities or other structures buried on the site. The national One Call (811) system is used to notify utility companies to set stakes or flags to mark underground utilities prior to excavation. Site markers are color-coded as shown in *Table 1*. Proposed excavations should be marked with white flags prior to the utility survey.

Table 1 Site Marker Color Codes

Color	Meaning
Red	Electric power lines, cables, conduit, and lighting cables
Yellow	Gas, oil, steam, petroleum, or gaseous material
Orange	Communication, alarm or signal lines, cables, or conduits
Blue	Potable water
Green	Sewers and drain lines
Purple	Reclaimed water, irrigation, and slurry lines
White	Proposed excavation
Pink	Temporary survey markings

2.1.2 Trench Protection Methods

When earth is removed from a trench, extreme pressures may be generated on the trench walls, resulting in collapse. OSHA requires that any trenches 5' deep or deeper include a protective system unless the excavation is made entirely in stable rock. If a trench is less than 5' deep, an OSHA-approved **competent person** may determine whether a protective system is required.

In addition to providing a safe environment for the workers in a trench, protective systems restrain the movement of trench walls to protect adjacent soil that may contain buried utilities or support nearby building foundations or other structures.

The type of protective system required is determined by the trench depth, trench width, and type of soil. Average soils are usually a mixture of sand, silt, and clay and can behave in very different ways depending on the percentage of each type. Soils with a higher percentage of clay tend to be more stable than those with a higher percentage of sand. A project can include varying soil conditions and require several types of protective systems on the same site. Methods of trench protection include **shoring**, **shielding**, and **sloping**.

Shoring Systems

Shoring systems are wood or metal structures that provide a framework to support the walls of an excavation. Shoring systems can be pneumatic, hydraulic, or mechanical. Pneumatic and hydraulic shoring systems use pistons that are pumped outward until they press up against the trench walls (*Figure 21*). Mechanical shoring uses **uprights**, **wales**, and **cross braces** to support the trench walls. Uprights placed so that individual members are closely spaced or interconnected are often called *sheeting* and are used in tight excavations against structures.

Figure 21 Shoring system.

To avoid accidents and injury, special safety rules must be followed when shoring an excavation:

- Follow the manufacturer's safety and installation instructions.
- Shoring must extend a minimum of 1' above the trench edge to prevent soil from falling back into the trench.
- If the trench is deeper than 20', the shoring must be designed by a professional engineer.
- Never enter an excavation before the shoring is in place.
- Do not install the shoring from inside the trench. All shoring must be installed from the top of the trench.
- The cross braces must be level across the trench and exert equal pressure on each side of the trench.
- The vertical uprights must be flat against the excavation wall.
- All materials used for shoring must be thoroughly inspected before use and must be in good condition.
- Shoring is removed by starting at the bottom of the excavation and going up.
- Vertical supports are pulled out of the trench from above.
- Every excavation must be backfilled immediately after the support system has been removed.

Shielding Systems

A trench shielding system is a structure designed to withstand a cave-in and protect employees within the structure. Shields can be permanent structures, or they can be portable and moved along as work progresses. A portable trench shield is also called a **trench box** (*Figure 22*). If the trench will not stand long enough to excavate for the shield, the shield can be placed high and pushed down as material is excavated.

The excavated area between the outside of the trench box and the face of the trench should be as narrow as possible. The space between the trench box and the excavation walls must be backfilled to prevent side-to-side movement of the trench box.

Figure 22 Trench box.

Follow these safety guidelines when using a trench box:

- Read and follow the manufacturer's safety and installation instructions.
- Be sure that the vertical walls of a trench box extend at least 18" above the lowest point where the excavation walls begin to slope.
- Never enter a trench box during installation or removal.
- Never enter an excavation behind a trench box after the trench box has been moved.
- Backfill the excavation as soon as the trench box has been removed.

Sloping Systems

Sloping an excavation means cutting the walls of the excavation to a specific angle at which the soil will rest without sliding. **Benching**, also known as *step-back*, is another method of sloping that uses a series of steps rather than a continuous slope. The design of a sloping system must be completed by a registered professional engineer. Design considerations include the soil type, trench depth, amount of time the trench is to remain open, environmental conditions, nearby loads, and other factors. At least one copy of the design must be maintained at the jobsite while the slope is being constructed. After construction, the design must be made available upon request.

2.1.3 Ladders

There must be at least one method of entering and exiting all excavations over 4' deep. Ladders are generally used for this purpose (*Figure 23*). Ladders must be placed within 25' of each worker.

When ladders are used, there are several requirements that must be met:

- Ladder side rails must extend a minimum of 3' above the landing.
- Ladders must have nonconductive side rails if work will be performed near equipment or systems using electricity.
- Two or more ladders must be used where 25 or more workers are working in an excavation in which ladders serve as the primary means of entry and exit, or where ladders are used for two-way traffic in and out of the trench.
- All ladders must be inspected before each use for signs of damage or defects. Damaged ladders should be labeled "DO NOT USE" and removed from service until repaired.
- Use ladders only on stable or level surfaces.
- Secure ladders when they are used in any location where they can be displaced by excavation activities or traffic.
- While on a ladder, do not carry any object or load that could cause you to lose your balance.

2.2.0 Trench Safety

Work within a trench is dangerous. The collapse of unsupported trench walls can instantly crush and bury workers. One cubic yard of earth weighs approximately 3,000 pounds, which is the weight of a small car, and more than enough to seriously injure or kill a worker. Each year in the United States, more than 100 people are killed and many more are seriously injured in cave-in accidents.

Some of the hazards you may encounter while working in an excavation include cave-ins, falls, flooding, electric shock, toxic liquids or gases, and falling or flying rocks or dirt from excavation equipment.

Figure 23 Ladder in a trench.

Many conditions on a construction site can lead to trench failure. If you notice any of these conditions, immediately inform your supervisor:

- Trench walls containing **disturbed soil** from previously excavated ground
- Trench intersections where large corners of earth can break away
- A narrow right-of-way causing heavy equipment to travel too close to the edge of the trench
- Vibrations from construction equipment, nearby traffic, or trains
- Increased subsurface water that causes soil to become saturated and unstable
- Drying of exposed trench walls, potentially causing them to crumble and collapse
- Inclined layers of soil dipping into the trench, creating a potential for a slide

Stay alert for developing cracks, moisture, or small movements in the trench material. Changes in trench walls may be an early indication of trench failure. See *Figure 24*.

Tension cracks usually form at a depth of one quarter to one half of the way down from the top of a trench. Sliding or slipping may occur as a result of tension cracks. In addition to sliding, tension cracks can cause toppling. Toppling occurs when the trench's vertical face shears along the tension crack line and topples into the excavation. An unsupported excavation can create an unbalanced stress in the soil, which in turn causes **subsidence** (sinking) at the surface and bulging of the vertical face of the trench. If uncorrected, this condition can cause wall failure and trap workers in the trench, or greatly stress any protective system. Bottom heaving is caused by downward pressure created by the weight of adjoining soil. This pressure causes a bulge in the bottom of the cut. Heaving and squeezing can occur even when protective systems are properly installed.

Another indication of an unstable trench is boiling. Boiling occurs when water flows upward into the bottom of the cut, flooding and destabilizing the trench. A high water table is one of the causes of boiling. Boiling can happen quickly and can occur even when shoring or shielding are used. If boiling starts, stop what you are doing and leave the trench immediately.

Failure of unsupported trench walls is not the only cause of worker burial. Tons of dirt can be dumped on workers if the excavated earth slides into the trench. Such slides occur when the spoil pile is placed too close to the edge of the trench, or when the ground beneath the pile gives way. There must be a minimum of 2' between the trench wall and the spoil pile. This area must also be kept free of any tools or materials.

The following guidelines must be enforced to ensure the safety of all workers in or around an excavation:

- Never enter an excavation until it has been inspected and you have the approval of the OSHA-approved competent person on site. Some sites require excavation entry permits.
- Inspect trenches at the start of each shift for changes that may have resulted from rain, frost, or equipment vibration.
- Test for atmospheric hazards such as low oxygen, hazardous fumes, and toxic gases prior to entry. Some trenches may require the use of respiratory protection.
- Wear all appropriate personal protective equipment, including high-visibility safety vests.
- Keep heavy equipment away from trench edges.
- Know where underground utilities are located.
- Do not work under raised loads.
- If water starts to accumulate in the trench, get out of the trench immediately.
- Do not walk under loads being handled by power shovels, cranes, or hoists.
- Stay clear of any vehicle that is being loaded.
- Do not work above or below a coworker on a sloped or benched excavation wall.
- Ladders and ramps used as exits must be located every 25' in any trench that is over 4' deep.
- Make sure someone is on top to watch the walls when you enter a trench.
- Make sure you are never alone in a trench. Two people can cover each other's blind spots.
- Keep tools, equipment, and excavated dirt at least 2' from the edge of the excavation.
- Make sure shoring, shielding, or sloping are used for excavations and trenches over 5' deep.
- Stop work immediately if there is any potential for a cave-in. Make sure any problems are corrected before starting work again.
- Barricade access to excavations to protect pedestrians and vehicles (*Figure 25*).

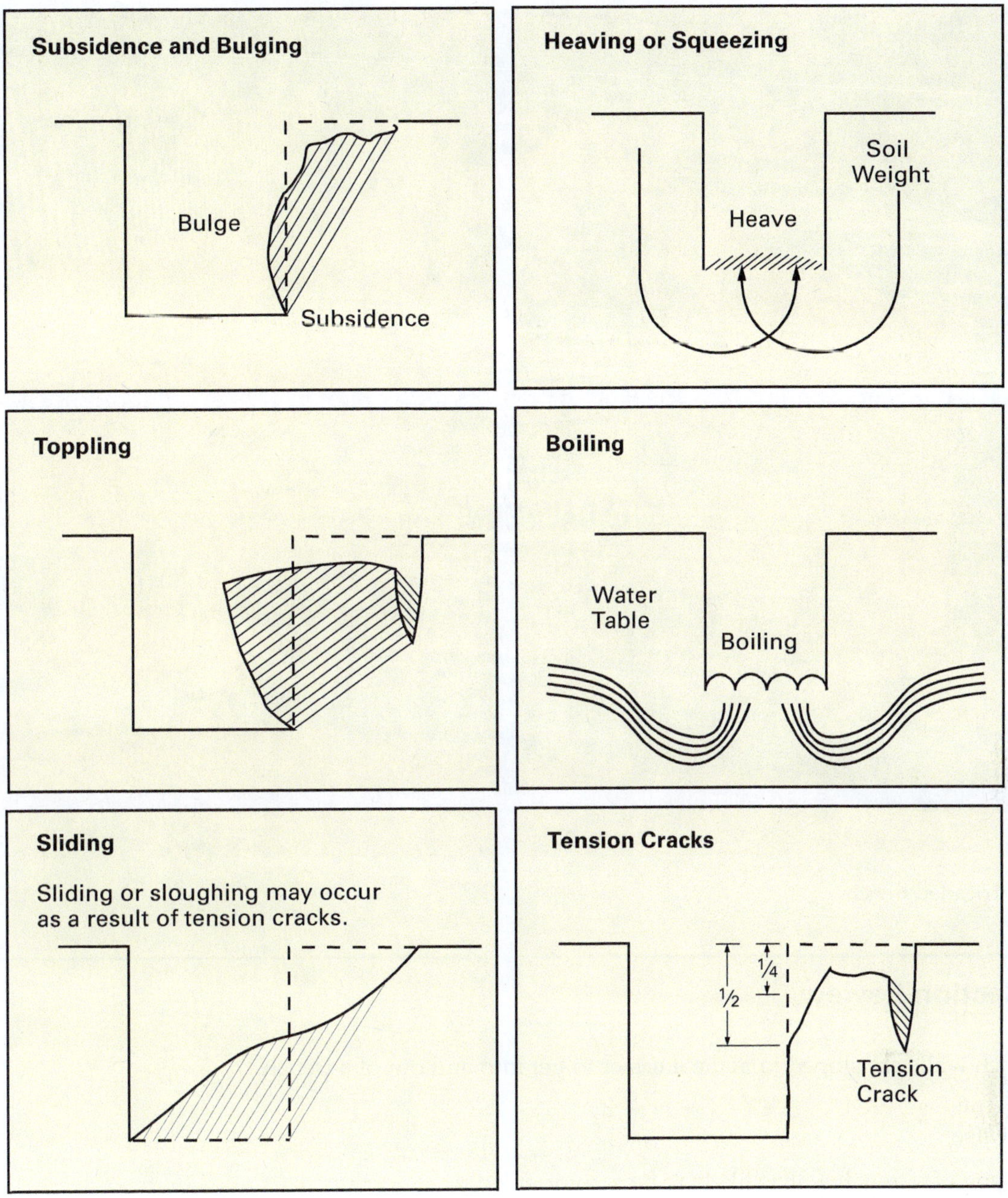

Figure 24 Indications of an unstable trench.

Figure 25 Barricaded trench.

2.0.0 Section Review

1. Trench walls are sloped to make it easier to get into and out of a trench.
   a. True
   b. False
2. Bulging of a trench wall is likely to be accompanied by _____.
   a. subsidence at the surface
   b. boiling of the trench cut
   c. heaving of the trench floor
   d. tension cracks

SECTION THREE

3.0.0 INSTALLING UNDERGROUND PIPING SYSTEMS

Objective

Describe the installation of underground piping systems.

a. Explain how bedding is placed.
b. Describe the methods for cutting and joining underground pipe.
c. Explain how to install thrust blocks and restraints.
d. Identify methods of corrosion protection for pipe and fittings.
e. Explain how to install tracer wire.
f. Describe how a trench is backfilled.
g. Identify the testing, inspection, and flushing requirements for underground piping systems.

Performance Task

1. Complete a Contractor's Material and Test Certificate for Underground Piping.

Trade Terms

Bedding: The surface or foundation on which a pipe rests in a trench.

Deadman: A block of concrete in which tie rods are secured.

Ditch breakers: Sandbags placed under, beside, and over the pipe in a trench to prevent erosion of the pipe bedding material caused by groundwater flow along the pipeline.

Haunching: The process of placing and consolidating material in areas that may remain void or loose when the rest of the backfill is added.

Proctor curve: A test performed in a laboratory that determines the moisture content, cohesion factor, and soil gradation of a particular soil.

A sound pipe installation begins with a stable foundation and ends with returning the surface to its pre-excavation condition. The shop drawings will indicate the specified laying conditions, joining methods, and backfill requirements for a particular installation. Follow these instructions. If laying conditions are not specified, do not proceed until they are known.

3.1.0 Placing Bedding

All pipe foundations require a layer of **bedding** material to cushion and support the pipe. Certain soil conditions may also require a layer of stabilization material to support the pipe bedding.

3.1.1 Stabilization

Before installing bedding, inspect the bottom of the trench to determine the need for stabilization. Walk in the trench, shifting weight from one foot to the other to determine if the soil underneath moves, shifts, or gives. If it does not feel solid, stabilization may be needed. Crushed stone is often used, in sizes ranging from less than 1" to large rock (riprap). The material is poured into the trench and spread to a uniform depth to provide a solid foundation.

3.1.2 Bedding

Pipe bedding is placed on top of the stabilization layer, if present, and is the material on which the pipe is laid. Bedding requirements can be found in the shop drawings and depend on the location, soil conditions, and type of pipe. Bedding should be sufficient to provide uniform, firm support for the pipe and to maintain the pipe grade (typically 4" to 6").

Bedding materials include native soil or outside materials such as sand, gravel, or other specified fill. The material is poured into the trench and leveled to grade along the full length of the trench. Lasers are often used to check the grade.

3.1.3 Laying Pipe in Trenches

PVC is relatively light and easily laid by hand in most pipe sizes. Due to its weight, ductile iron pipe must be laid with the use of heavy machinery. The best machinery for this is a boom or crane truck and Listed nylon slings to avoid pipe damage.

WARNING!

Heavy equipment may only be operated by qualified individuals trained in its safe use.

The pipe should lie level in the trench. When installing bell-and-spigot pipe, the ground must be dug out beneath each bell to prevent the bell from supporting the pipe. If the pipe is joined by means of couplings, the soil must also be dug out underneath the couplings to keep the weight of the pipe off the couplings. *Figure 26* shows a proper laying method.

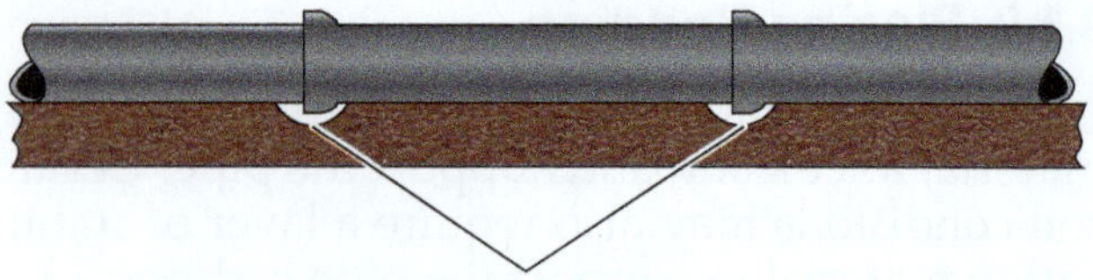

Figure 26 Proper laying method.

3.2.0 Cutting and Joining Pipe

Creating strong, leakproof joints is critical to a safe and long-lasting installation. If the joints are not properly cut, prepared, and fitted, leaks may occur and require rework. Leaks can also cause piping to shift and trigger dangerous blow-outs.

3.2.1 Cutting Pipe

There are many occasions when cutting pipe in the field is required. The type of tool selected depends on the type of pipe.

Cutting Ductile Iron Pipe

Ductile iron pipe is cut using a power saw with an abrasive wheel. Hinged four-wheel cutters with appropriate cutter wheels can also be used in close quarters. Before cutting, the pipe is marked to the correct length and a line drawn around the pipe circumference using chalk or a marking crayon to ensure a square cut. Note that for pipe 12" and smaller, no cuts should be made within 24" of the bell. The cut end must be beveled at a 30-degree angle to facilitate pipe entry and remove any rough edges that might damage the gasket. This can be done using a coarse file or angle grinder.

WARNING!

Never use a saw unless all guards are in place and you have been trained in its safe use. Cutting ductile iron produces flying sparks and metal shavings. Always wear appropriate personal protective equipment, including eye protection.

Cutting PVC Pipe

PVC pipe is cut using a fine-toothed hacksaw, handsaw, or portable power saw. If a portable power saw is used, use a steel blade or abrasive wheel. As with ductile iron pipe, measure the desired length and mark the pipe circumference to ensure a square cut. PVC must also be beveled on the end, typically using a PVC beveling tool. A coarse file or rasp, portable sander, or abrasive disc can also be used. The angle can be determined by using an end beveled at the factory as a guide.

3.2.2 Making Pipe Joints

Most underground pipe is connected using a push-on joint. A push-on joint consists of a bell end containing a rubber gasket and a beveled spigot end. The joint is assembled by pushing the spigot end into the bell end, which compresses the gasket and forms a watertight seal.

Because underground piping is large and can be difficult to maneuver, different methods of leverage are used to aid in pipe insertion. The first is the bar-and-block method (*Figure 27*). This method can be used as long as the soil in the trench is firm. A wooden block is placed against the bell end of the pipe, opposite the end being joined. The pry bar is then driven into the soil in front of the pipe and pushed forward against the block to drive the pipe in. The wooden block acts as a softener to protect the pipe end from damage and provides a straight center-line push.

Lever pullers can also be used to assist in making pipe joints (*Figure 28*). One end of the puller chain is attached to the bell end of the pipe and the other end is placed on the spigot side. The handle is pushed back to pull the spigot end in and create a tight connection.

Joining Ductile Iron Pipe

To join bell-and-spigot ductile iron pipe, proceed as follows:

Step 1 Inspect the bell and gasket to be certain they are clean. The gasket will not seal properly if foreign material is present. If necessary, wipe the gasket and bell with a clean, dry cloth.

Step 2 Loop the gasket and place it into the bell end socket. Inspect to ensure that the gasket is evenly seated around the retainer ring.

NOTE

Gaskets must be kept at a temperature of at least 40°F. In cold weather, keep gaskets in a heated area before use or immerse them in a tank of warm water. Gaskets immersed in water must be dried before being placed in the pipe socket.

Step 3 Brush a thin layer of manufacturer-approved lubricant on the inside surface of the gasket that will come in contact with the spigot end of the pipe.

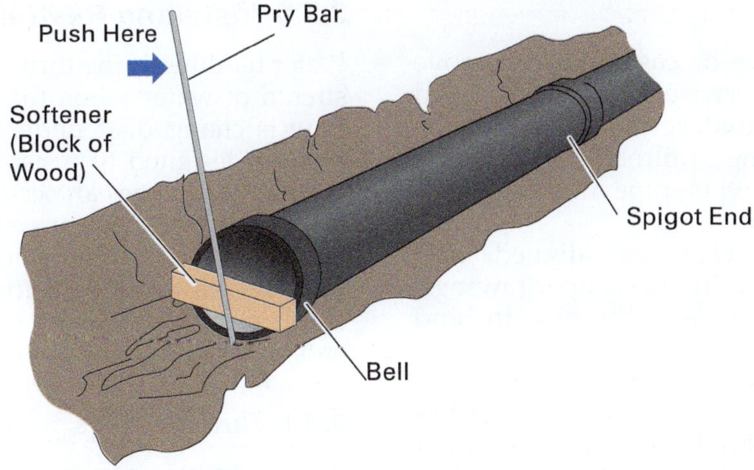

Figure 27 Bar-and-block method.

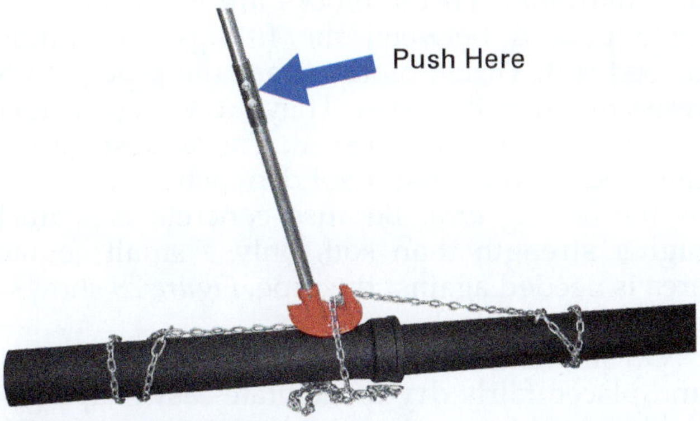

Figure 28 Lever puller.

Step 4 Clean the spigot end of the pipe from the end of the pipe to the stripe.

> **CAUTION**
> In cold weather, frozen material may cling to the pipe and must be removed before joining. In some cases, it is desirable to apply a thin layer of Listed lubricant to the outside of the pipe about 3" back from the end of the pipe. Do not allow the pipe to touch the ground after lubricating since foreign material may adhere to the lubricant and result in a leak.

Step 5 Insert the spigot end of the pipe into the socket until it makes contact with the bottom of the socket. Tighten the joint using a bar and block or lever puller.

Joining PVC Pipe

To join bell-and-spigot PVC pipe, proceed as follows:

Step 1 Inspect the bell and gasket to be certain they are clean. The gasket will not seal properly if foreign material is present. If necessary, wipe the gasket and bell with a clean, dry cloth.

Step 2 Clean the pipe end around the entire circumference from the end of the spigot to 1" beyond the reference mark.

Step 3 Lubricate the spigot end of the pipe using a cloth, pad, sponge, or gloved hand. Cover the entire circumference, paying particular attention to the beveled end.

> **CAUTION**
> Do not lubricate the gasket or bell end as it may cause the gasket to be displaced when the pipe is joined. After lubricating the spigot end of the pipe, take precautions to keep it clean. If it gets dirty, it will have to be cleaned and then relubricated.

Step 4 While bracing the bell end, insert the spigot end so that it makes contact with the gasket. Push the spigot end in until the reference mark on the spigot end is flush with the end of the bell. Do not insert beyond the reference mark. If necessary, carefully tighten the joint using a bar and block or lever puller.

Installing Sleeving

The purpose of a sleeve or casing is to remove physical loads that otherwise would be applied to the pipe. Sleeving of underground pipe may be used at roadway crossings, railroad crossings, for seismic protection, and when pipe must be laid in shifting soil.

Sleeving should always be aligned and positioned as called for in the shop drawings. Improper alignment can cause the pipe to bind when passing through, making the load condition worse rather than better. Under roadways and railroad crossings, where vehicular load is the problem, the pipe should lie in the bottom of the sleeve. Where seismic protection or protection against shifting soil is the objective, the pipe should be centered in the sleeve.

The most common material for sleeving is steel pipe. In soil-covered underground applications, the sleeve material class should at least equal that of the pipe being protected. Where the sleeve function is primarily a form for concrete, other materials may be used. These include lightweight pipe, sleeves formed from sheet metal, conduit, and treated fiberboard. Cost, availability, specifications, and local codes will govern this decision.

The seismic protection requirements in NFPA 24 call for at least 1" nominal clearance all around for pipes 1" through $3^1/_2$" and 2" for pipes 4" and larger. If the sleeve is made of pipe, a nominal pipe size 2" larger than the pipe being protected may be used for pipe 1" through $3^1/_2$". For pipes 4" and larger, a nominal pipe size 4" larger than the pipe must be used.

When passing through floor and grade beams, sleeves should always be located from a fixed object such as the center line of a beam, a column, or a fixed point on the face of a wall. The size of the pipe to go through the sleeve, as well as the size and type of sleeve, must be specified.

It is necessary to install a watertight seal between the pipe passing through the sleeve and the inside of the sleeve when it penetrates a building. This prevents water from passing through the sleeve from one area to another. With soil cover, soil intrusion must be prevented in a similar manner. This reduces the potential for pipe fractures caused by stresses or minor movement of the soil.

3.3.0 Installing Restraints

Water hammer is the thrust produced by a flowing stream of water when the direction or volume of flow is changed. If underground piping systems are not designed to resist the force of thrust, tremendous damage can occur, not only to the piping system, but also to streets, parking lots, lawns, buildings, and building foundations. Mechanical push-on joints are held together by restraints, including concrete thrust blocks, deadman blocks and rodding, and mechanical restraints.

3.3.1 Thrust Blocks

Water hammer occurs at every place that a pipeline changes direction; therefore, thrust blocks must be placed at all bends, tees, caps, valves, and hydrants. Thrust blocks are masses of concrete poured between the fittings and undisturbed soil. Thrust blocks brace the pipe joint to resist the force of thrust. They are wedge-shaped for efficient use of material. The widest part of the wedge is against a solid trench wall known as the *bearing area*. Because concrete has much higher strength than soil, only a small contact area is needed against the pipe. *Figure 29* shows a typical thrust block.

On smaller sizes of pipe, the concrete is mixed and placed fairly dry to facilitate easy shaping of the block into a wedge. For larger pipe sizes (12" or more) and greater pressures, the composition of the block should be specified by an engineer to ensure adequate strength.

Did You Know?

The phrase *fire plug* dates from the 1600s when water mains were made from hollowed out logs that ran beneath the cobblestone streets. In the event of a fire, the fire brigade would have to travel to the site, remove the cobbles, dig down to the main, and then bore a hole in the log. This flooded the excavation, which was then pumped out to the fire using hand pumps. Afterwards, a wooden plug was inserted to close the main, hence the term *fire plug*. This was a very time-consuming process that resulted in many losses. Following the Great London Fire of 1666, which destroyed more than 13,000 homes, the city of London installed new mains and plugs above ground level.

Source: www.firehydrant.org

Thrust Block Sizing

The size of a thrust block is determined by four factors:

- *Soil conditions* — The stability of the soil under a load varies with the kind of soil. Hard soils can take higher loads than soft soils. The safe bearing load of a particular soil is usually determined on a local basis by an engineer. There are soil bearing tests that can be used to do this accurately. *Table 2* shows the approximate safe bearing loads of different soils.
- *Pipe size* — The larger the pipe, the more force or thrust the moving water will have.
- *Type of fitting* — The type of fitting determines how much of the thrust will be intercepted and the direction in which it will be applied.
- *Maximum pressure* — The magnitude of the thrust is also affected by the pressure in the line. Higher pressures produce greater thrust.

Thrust at Fittings

Table 3 lists the approximate amount of thrust at fittings, in pounds. The figures are based on a water pressure of 100 psi.

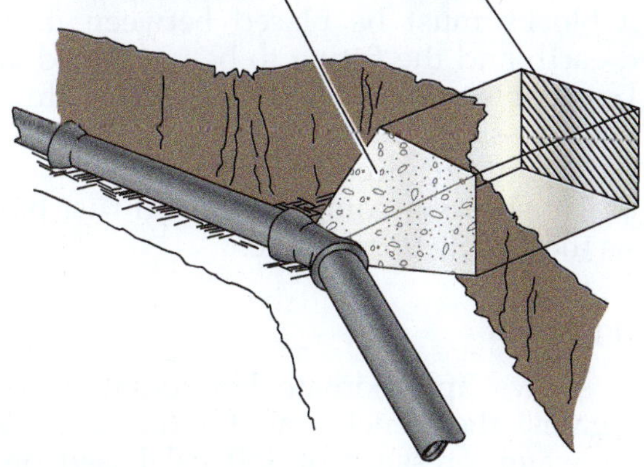

Figure 29 Typical thrust block.

Table 2 Horizontal Bearing Strengths [Data from NFPA 13, *Table A.6.6.1(c)*]

Type of Soil	Bearing Strength, S_b (lb/ft^2)
Muck	0
Soft Clay	1,000
Silt	1,500
Sandy Silt	3,000
Sand	4,000
Sandy Clay	6,000
Hard Clay	9,000

Note: Although the bearing strength values in this table have been used successfully in the design of thrust blocks and are considered to be conservative, their accuracy is totally dependent on accurate soil identification and evaluation. The ultimate responsibility for selecting the proper bearing strength of a particular soil type must rest with the design engineer.

Table 3 Approximate Thrust at Fittings at 100 psi [Data from NFPA 13, *Table A.6.6.1(a)*]

Nominal Pipe Diameter (in)	Total Pounds (lb)			
	Dead End	90° Elbow	45° Elbow	22½° Elbow
4	1,810	2,559	1,385	706
6	3,739	5,288	2,862	1,459
8	6,433	9,097	4,923	2,510
10	9,677	13,685	7,406	3,776
12	13,685	19,353	10,474	5,340

Note: To determine thrust at pressures other than 100 psi, multiply the thrust obtained in the table by the ratio of the pressure to 100 psi. For example, the thrust on a 12", 90-degree bend at 125 psi is 19,353 × 125/100 = 24,191 lb.

Using this information, thrust blocks can be calculated using a formula presented in NFPA 13 and NFPA 24. However, that calculation is normally performed by a sprinkler system layout technician.

NFPA 13 requires that thrust blocks consist of a concrete mix not leaner than one part cement, two and one-half parts sand, and five parts stone. Thrust blocks must be placed between undisturbed earth and the fitting to be restrained and must be capable of resisting the calculated thrust. Wherever possible, thrust blocks must be placed so that the joints are accessible for repair.

Figure 30 shows a comparison of thrust block areas as they relate to pipe size.

Bearing Areas

Table 4 shows approximate horizontal bearing areas against the trench wall for thrust blocks, using a water pressure of 100 psi based on a 90-degree horizontal bend and ductile iron pipe outside diameters, and a soil horizontal bearing strength of 1,000 lb/ft^2.

Thrust Block Locations

The location of thrust blocks is determined by the direction of thrust and the type of fitting being used. *Figure 31* shows thrust block locations.

- *Installation on inclines* — When pipe must be laid on an incline, it is very important to prevent any downward movement of the pipe. The weight of the pipe can cause the joints to loosen. The pipe must be anchored at the bottom of the incline and then at all changes in direction. Thrust blocks should also be placed every 48' or so on a straight run. In addition, pipe going up or down a slope may require a **deadman** and rodding. A deadman is simply a block of concrete in which tie rods are secured. When installing bell-and-spigot pipe on an incline, the bell must always face uphill. *Figure 32* shows a concrete block deadman and rodding.

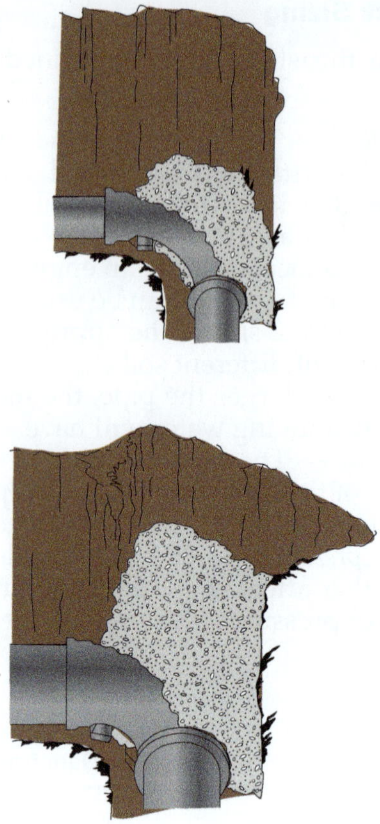

Figure 30 Thrust block areas for different pipe sizes.

Table 4 Required Horizontal Bearing Block Area [Data from NFPA 13, ***Table A.6.6.1(b)***]

Nominal Pipe Diameter (in)	Bearing Block Area (ft^2)
3	2.6
4	3.8
6	7.9
8	13.6
10	20.5
12	29.0

Note: Values listed are based on a 90-degree horizontal bend, an internal pressure of 100 psi, a soil horizontal bearing strength of 1,000 lb/ft^2, a safety factor of 1.5, and ductile iron pipe outside diameters.

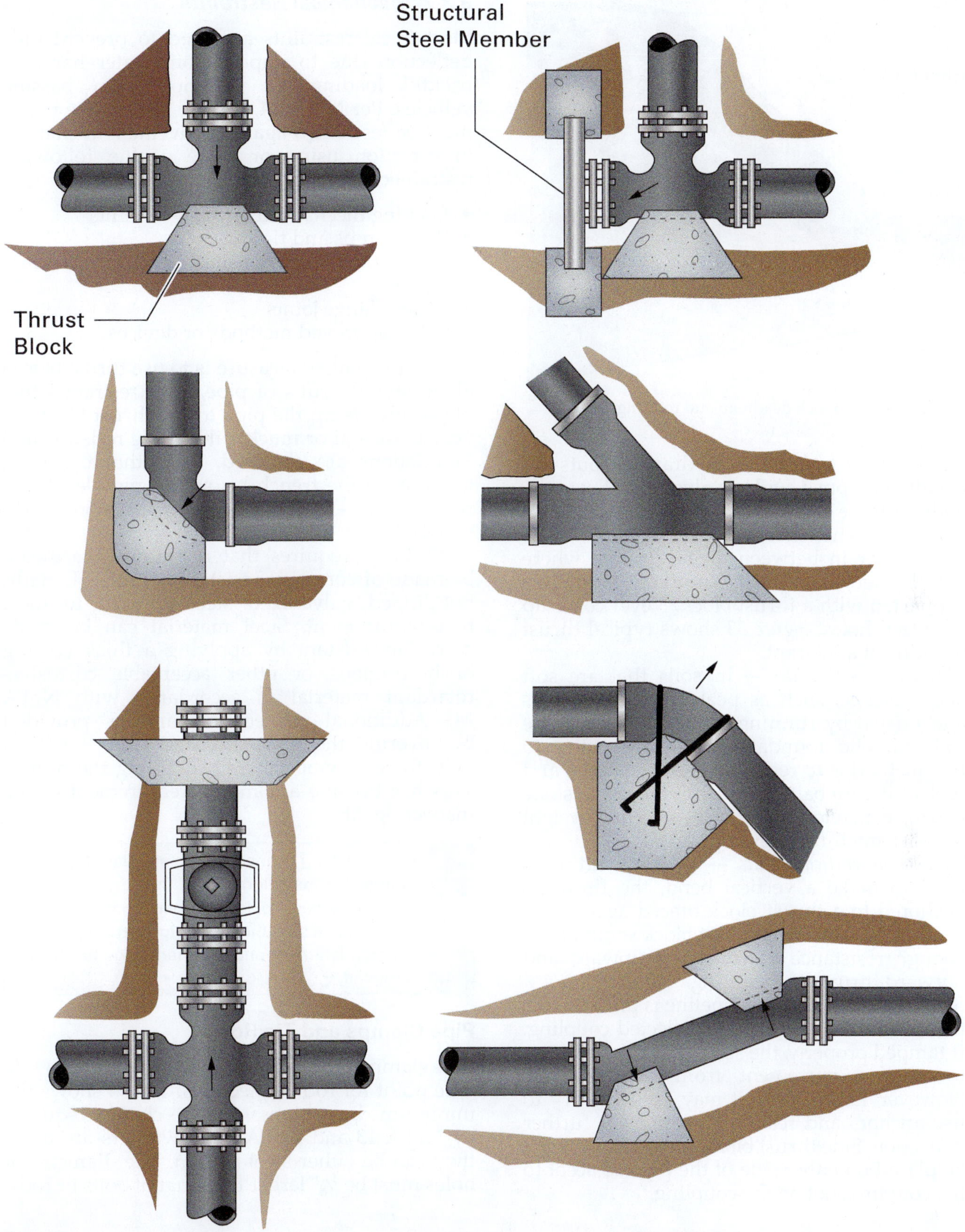

Figure 31 Thrust block locations.

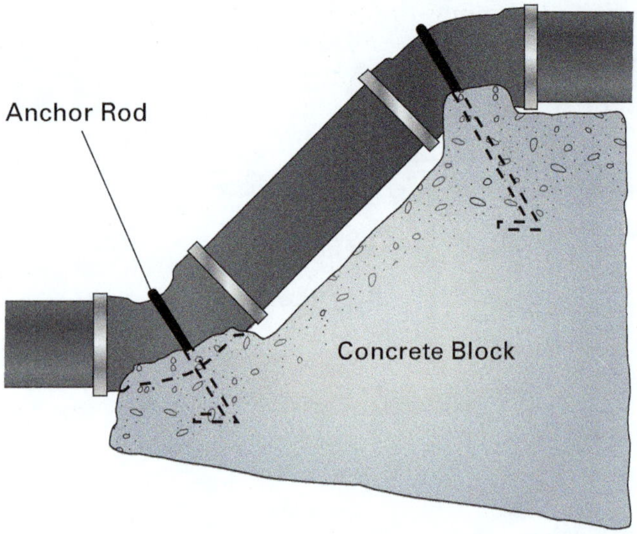

Figure 32 Concrete block deadman and rodding.

- *Installation at hydrants* — Hydrant runouts are usually equipped with a valve that allows the hydrant to be serviced without shutting down the supply. The valve can be at the main, at the hydrant, or in between. Regardless of where it is located, each fitting must be separately supported with a thrust block. Never cover up the weep holes. *Figure 33* shows typical thrust blocking at a hydrant.
- *Thrusts in soft soils* — In soils that are soft and unstable, such as peat or muck, thrusts are resisted by running corrosion-resistant tie rods to solid foundations. An alternative to this method is to remove the soft material and replace it with ballast, which is a mass of stone and/or concrete, of sufficient size and weight to resist any thrust developed.
- *Thrusts in vertical bends* — Where a fitting is used to make a vertical bend, the fitting is anchored to a thrust block braced against the undisturbed soil. The thrust block should have enough resistance to withstand upward and outward thrusts at the fitting.
- *Side thrust* — When the pipeline is pressurized, there is side thrust at each deflected coupling. If tamped properly, the soil is usually sufficient to prevent movement from side to side. However, in soft soils it may be necessary to use anchors and restraints to prevent further deflection. Side thrust blocks or restraints must be placed on either side of the pipe adjacent to the coupling, not on the coupling.

3.3.2 Mechanical Restraints

Mechanical restraints are used to prevent pipe deflection due to impact from water hammer, backfill loading, or vibration from passing vehicles. Per NFPA 24, private fire service mains must be restrained against movement at changes in direction using or more of the following restrained-joint systems:

- Locking mechanical or push-on joints
- Pipe clamps and tie rods
- Mechanical joints using setscrew retainer glands
- Bolted flange joints
- Other approved methods or devices

One preventive measure is to use thrust blocks along straight runs of pipe, with restraints that physically clamp the pipe to the thrust block. In very loose soil or muck, piles or tie rods to solid foundations are required. The other option is to excavate the trench below the pipe level and replace the removed material with more stable material.

NFPA 24 requires that restraining hardware be made of corrosion-resistant material. Usually, hot-dipped galvanized steel is used to meet this requirement. Steel material can be made corrosion-resistant by applying a thick coating of bituminous or other acceptable corrosive-retardant material in accordance with NFPA 24. Additional protection can be provided by covering the pipe, rod, and clamps with a polyethylene wrap before backfilling the trench. This has become a common requirement of the inspecting AHJ.

NOTE

When making connections to standard pipe fittings, cut the spigot end of the pipe to remove the factory bevel before connecting it to the fitting. Make sure the pipe spigot end is cut square, deburred, and the sharp edge has been removed.

Pipe Clamps and Tie Rods

Pipe clamps are used to grip the pipe and provide a tie point for rod connections. *Table 5* shows the minimum dimensions for pipe clamps required by NFPA 13 and NFPA 24. If washers are used, they can be either steel or iron. The diameter of holes must be $\frac{1}{8}$" larger than that of bolts or rods.

Figure 33 Thrust block at a hydrant.

Figure 34 shows pipe clamps used with tie rods. The tie rods connect to the pipe clamps on one end and special lugs with eye bolts attached to the fitting on the opposite end. The lugs and eye bolts are used to offset the threaded rod from the mechanical joint to align with the pipe clamp.

Tie rods allow for balanced force on both sides of the clamp. NFPA 24 states that the minimum rod size is $\frac{5}{8}$". Tie rods must be corrosion resistant and meet minimum size criteria. NFPA 24 further states that if clamps are used, rods must be used in pairs, two to a clamp.

Mechanical Joints Using Setscrew Retainer Glands

Mechanical joints are made between a plain pipe end and a gasketed fitting using setscrew retainer glands. The joint consists of multiple gripping wedges incorporated into a ductile iron follower gland, a rubber gasket, and necessary bolts and nuts (*Figure 35*).

Table 5 Minimum Dimensions for Pipe Clamps

Nominal Pipe Diameter	Clamp Dimensions	Bolt Size
4	$\frac{1}{2} \times 2$	$\frac{5}{8}$
6	$\frac{1}{2} \times 2$	$\frac{5}{8}$
8	$\frac{5}{8} \times 2\frac{1}{2}$	$\frac{5}{8}$
10	$\frac{5}{8} \times 2\frac{1}{2}$	$\frac{3}{4}$
12	$\frac{5}{8} \times 3$	$\frac{7}{8}$

Note: All dimensions in inches.

The following procedure is for the mechanical joint shown in *Figure 35*. Other mechanical joints are assembled in a similar way. Always follow the manufacturer's instructions for the product in use.

Step 1 Clean the gasket and the pipe end with soapy water, then brush with an approved pipe lubricant.

Step 2 Place the ductile iron gland on the spigot end followed by the gasket. Press the gasket into position, ensuring that it is evenly seated around the joint.

Step 3 Insert the pipe into the fitting socket and press the gasket firmly and evenly into the gasket recess. Maintain straight alignment during the assembly.

Step 4 Insert the bolts and tighten them finger-tight.

Step 5 Tighten all bolts in alternating order to the torque recommended by the fitting manufacturer.

Step 6 Tighten the torque-limiting twist-off nuts clockwise until the wedges are all in firm contact with the pipe surface. Continue tightening in an alternating manner until all of the nuts have been twisted off.

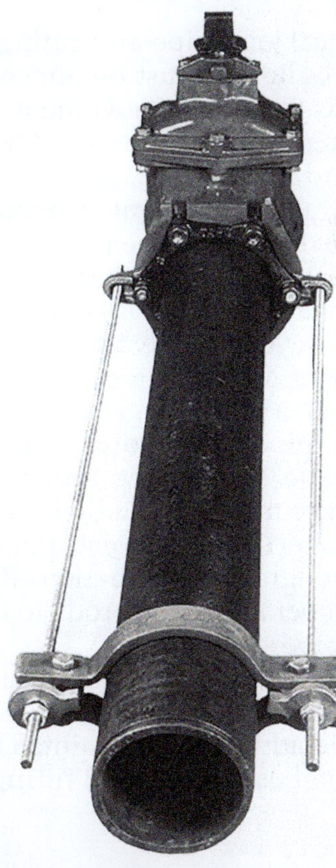

Figure 34 Pipe clamps and tie rods.

Figure 35 Mechanical joint in ductile iron pipe.

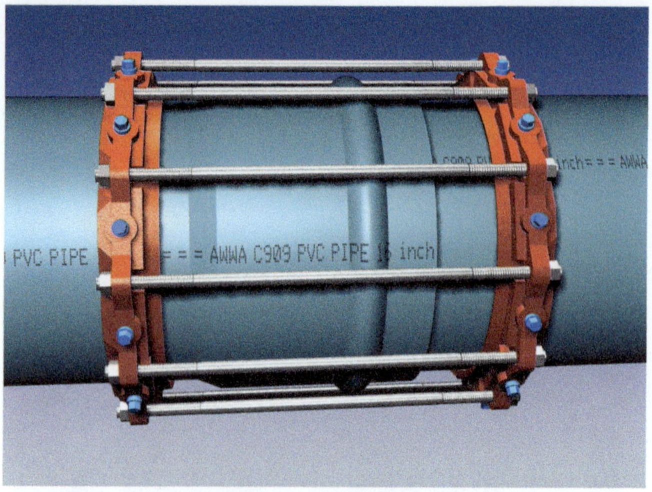

Figure 36 Wedge restraint on PVC pipe.

After assembly, all bolted metal parts must be cleaned and coated with bituminous or other acceptable corrosive-retardant material in accordance with NFPA 24.

A similar type of wedge restraint can also be used to brace straight sections of either ductile iron or PVC (*Figure 36*). In this case, the wedges in the follower gland bite into and displace the pipe material to hold it securely in place.

Flange Spigot Piece Restraint

The mechanical joint pipe-and-fitting fire-line entry into the building must be anchored with restraining rods to prevent movement. The number of rods is determined by the size of the pipe being restrained and its diameter. The use of corrosion-resistant restraining hardware is required.

The rods extend from the interior flange to the base ell beneath the flange spigot piece, and/or to an anchor (one or two underground pipe clamps as required) beyond the exterior of the building grade beam or wall. In the case of horizontal penetration of a wall or grade beam, adequate clearance of pipe and rods must be provided through a pipe sleeve. The space between the pipe and sleeve must be filled with a waterproofing filler material to prevent flooding of the building through the penetration. Refer to *Table 6* for the number and size of rods to be used with the various pipe sizes.

In the event the rods become loose or broken, a concrete thrust block may be installed at the base ell to provide additional restraint. This will prevent separation of the pipe and fitting. The use of retaining glands with mechanical joint fittings in lieu of rods and clamps may be required by some AHJs.

3.4.0 Corrosion Protection of Pipe and Fittings

Ductile iron pipe and fittings may require protection from corrosive soil conditions. Two factors must be considered when selecting corrosion protection: chemical attack and electrolysis.

A major factor in corrosion is the chemical attack that can take place with highly active soils. These are usually alkaline soils, but this can also occur with highly acidic soils, or soils with a high salt content like those found in coastal areas. A second factor in corrosion is electrical attack by earth currents (electrolysis). This is a galvanic action similar to electroplating and is not necessarily related to the chemistry of the soil. In either case, the basic goal is to provide a separation between the metal parts and the soil, so any chemical action is blocked, and electrical current does not flow.

Corrosion protection can be provided by selecting zinc-coated piping and/or encasing the piping and fittings in polyethylene tubing (*Figure 37*). Encasing the pipe and fittings in polyethylene not only blocks chemical action, but also blocks the stray electrical currents at most levels encountered in the field. Eight mil polyethylene tube is available for this purpose. Material and installation procedures are specified in ANSI/AWWA C105/A21.5, *Polyethylene Encasement for Ductile-Iron Pipe Systems*.

Table 6 Number and Size of Rods (Data from NFPA 13, *Table 6.6.2.1.2.2*)

Nominal Pipe Size	Rod Number — Diameter Combination			
	$\frac{5}{8}$	$\frac{3}{4}$	$\frac{7}{8}$	1
4	2	–	–	–
6	2	–	–	–
8	3	2	–	–
10	4	3	2	–
12	6	4	3	2

Note: All dimensions in inches. This table has been derived using a pressure of 225 psi and design stress of 25,000 psi.

Figure 37 Pipe encased in polyethylene tubing.

Cathodic protection is a means of preventing corrosion in underground piping where the earth's electrical currents have an effect in certain soil conditions. This is often done by using metal straps that bridge across gasketed joints and fittings in order to provide continuity along the entire length of the underground piping. An anode is bonded to the pipe in a convenient location. Wiring is then run from the anode to an accessible location where the current dissipates.

Bolts and nuts are particularly prone to corrosion. These fasteners should be completely covered with bituminous mastic to avoid any moisture contact. Highly active soil can destroy fasteners in six months or less.

3.5.0 Installing Tracer Wire

Some jobs and/or jurisdictions require that nonmetallic underground piping be equipped with a conductive locating wire, also known as *tracer wire*, as shown in *Figure 38*. This allows the pipe to be located using a receiver if a leak occurs or an excavation is required for other purposes.

Tracer wires must have a jacket approved for direct burial and are typically colored blue for potable water connections. The wire is affixed to the top of the pipe and secured at specified intervals per the shop drawings (typically every 5' to 10'). The tracer wire must be continuous. Where splicing is required, only Listed waterproof connectors may be used.

Figure 38 Pipeline tracer wire.

> **CAUTION**
>
> Do not install tracer wire on metal piping or components as it can result in galvanic corrosion over time.

Each end of the tracer wire is attached to a ground rod. When installing tracer wire, include an additional 6' or more coiled at the end of the run to allow for future connections. Each end of the tracer wire is attached to a ground rod. Be sure to protect the wire from damage when backfilling and install warning tape about a foot above the traced pipeline. This will serve as an indicator to future excavators that the pipeline has been located and only hand digging should be used beyond that point.

3.6.0 Backfilling

Adequate fill material must be provided under the pipe as well as over it. The spoil bank is the main source for backfill material, but loss through excavation may require additional material. The backfill is laid around the pipe in such a manner that it supports the pipe evenly, reducing stresses that may later cause pipe damage. The bedding, side fill, and cover should provide a uniform, even blanket around the pipe.

Before backfilling, skids must be removed. This is only done while other support is being provided under the pipe. In rocky, rough soils, padding of earth, sandbags, or other approved supports may be laid under the pipe. Synthetic materials are often used to provide a padded surface on the bottom of the ditch.

When filling is completed, the surface should be restored as closely as possible to its original condition, including the replacement of topsoil.

> **NOTE**
>
> In some jurisdictions, the AHJ may require inspection before a trench is backfilled. Note that the AHJ responsible for inspecting bedding, backfilling, and compaction may not be the Fire Department. It could be the Water Authority or Public Works Department.

3.6.1 Initial Backfill

After laying the pipe on the bedding, the initial backfill is added. This protects the pipe by cradling it in compacted soil to prevent movement, settling, or floating, and to protect it from any surface loads. In certain geographic areas, and especially where the municipal utility piping can be directly affected, the municipality determines the material to be used for the initial backfill. Always follow the shop drawings and local requirements when installing backfill. The pipe laying conditions identify the type and thickness of material upon which the pipe rests.

Figure 39 identifies standard laying conditions for underground ductile iron pipe, as defined in ANSI/AWWA C150/A21.50, *American National Standard for the Thickness Design of Ductile Iron Pipe*:

- *Type 1* — Type 1 is a flat-bottomed trench, which means that the soil on which the pipe rests is undisturbed (not backfilled). This method is not recommended when laying pipe larger than 30" in diameter.
- *Type 2* — In Type 2, the pipe rests on undisturbed, flat-bottomed ground and is surrounded to its center line with backfill that is lightly tamped.
- *Type 3* — Type 3 shows the pipe laying on at least 4" of loose soil that is leveled to a given grade. Backfill is then placed to the top of the pipe and lightly tamped.
- *Type 4* — In Type 4, the pipe is placed on at least 4" of sand, gravel, or crushed stone, depending on the specifications. Backfill is placed to the top of the pipe and compacted to about 80 percent of the density of the undisturbed soil.
- *Type 5* — In Type 5, the pipe is placed on a bed of at least 4" of compacted granular material. The material is compacted to the center line of the pipe and then backfilled to the top of the pipe. Backfill should be granular material or soil from the original excavation that is free of rocks, frozen earth, or other foreign matter. The backfill is compacted to 90 percent of the undisturbed soil.

Laying Condition	Description
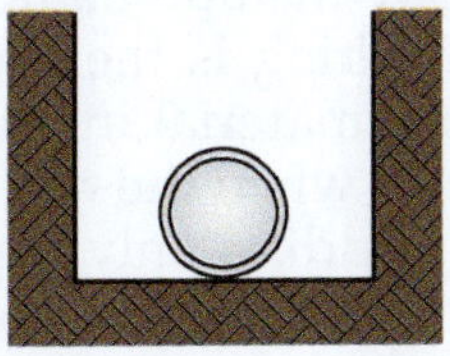 Type 1	Flat-bottom trench. Loose backfill.
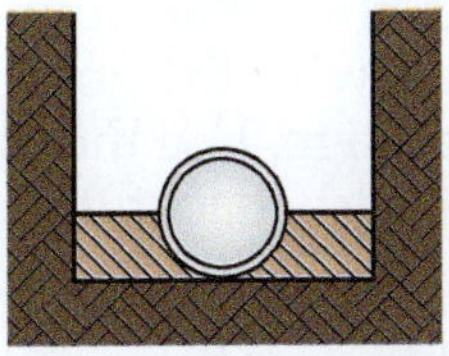 Type 2	Flat-bottom trench. Backfill lightly. Consolidated to center line of pipe.
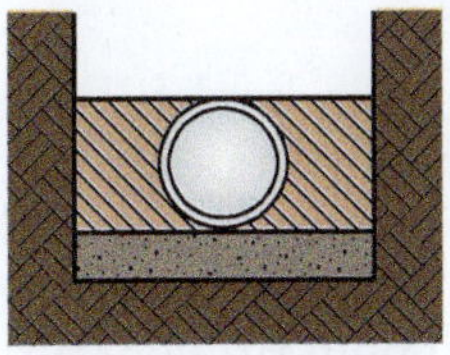 Type 3	Pipe bedded in 4" minimum loose soil or bedding material. Backfill lightly consolidated to top of pipe.
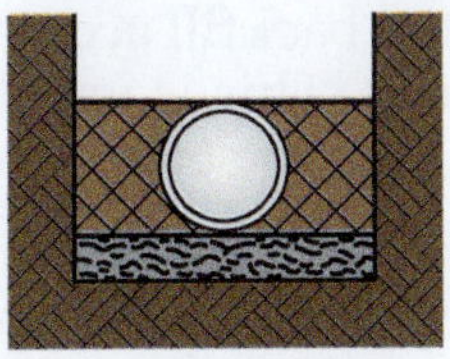 Type 4	Pipe bedded in sand, gravel, or crushed stone to depth of 1/8 pipe diameter, 4" minimum. Backfill compacted to top of pipe.
 Type 5	Pipe bedded to its center line in compacted granular material, 4" minimum under pipe. Compacted granular or select material to top of pipe.

NOTE: These options vary with job specifications and locations.

Figure 39 Different pipe laying conditions.

The defined laying condition is influenced by depth of burial, soil stability, and whether or not the piping run is under paved areas. Specific compaction requirements may also be included.

Buried piping must be located below the frost line to prevent freezing during the winter months. The piping must also be located deep enough to be protected from surface loads that could cause mechanical damage, including driveways, roads, and railroad tracks. Laying condition requirements directly affect the maximum depth of cover. Other factors include the size and type of pipe, wall thickness, and type of soil. *Figure 40* shows the effect of the laying condition on the maximum depth of cover for 12" ductile iron pipe with cement lining. Note that pipe that is fully embedded in a compacted granular material (Type 5) can withstand burial to more than twice the depth of pipe laying on a flat-bottom trench (Type 1).

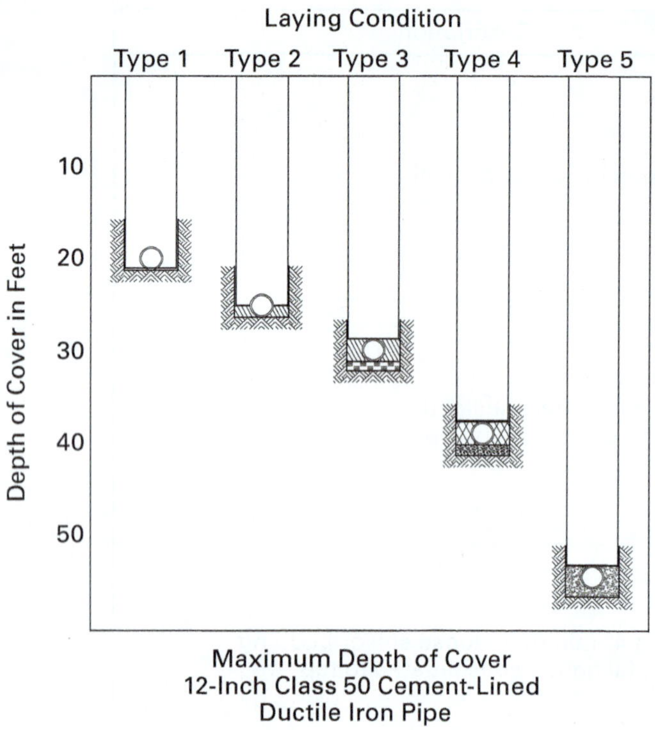

Figure 40 Effect of laying condition on maximum depth of cover.

3.6.2 Preparing Fill

When installing backfill, use only clean fill material. Any rocks, stumps, scrap, and foreign objects left in the ditch or in the backfill material are potential hazards to the piping. Foreign material such as cans, tools, clamps, or scrap metal can weaken corrosion protection or damage the pipe in that area.

Fill dirt, particularly that used for bedding and close to the pipe, should be sifted to remove rocks, large chunks, and other debris. The spoil from a new ditch made through rock cannot be used as fill because of the amount of rock debris. In addition, frozen soil can be as hard as rock and pose a hazard to the pipe. It may be used if it can be broken up sufficiently; however, frozen soil is difficult to distribute evenly and may result in areas of unsupported pipe or uneven fill after it thaws. If the spoil is unusable, new fill dirt must be brought in to pad and fill most of the ditch. Sometimes rocky soil may be used in the upper part of the ditch, as long as the rocks are small, well away from the pipe, and only used up to about 1' from the surface. The final layer must be clean fill dirt or topsoil.

3.6.3 Steps for Backfilling

Initial backfill may include **haunching**, tamping, and filling. Haunching is the process of placing and consolidating material in areas that may remain void or loose when the rest of the backfill is added. Because bedding only supports a narrow strip along the bottom of the pipe, sand or dirt is packed in along the bottom sides or haunches of the pipe. This material is compacted with a hand tamp or similar tool. Backfilling a trench is a delicate operation with the goal of protecting the pipeline from damage. Use the following steps as a checklist to properly backfill a trench:

Step 1 Perform a visual inspection of backfill material to identify and remove any material or objects that could damage the pipeline.

Step 2 Use suitable backfill material within 6" of the pipe, or per company policy. Fill equally along the haunches and up both sides of the pipe until the pipe is covered. This will help prevent pipe movement and provide additional support for the pipe.

Step 3 Crown the backfill according to company procedures. This is usually performed to compensate for settlement of backfill.

Step 4 Compact the soil as appropriate for proper pipe support, size of support, and spacing of support. Use a tamping tool or other appropriate tool. Be sure that you know when the soil is properly compacted.

NOTE

Provide extra cover and compaction at road crossings or other locations with heavy surface loads. This provides extra protection for pipe. Less settlement also means there is less chance of third-party damage.

3.6.4 Performing Soil Compaction

After the water line has passed inspection, the final backfilling and compaction can begin. Some municipalities enforce compaction measurements on the backfill. The replaced soil needs to be compacted to avoid a depression caused by later settlement and because loose earth is subject to erosion by rainwater. By consolidating the soil particles in the fill, compaction also strengthens the pipeline cover and increases the protection of the blanket of fill around the pipe.

Compaction can be accomplished by having a trained, qualified operator drive a heavy crawler-type tractor over the backfilled ditch. Extra cover is often used at road crossings, ditches, and streams. Extra compaction is allowed at road crossings and other locations where heavy loads are expected to cross the pipeline.

Backfill is normally crowned after compaction to allow for some settlement. **Ditch breakers** (sandbags) may be required on the slope of a ditch, diversion level, or surface to prevent backfill from washing away.

Generally, there is a specification on achieving 90–95 percent of the maximum attainable density in the soil. A typical municipality might require a minimum of 90 percent compaction under all circumstances, with a minimum compaction of 95 percent under streets. This means 90–95 percent of the volume is soil or sand and 5–10 percent is air. Obviously, coarse material will be more difficult than fine material to compact to these high levels.

The compaction process begins with the first lift or layer placed and tamped in the trench. Each lift should be 8" to 12" deep. Most trenches are tamped using vibratory compacting equipment or tamping rammers. *Figure 41* shows a tamping rammer, also referred to as a *jumping jack*.

After compacting the first lift, the contractor requests a soil test from an ASTM-approved testing laboratory. The laboratory tests the soil moisture content and issues a **proctor curve**. A proctor curve is a moisture density curve. This determines what, if anything, needs to be added to the soil in order to meet the compaction criteria that may be required on the job. If the soil from the first test passes inspection, subsequent lifts can then be added. Tests are made for each lift.

3.7.0 Testing, Inspection, and Flushing

Many installations require the sprinkler fitter to perform testing, inspection, flushing, and chlorinating of underground pipe system installations.

> **NOTE**
> It is a good practice to test underground water lines as they are being installed. It is much easier to correct errors while the work is in process than it is to discover them after the entire system has been installed.

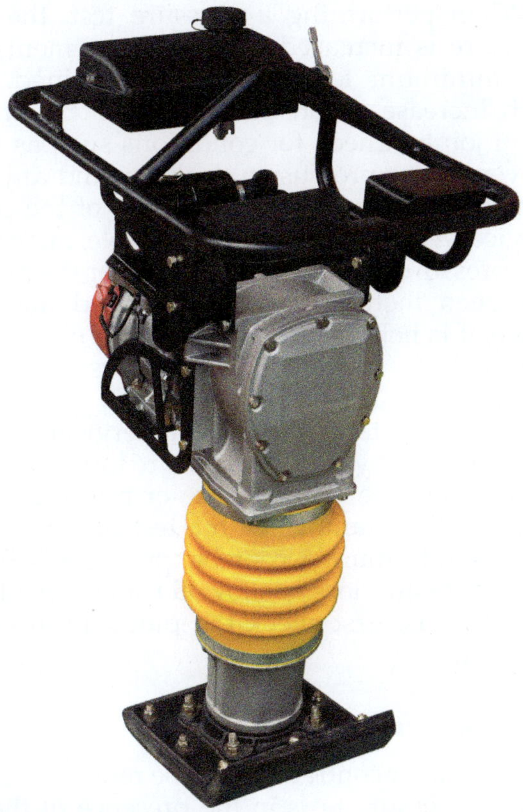

Figure 41 Tamping rammer.

3.7.1 Testing

A properly conducted pressure test should detect any leaks, loose fittings, or damaged pipe. NFPA 24 requires that water service piping tests be performed prior to covering or concealment. The trench must be backfilled between joints before testing to prevent the pipe from moving during testing, but the joints must remain exposed so that any leaks can be detected. All thrust blocks and restraints must be in place prior to pressure testing. Thrust blocks must be allowed to cure for at least 24 hours before the test.

Testing is done by closing all valves and openings and pressurizing the line with water. Additional pressure beyond that available in the local system must be supplied by means of a hydrostatic pump equipped with a pressure gauge. Testing equipment must be maintained properly and should be drained after each use.

Test pressures must be at least 50 psi greater than the normal maximum supply pressure. The minimum allowable test pressure is 200 psi. Water should be used unless the weather will not allow a water test without freezing.

When performing a pressure test, the water pressure is increased in 50-psi increments until the minimum test pressure is attained. After each increase in pressure, observe the stability of all joints. Check for conditions such as gasket movement or protrusion, leakage, and any other condition likely to impact the use of the pipe in service. When all joints appear stable, increase the test pressure by another 50 psi. After the pressure has been increased to the required maximum value, it is held for two hours while observations are made for leakage and the pressure readings checked.

Acceptable test results for underground piping are indicated by either a pressure loss of less than 5 psi below the test pressure or no visible water leakage. For example, with a test pressure of 200 psi, the minimum acceptable pressure is 195 psi. If the pressure loss is above 5 psi and the leak is not obvious, first check the pump itself and its connections.

3.7.2 Inspection

The sprinkler contractor may be required to make formal pressure tests in the presence of the AHJ or the owner's representative. Before these tests are made for official acceptance, tests should be made privately to make sure the installation will pass when the formal test takes place. The contractor will be responsible for completing all required acceptance tests and, subsequently, the Contractor's Material and Test Certificate for Underground Piping (*Figure 42A* and *Figure 42B*). Witnesses for both the contractor and the property owner must sign the completed test certificate.

As shown in the test description in *Figure 42A*, NFPA 13 provides a calculation for determining the hydrostatic testing allowance during the inspection. This calculation may be applied in large or complex installations and takes into account additional system criteria. It is typically performed by a layout technician, if required. Most inspections rely on the 5-psi leakage rate to determine acceptance.

3.7.3 Flushing

After the underground pipe has been tested and inspected, the entire system must be thoroughly flushed before being connected to the sprinkler piping and the process documented. If the system is flushed by others, it is important to obtain documentation. If the proper documentation is not obtained in a timely manner, the flushing will have to be repeated for that purpose.

Flushing is made easy if the inside of the piping is kept clear of foreign matter as the pipe is being laid, and if the open end is kept closed when not attended. Flushing is done under supply pressure through the flanged spigot, hydrants, or other outlets. Care must be taken to section off connected sprinkler or standpipe equipment using valves or blind flanges to avoid lodging foreign material in these systems.

The system must be flushed until all foreign material is ejected and the water leaving the system runs clear. Mains supplying sprinkler systems must be flushed at the minimum rates specified by NFPA 24, or at the hydraulically calculated water demand rate of the system, whichever is greater. *Table 7* shows the flow required to produce a velocity of 10' per second as given in NFPA 13. If the flow rate cannot be verified or met, refer to NFPA 24 for exceptions.

CAUTION

Flushing involves a large release of water. Make certain that it is properly diverted to avoid damage to landscaping and other property.

If the underground pipe supplies a fire pump system, the required velocity is 15' per second per NFPA 20, *Standard for the Installation of Stationary Pumps for Fire Protection* (*Table 8*). Per NFPA 20, flushing of a fire pump system must occur before hydrostatic testing.

CAUTION

Do not proceed unless the flushing has been completed, approved, and documented. Failure to do so can result in having to completely disassemble the entire system.

3.7.4 Chlorinating

The water mains of municipalities supply potable water. The term *potable* water means water that is suitable for drinking, cooking, and domestic purposes. To be potable, it must meet the requirements of the Water or Health Authorities.

Local ordinances and/or specifications may require chlorinating of underground systems by the sprinkler contractor. In those instances, the potable water supply lines must be disinfected (sterilized) by chlorinating after passing the pressure test. The local Water Authority determines the requirements for chlorination.

Contractor's Material and Test Certificate for Underground Piping

PROCEDURE

Upon completion of work, inspection and tests shall be made by the contractor's representative and witnessed by an owner's representative. All defects shall be corrected and system left in service before contractor's personnel finally leave the job.

A certificate shall be filled out and signed by both representatives. Copies shall be prepared for approving authorities, owners, and contractor. It is understood the owner's representative's signature in no way prejudices any claim against contractor for faulty material, poor workmanship, or failure to comply with approving authority's requirements or local ordinances.

Property name	Date
Property address	

Plans	Accepted by approving authorities (names)		
	Address		
	Installation conforms to accepted plans	❑ Yes	❑ No
	Equipment used is approved If no, state deviations	❑ Yes	❑ No
Instructions	Has person in charge of fire equipment been instructed as to location of control valves and care and maintenance of this new equipment? If no, explain	❑ Yes	❑ No
	Have copies of appropriate instructions and care and maintenance charts been provided to the owner or owner's representative? If no, explain	❑ Yes	❑ No
Location	Supplies buildings		
Underground pipes and joints	Pipe types and class / Type joint		
	Pipe conforms to ____________ standard	❑ Yes	❑ No
	Fittings conform to ____________ standard If no, explain	❑ Yes	❑ No
	Joints needing anchorage clamped, strapped, or blocked in accordance with ____________ standard If no, explain	❑ Yes	❑ No

Test description

Flushing: Flow the required rate until water is verified to be clear of debris at outlets such as hydrants and blow-offs. Flush at one of the flow rates as specified in 6.10.2.1.3.

Hydrostatic: All piping and attached appurtenances subjected to system working pressure shall be hydrostatically tested at 200 psi (13.8 bar) or 50 psi (3.4 bar) in excess of the system working pressure, whichever is greater, and shall maintain that pressure ±5 psi (0.34 bar) for 2 hours.

Hydrostatic Testing Allowance: Where additional water is added to the system to maintain the test pressures required by 6.10.2.2.1, the amount of water shall be measured and shall not exceed the limits of the following equation (for metric equation, see 6.10.2.2.6):

$$L = \frac{SD\sqrt{P}}{148{,}000}$$

L = testing allowance (makeup water), in gallons per hour (lpm)
S = length of pipe tested, in feet (m)
D = nominal diameter of the pipe, in inches (mm)
P = average test pressure during the hydrostatic test, in pounds per square inch (gauge) (bar)

Flushing tests

New underground piping flushed according to ____________ standard by (company) If no, explain	❑ Yes	❑ No
How flushing flow was obtained: ❑ Public water ❑ Tank or reservoir ❑ Fire pump	Through what type opening: ❑ Hydrant butt	❑ Open pipe
Lead-ins flushed according to ____________ standard by (company) If no, explain	❑ Yes	❑ No
How flushing flow was obtained: ❑ Public water ❑ Tank or reservoir ❑ Fire pump	Through what type opening: ❑ Y connection to flange and spigot	❑ Open pipe

 NFPA 13 (p. 1 of 2)

Figure 42A Contractor's Material and Test Certificate for Underground Piping (NFPA 13, *Figure 6.10.1*) (1 of 2).

Hydrostatic test	All new underground piping hydrostatically tested at ______ psi (bar) for ______ hours		Joints covered ❑ Yes ❑ No
Leakage test	Total amount of leakage measured ______ gallons (liters) ______ hours		
	Allowable leakage ______ gallons (liters) ______ hours		
Forward flow test of backflow preventer	Forward flow test performed in accordance with 6.10.2.5.2:		❑ Yes ❑ No
Hydrants	Number installed	Type and make	All operate satisfactorily ❑ Yes ❑ No
Control valves	Water control valves left wide open If no, state reason		❑ Yes ❑ No
	Hose threads of fire department connections and hydrants interchangeable with those of fire department answering alarm		❑ Yes ❑ No
Remarks	Date left in service		
Signatures	Name of installing contractor		
	Tests witnessed by		
	For property owner (signed)	Title	Date
	For installing contractor (signed)	Title	Date

Additional explanation and notes

NFPA 13 (p. 2 of 2)

Figure 42B Contractor's Material and Test Certificate for Underground Piping (NFPA 13, ***Figure 6.10.1***) (2 of 2).

Table 7 Minimum Flow Rates for Flushing Underground Piping (Data from NFPA 13, *Table 6.10.2.1.3*)

Nominal Pipe Size (in)	Flow Rate (gpm)
2	100
$2\frac{1}{2}$	150
3	220
4	390
6	880
8	1,560
10	2,440
12	3,520

Table 8 Minimum Flow Rates for Flushing Suction Piping (Data from NFPA 20, *Table 14.1.1.1*)

Nominal Pipe Size (in)	Flow Rate (gpm)
2	150
$2\frac{1}{2}$	229
3	330
4	590
6	1,360
8	2,350
10	3,670
12	5,290

3.0.0 Section Review

1. Which of the following is *most* likely to be used as a bedding material?
   a. Mulch
   b. Muck
   c. Sand
   d. Straw
2. The same tools are used to cut both ductile iron pipe and PVC pipe.
   a. True
   b. False
3. The horizontal bearing strength of hard clay is _____.
   a. 3,000 lb/ft^2
   b. 4,000 lb/ft^2
   c. 6,000 lb/ft^2
   d. 9,000 lb/ft^2
4. The underground piping components at the greatest risk of corrosion are the _____.
   a. fasteners
   b. fittings
   c. spigot ends
   d. bell ends
5. Each end of a tracer wire is attached to a(n) _____.
   a. supervisory switch
   b. ground rod
   c. amplifier
   d. pressure switch
6. Tests are made for each lift in the compaction process (every 8" to 12").
   a. True
   b. False
7. The *minimum* flow rate for flushing 6" underground piping is _____.
   a. 390 gpm
   b. 880 gpm
   c. 1,560 gpm
   d. 2,440 gpm

Review Questions

1. A device used to protect against cross-connections in the potable water supply is a _____.
   a. PIV
   b. backflow preventer assembly
   c. gate valve
   d. supervisory switch

2. Which of the following devices is *most* likely to be used in a residential application?
   a. Reduced-pressure zone (RPZ) backflow preventer assembly
   b. Double-check valve backflow preventer assembly (DCV)
   c. Dual-check valve backflow preventer
   d. Double-check detector backflow preventer assembly

3. The type of hydrant *most* likely to be used in areas with freezing temperatures is a _____.
   a. dry barrel hydrant
   b. California hydrant
   c. wet barrel hydrant
   d. bonnet valve hydrant

Figure RQ01

4. Refer to the hydrant in *Figure RQ01*. Assuming that it follows standard color codes, determine whether it is a public or private hydrant and its available flow rate.
   a. Private with a flow rate of less than 500 gpm
   b. Private with a flow rate of 500 to 1,499 gpm
   c. Public with a flow rate of less than 500 gpm
   d. Public with a flow rate of over 1,500 gpm

5. Average soils are usually a mixture of _____.
   a. sand, iron ore, and water
   b. sand, silt, and clay
   c. sand, bark, and lime
   d. clay, silt, and water

6. In any trench that is over 4' deep, ladders must be placed every _____.
   a. 15'
   b. 25'
   c. 50'
   d. 75'

7. A cubic yard of soil weighs _____.
   a. 300 pounds
   b. 500 pounds
   c. 1,000 pounds
   d. 3,000 pounds

8. For ductile iron pipe 12" and smaller, no cuts should be made within _____.
   a. 24" of the bell
   b. 30" of the bell
   c. 36" of the bell
   d. 42" of the bell

9. For the seismic protection of sleeved pipe 4" and larger, the required clearance between the sleeve and the pipe is _____.
   a. 1"
   b. 2"
   c. 3"
   d. 4"

10. A 2" pipe would require a sleeve with a diameter of _____.
   a. 1"
   b. 2"
   c. 3"
   d. 4"

11. The horizontal bearing strength of silt is _____.
   a. 1,000 lb/ft^2
   b. 1,500 lb/ft^2
   c. 3,000 lb/ft^2
   d. 4,000 lb/ft^2

12. How many pounds of thrust is developed on an 8" 90-degree elbow?
   a. 9,097 pounds
   b. 10,474 pounds
   c. 13,685 pounds
   d. 19,353 pounds

13. Thrust block calculations are typically done by the _____.
   a. sprinkler fitter
   b. shop foreman
   c. layout technician
   d. AHJ

14. The spoil from a new ditch *cannot* be used as fill if it is cut through _____.
   a. rock
   b. sand
   c. silt
   d. loam

15. The *minimum* flow rate for flushing 8" suction piping on a fire pump system is _____.
   a. 1,360 gpm
   b. 2,350 gpm
   c. 3,670 gpm
   d. 5,290 gpm

Trade Terms Quiz

Fill in the blank with the correct term learned from your study of this module.

1. A(n) _______________ is a narrow excavation (relative to its length) made below the surface of the ground.
2. A method of protecting employees working in excavations from cave-ins by cutting the excavation walls to a specific angle at which the soil will rest without sliding is called _______________.
3. A(n) _______________ uses two spring-loaded check valves plus a hydraulic spring-loaded pressure differential relief valve.
4. A direct link between the potable water supply and water of unknown or questionable quality is a(n) _______________.
5. Horizontal members of a shoring system placed parallel to the excavation face whose sides bear against the vertical members of the shoring system are referred to as _______________.
6. _______________ is a structure such as a metal, hydraulic, mechanical, or timber system that supports the sides of an excavation and is designed to prevent cave-ins.
7. Valves in a backflow preventer assembly that permit testing of individual pressure zones are called _______________.
8. A simple backflow preventer that does not have a test cock and that uses two spring-loaded check valves to prevent back siphonage and back pressure is a(n) _______________.
9. A(n) _______________ is a pre-manufactured steel box that provides a shielding system by allowing workers to safely lay pipe within the box while the box is inside the excavation.
10. Any manmade cut, cavity, trench, or depression in an earth surface, formed through earth removal is a(n) _______________.
11. _______________ is a structure that is able to withstand the forces imposed on it by a cave-in and thereby protect employees within the structure.
12. A(n) _______________ is installed on a reduced-pressure zone backflow preventer assembly to prevent back pressure and back siphonage.
13. _______________ is an undesirable condition that results when non-potable liquids enter the potable water supply by reverse flow through a cross-connection.
14. The horizontal members of a shoring system installed perpendicular to the sides of the excavation, the ends of which bear against either uprights or wales, are called _______________.
15. _______________ is a method of protecting workers from cave-ins by excavating the sides of a trench to form one or a series of horizontal levels or steps, usually with vertical or near-vertical surfaces between them.

16. _______________ are vertical members of a trench shoring system placed in contact with the earth and usually positioned so that individual members do not contact each other.

17. The process of placing and consolidating material in areas that may remain void or loose when the rest of the backfill is added is known as _______________.

18. A backflow preventer assembly that includes test cocks and prevents back pressure and back siphonage through the use of two spring-loaded check valves is a(n) _______________.

19. _______________ is soil that has been previously backfilled or removed from the earth and replaced by any means.

20. The surface or foundation on which a pipe rests in a trench is called _______________.

21. _______________ is a form of backflow caused by lower-than-normal upstream pressure in the potable water line.

22. A(n) _______________ is capable of identifying existing and predictable hazards in the area, or pointing out working conditions that are unsanitary, hazardous, or dangerous to employees, and who has authority to take prompt corrective measures.

23. A laboratory test that determines the moisture content, cohesion factor, and soil gradation of a particular soil is a(n) _______________.

24. _______________ is a depression in the earth that is caused by unbalanced stresses in the soil surrounding an excavation.

25. _______________ are sandbags placed under, beside, and over the pipe in a trench to prevent erosion of the pipe bedding material caused by groundwater flow along the pipeline.

26. A block of concrete in which tie rods are secured is known as a(n) _______________.

Trade Terms

Back siphonage
Backflow
Bedding
Benching
Competent person
Cross braces
Cross-connection
Deadman
Disturbed soil
Ditch breakers
Double-check valve backflow preventer assembly (DCV)
Dual-check valve backflow preventer
Excavation
Haunching
Manufactured air gap
Proctor curve
Reduced-pressure zone (RPZ) backflow preventer assembly
Shielding
Shoring
Sloping
Subsidence
Test cocks
Trench
Trench box
Uprights
Wales

Cornerstone of Craftsmanship

Byron Weisz
CFO
Cen-Cal Fire Systems, Inc.

How did you choose a career in the industry?

I actually fell into this career. I wanted to go to school to become an architect, but after attending junior college for prep classes, I lost interest and wanted to get to work. I was doing odd jobs, and then I received a call from my brother. He worked for his father-in-law who owned a fire sprinkler contracting company, and they were looking for someone to train as a designer. So, I interviewed, was offered the position, and from my first day I found what God intended me to do.

Who inspired you to enter the industry?

I'm not sure anyone inspired me to enter the industry. It was more like an opportunity. I was offered a position and decided to commit to becoming a fire sprinkler designer.

What types of training have you been through?

I've always compared learning to design fire protection systems to becoming an apprentice and working in the field. It started with me simply taking NFPA 13 home every evening, reading and studying the code. My employer expected me to study on my own time and apply what I was learning during the day.

In my mind, you learn by making mistakes once and using that experience to make sure you get it right the second time. I took correspondence courses through the University of Oklahoma on my own, and I studied all applicable codes, including NFPA standards and building codes. Then, I applied and tested for NICET certification. The University of California at Davis offered a certification program in Fire Protection Engineering that took two years of night classes, and I enrolled in and was certified in that program, too. All major manufacturers offer training, and I was like a sponge. Overall, I believe the best training is the opportunity to design and manage work.

How important is education and training in construction?

Training is very important. Materials, processes, and tools of the trade evolve. There are always improvements and advancements in how work is done. You must be willing to listen and learn, whatever your position might be. As someone who is committed to the trade, you either need to decide what you'll take the lead on and keep up or you'll be left behind.

How important are NCCER credentials to your career?

Having an NCCER credential is very important. Some say they are fire sprinkler fitters and others have something to show for it. But having a credential doesn't make you special. Taking that credential and applying yourself to be the best you can be at what you do is what sets you apart.

How has training/construction impacted your life?

Very simply put, it has provided me a great life and some awesome experiences over the last 40 years.

What kinds of work have you done in your career?

I have been involved in design, fabrication, and project management for all types of fire protection systems and applications.

Tell us about your present job.

Currently, I co-own a fire sprinkler company that my brother and I started in 1986, and I recently purchased a partnership in a second company. I consult, work with our estimators, oversee two design departments, and co-manage the day-to-day business operations of two companies employing over 200 people.

What do you enjoy most about your job?

Every job I'm involved with is unique. I enjoy driving up to a dirt field and being part of a construction project that becomes something. Working through the planning, designing, getting approvals through the local AHJ, and being part of a team building something is exciting stuff.

What factors have contributed most to your success?

I credit my parents for being an amazing example—both showing and teaching me what it means to have a good work ethic. It starts with getting up, staying committed, and wanting to do the best you can every day.

Would you suggest construction as a career to others? Why?

I would recommend a career in fire sprinkler installation. In my career, those who commit and are willing to work have never missed a day of work because there wasn't something to do.

What advice would you give to those new to the field?

Commit to the opportunities you've been given and be willing to work.

Interesting career-related fact or accomplishment?

After being is this trade for almost 40 years it gives me a sense of accomplishment to go by projects I have been involved with over the years and know that at one time it was nothing more than a dirt field. Knowing that I had a small part in making a project happen and being able to see what it has become is pretty rewarding.

How do you define craftsmanship?

Being able to actually do something and not simply talking about it. Follow through on what you tell someone you can do and let the quality of your work speak for you.

Cornerstone of Craftsmanship

Matt Rouleau
Technical Services Manager
Reliable Automatic Sprinkler Company, Inc.

How did you choose a career in the industry?

I started in the fire protection industry by holding a second job at a fire protection contracting company as a material fabricator. This was while I was attending college pursuing a nursing degree and working part time as a certified nursing assistant at a local hospital.

Who inspired you to enter the industry?

The grandfather of my girlfriend at the time (who has now been my wife since 1993), was the owner of a fire protection contracting company as well as the former owner of Globe Fire Sprinkler Corporation. Through my wife I was introduced into the world of automatic fire suppression, both contracting and manufacturing.

What types of training have you been through?

Training is a never-ending process with fire protection. I have been through first aid, C.P.R., OSHA, and many other safety training classes throughout my career. I have attended manufacturer training sessions from most of the players in the industry. I have attended "Fire Tech Productions" training courses. I am also a UA journeyman sprinkler fitter, and NICET-certified S.E.T. in Inspection, Testing, and Maintenance and Water-Based Systems Layout. To retain the NICET certification, you must have 90 hours of continuing education every three years. I have participated in more training sessions for that than I can remember.

How important is education and training in construction?

Education and training are critical to success in construction. Technology is advancing at a faster rate than ever before in history and code cycles are under constant updates. If you do not stay educated, and up to date on the current codes as well as technology, your chances for success are greatly hindered.

How important are NCCER credentials to your career?

I feel it is extremely important. My goal has always been to expand my personal knowledge in the industry. It also provides me with a new network of like-minded professionals to utilize as a resource, as well as an opportunity to share my knowledge and experiences with new people as well.

How has training/construction impacted your life?

Through training, I am where I am today. I was trained to fabricate at the beginning of my career, which led me to being trained on system design. System design provided me with the level of expertise that then allowed me to perform project estimation. The incremental steps continually build your abilities to advance in the industry.

What kinds of work have you done in your career?

I think I have worn every hat in the industry. I started in material fabrication, then progressed into designing and estimating. In 2006, I became a journeyman sprinkler fitter. A brief stint away from contracting landed me in a Fire Protection Consultant role for three years with one of the top firms in the nation. I moved from consultation back into contracting where I became an operating partner in a contracting company. I sold my portion of the company and took a position in the Product Development Engineering team with Globe Fire Sprinkler for several years, and I am now a Technical Services Manager for Reliable Automatic Sprinkler Co., Inc.

Tell us about your present job.

I am a Technical Services Manager at Reliable Automatic Sprinkler Co., Inc. In this role, I offer product support to our existing and potentially new customers on our full line of valves, sprinklers, and system components. I provide product training for designers, installers, suppliers, architect/engineers, and Authorities Having Jurisdiction, both in person and over the internet. I am a direct interface with Product Development Engineering, where I provide input to the engineering team as it relates to service and maintenance of the products. This role to me is very important, as I provide suggestions and input to the product designs based on my personal field experience and previous manufacturing experience, in addition to the feedback I get from interfacing with end users.

What do you enjoy most about your job?

Saving lives and property! No greater job in my humble opinion.

What factors have contributed most to your success?

Goal setting, determination, willingness to ask questions, desire to never stop learning or to be challenged, and some of the greatest mentors in the industry.

Would you suggest construction as a career to others? Why?

Personally, I think it is an extremely gratifying career. You are always challenged one way or another in construction. The sky is the limit on how far you can climb the industry ladder. Life safety will never go away, it will only get more stringent as time passes. It is a niche market too. Fire protection is not a field students typically hear about or are informed of when they are considering what to do after high school.

What advice would you give to those new to the field?

Be safe. Pay attention. Ask questions. Network. Study your apprenticeship books! Get involved early with fire protection activities outside of work in addition to your work.

Interesting career-related fact or accomplishment?

During my time in contracting, in my role as a sprinkler fitter, I performed system inspections at the Michigan Governor's Summer Residence, also known as the Lawrence A. Young Cottage, which is a house located on Mackinac Island, Michigan. This structure was built in 1895. It was one of the neatest properties I had the opportunity to inspect and service.

How do you define craftsmanship?

To me, craftsmanship is a lifelong pursuit of excellence in a field or skill set. Always striving for expertise in your activities whatever they might be. Every journey starts with the first step. Becoming an apprentice and truly embracing the program's content and training is the first step to becoming a true craftsman sprinkler fitter.

Trade Terms Introduced in This Module

Back siphonage: A form of backflow caused by lower-than-normal upstream pressure in the potable water line.

Backflow: An undesirable condition that results when non-potable liquids enter the potable water supply by reverse flow through a cross-connection.

Bedding: The surface or foundation on which a pipe rests in a trench.

Benching: A method of protecting workers from cave-ins by excavating the sides of a trench to form one or a series of horizontal levels or steps, usually with vertical or near-vertical surfaces between levels.

Competent person: A person who is capable of identifying existing and predictable hazards in the area or working conditions that are unsanitary, hazardous, or dangerous to employees, and who has the authority to take prompt corrective measures to fix the problem.

Cross braces: The horizontal members of a shoring system installed perpendicular to the sides of the excavation, the ends of which bear against either uprights or wales.

Cross-connection: A direct link between the potable water supply and water of unknown or questionable quality.

Deadman: A block of concrete in which tie rods are secured.

Disturbed soil: Soil that has been previously backfilled or removed from the earth and replaced by any means.

Ditch breakers: Sandbags placed under, beside, and over the pipe in a trench to prevent erosion of the pipe bedding material caused by groundwater flow along the pipeline.

Double-check valve backflow preventer assembly (DCV): A backflow preventer assembly that prevents back pressure and back siphonage through the use of two spring-loaded check valves and includes test cocks for testing individual pressure zones.

Dual-check valve backflow preventer: A simple backflow preventer that uses two spring-loaded check valves to prevent back siphonage and back pressure but does not include test cocks.

Excavation: Any manmade cut, cavity, trench, or depression in an earth surface, formed through earth removal.

Haunching: The process of placing and consolidating material in areas that may remain void or loose when the rest of the backfill is added.

Manufactured air gap: An air gap installed on a reduced-pressure zone (RPZ) backflow preventer assembly to prevent back pressure and back siphonage.

Proctor curve: A test performed in a laboratory that determines the moisture content, cohesion factor, and soil gradation of a particular soil.

Reduced-pressure zone (RPZ) backflow preventer assembly: A backflow preventer assembly that uses two spring-loaded check valves plus a hydraulic, spring-loaded pressure differential relief valve to prevent back pressure and back siphonage.

Shielding: A structure that is able to withstand the forces imposed on it by a cave-in and thereby protect employees within the structure. Shielding can be permanent structures or designed to be portable and moved along as work progresses.

Shoring: A structure such as a metal hydraulic, mechanical, or timber system that supports the sides of an excavation and is designed to prevent cave-ins.

Sloping: A method of protecting employees working in excavations from cave-ins by cutting the excavation walls to a specific angle at which the soil will rest without sliding.

Subsidence: A depression in the earth that is caused by unbalanced stresses in the soil surrounding an excavation.

Test cocks: Valves in a backflow preventer assembly that permit the testing of individual pressure zones.

Trench: A narrow excavation (in relation to its length) made below the surface of the ground. In general, the depth is greater than the width, but the width of a trench (measured at the bottom) is not greater than 15' (4.6 m).

Trench box: A pre-manufactured steel box that provides a shielding system by allowing workers to safely lay pipe within the box while it is inside an excavation.

Uprights: The vertical members of a trench shoring system placed in contact with the earth and usually positioned so that individual members do not contact each other.

Wales: Horizontal members of a shoring system placed parallel to the excavation face whose sides bear against the vertical members of the shoring system or the earth.

Additional Resources

This module presents thorough resources for task training. The following reference material is suggested for further study.

Excavations. Washington, DC: OSHA Publications Office.

NFPA 13, *Standard for the Installation of Sprinkler Systems*, Latest Edition. Quincy, MA: National Fire Protection Association.

NFPA 20, *Standard for the Installation of Stationary Pumps for Fire Protection*, Latest Edition. Quincy, MA: National Fire Protection Association.

NFPA 24, *Standard for the Installation of Private Fire Service Mains and Their Appurtenances*, Latest Edition. Quincy, MA: National Fire Protection Association.

NFPA 25, *Standard for the Inspection, Testing, and Maintenance of Water-Based Fire Protection Systems*, Latest Edition. Quincy, MA: National Fire Protection Association.

NFPA 291, *Recommended Practice for Fire Flow Testing and Marking of Hydrants*, Latest Edition. Quincy, MA: National Fire Protection Association.

Figure Credits

Pro-Line Fittings, Module Opener

Courtesy of McWane Ductile, Figures 1, 37

iStock@Brenadette Faust, Figure 2

Courtesy of www.wattswater.com, Figures 3–9

John Denhardt, Figure 10

Courtesy of American AVK Co., Figures 11–12, RQ01

Courtesy of AMERICAN Flow Control, Figure 14

iStock@Songpin, Figure 16

Courtesy of FireResQ, Inc, Figure 17

Wholesale Fire & Rescue Ltd., Figure 18

Reprinted courtesy of Caterpillar Inc., Figure 20

Kundel Industries, Inc., Figure 21

iStock@Aksakalko, Figure 22

Mike Powers, Figure 23

iStock@ChrisVanLennepPhoto, Figure 25

Wheeler-Rex, Figure 28

Mac McGaffagan, Figures 33, 38

Trumbull Manufacturing, Inc., Figure 34

Courtesy of EBAA Iron, Inc., Figures 35–36

iStock@stefann11, Figure 41

Reproduced with permission of NFPA from NFPA 13, *Standard for the Installation of Sprinkler Systems*, 2019 edition. Copyright© 2018, National Fire Protection Association, Figures 42A and 42B

Reproduced with permission of NFPA from NFPA 13, *Standard for the Installation of Sprinkler Systems*, 2019 edition. Copyright© 2018, National Fire Protection Association. For a full copy of NFPA 13, please go to www.nfpa.org., Tables 2–4 and 6–8

Section Review Answer Key

Section 1.0.0

Answer	Section Reference	Objective
1. b	1.1.2	1a
2. d	1.2.3	1b

Section 2.0.0

Answer	Section Reference	Objective
1. b	2.1.2	2a
2. a	2.2.0	2b

Section 3.0.0

Answer	Section Reference	Objective
1. c	3.1.2	3a
2. b	3.2.1	3b
3. d	3.3.1; *Table 2*	3c
4. a	3.4.0	3d
5. b	3.5.0	3e
6. a	3.6.4	3f
7. b	3.7.3; *Table 7*	3g

NCCER CURRICULA — USER UPDATE

NCCER makes every effort to keep its textbooks up-to-date and free of technical errors. We appreciate your help in this process. If you find an error, a typographical mistake, or an inaccuracy in NCCER's curricula, please fill out this form (or a photocopy), or complete the online form at **www.nccer.org/olf**. Be sure to include the exact module ID number, page number, a detailed description, and your recommended correction. Your input will be brought to the attention of the Authoring Team. Thank you for your assistance.

Instructors – If you have an idea for improving this textbook, or have found that additional materials were necessary to teach this module effectively, please let us know so that we may present your suggestions to the Authoring Team.

NCCER Product Development and Revision
13614 Progress Blvd., Alachua, FL 32615

Email: curriculum@nccer.org
Online: www.nccer.org/olf

❑ Trainee Guide ❑ Lesson Plans ❑ Exam ❑ PowerPoints Other ______________________

Craft / Level: Copyright Date:

Module ID Number / Title:

Section Number(s):

Description:

Recommended Correction:

Your Name:

Address:

Email: Phone:

Glossary

Adopted standards: Standards that have been adopted through statutory or administrative laws in a specific geographic area. Once a document is adopted by reference in a state statute or local ordinance, it has the same force and effect as if published therein in its entirety.

Alloy: Any substance made up of two or more metals.

Ambient temperature: The surrounding air temperature.

American Fire Sprinkler Association (AFSA): A non-profit, international association representing merit shop sprinkler contractors, dedicated to the educational advancement of its members and promotion of the use of sprinkler systems.

Armover: A horizontal pipe that extends from a branch line to a single sprinkler.

Authority Having Jurisdiction (AHJ): An organization, office, or individual responsible for enforcing the requirements of a code or standard, or for approving equipment, materials, an installation, or a procedure.

Back siphonage: A form of backflow caused by lower-than-normal upstream pressure in the potable water line.

Backflow: An undesirable condition that results when non-potable liquids enter the potable water supply by reverse flow through a cross-connection.

Bedding: The surface or foundation on which a pipe rests in a trench.

Benching: A method of protecting workers from cave-ins by excavating the sides of a trench to form one or a series of horizontal levels or steps, usually with vertical or near-vertical surfaces between levels.

Beveling: Creating a slight angle on the end of a pipe to facilitate insertion into a fitting.

Blocked: Secured for storage in tiers.

Borates: Boric acid salts used as a flux.

Brazing: A method of fusing metals with a nonferrous filler metal using heat above 840°F, but below the melting point of the base metals being joined.

Burr: A sharp, raised edge or other metal protrusion produced on the pipe end when cutting a pipe.

Bushing: A pipe fitting with both internal and external threads that can be used to connect a pipe or fitting with another pipe or fitting of a different size.

Capillary action: The spontaneous flow of a liquid into a narrow space between two solid surfaces.

Cast iron: An alloy of iron and carbon that is cast in a mold. Cast iron fittings have a classification of either 125 or 250.

C-factor: Refers to the roughness of the inside of the sprinkler pipe. The higher the C-factor, the smoother the pipe.

Check valve: A valve that allows water to flow in only one direction.

Chlorinated polyvinyl chloride (CPVC): A common thermoplastic pipe used in the sprinkler industry.

Close nipple: A nipple that is about twice the length of a standard thread that is threaded from each end with no shoulder.

Companion flange: A flange that is used to connect a fitting or flange directly to threaded pipe.

Competent person: A person who is capable of identifying existing and predictable hazards in the area or working conditions that are unsanitary, hazardous, or dangerous to employees, and who has the authority to take prompt corrective measures to fix the problem.

Compressed gas: Gas stored under pressure in cylinders.

Concentric reducing coupling: A reducing coupling that maintains the same center line between the two pipes that it joins.

Control valve: Controls the water supply to a sprinkler system.

Coupon: The metal disc removed from a steel pipe when using a hole cutter.

CPVC compound: An orange thermoplastic material used to produce sprinkler pipe and fittings.

CPVC resin: A white thermoplastic material that accounts for 85 percent of the weight of finished CPVC products.

Cross braces: The horizontal members of a shoring system installed perpendicular to the sides of the excavation, the ends of which bear against either uprights or wales.

Cross-connection: A direct link between the potable water supply and water of unknown or questionable quality.

Cut groove: A groove that is formed around pipe by cutting away the pipe material.

Deadman: A block of concrete in which tie rods are secured.

Deluge sprinkler system: A type of sprinkler system that uses open sprinklers rather than closed sprinklers. When heat triggers the detection system, the deluge valve operates, producing immediate water flow through all the sprinklers in a given area.

Dielectric: A nonconductor of electricity.

Dimple: A slight surface depression, or the act of creating one.

Disturbed soil: Soil that has been previously backfilled or removed from the earth and replaced by any means.

Ditch breakers: Sandbags placed under, beside, and over the pipe in a trench to prevent erosion of the pipe bedding material caused by groundwater flow along the pipeline.

Double-check valve backflow preventer assembly (DCV): A backflow preventer assembly that prevents back pressure and back siphonage through the use of two spring-loaded check valves and includes test cocks for testing individual pressure zones.

Drop/suspended ceiling: A ceiling that is not part of the structural framework of a building. This ceiling is installed by suspending it from the floor above, or from the roof, and is commonly used to provide space for services such as cables, recessed lighting, and piping.

Dry pipe sprinkler system: Dry pipe sprinkler systems are filled with pressurized air or nitrogen. Water flows into the system only when a sprinkler is activated.

Dual-check valve backflow preventer: A simple backflow preventer that uses two spring-loaded check valves to prevent back siphonage and back pressure but does not include test cocks.

Ductile iron: A type of cast iron in which magnesium is added to the molten gray iron to reduce brittleness.

Eccentric reducing coupling: A reducing coupling that displaces the center line of the smaller of the two joining pipes to one side.

Elbow: A fitting that makes an angle between adjacent pipes. An elbow is always 90 degrees unless another angle is stated on the fitting or in the drawing specifications. Also known as an *ELL*.

Electrolysis: The decomposition of a substance by the passage of electricity through it.

Excavation: Any manmade cut, cavity, trench, or depression in an earth surface, formed through earth removal.

Fin: A sharp, flared metal protrusion produced on a pipe end when cutting it. A fin juts out at an angle to the pipe.

Flexible sprinkler hose fittings: Flexible fittings designed to span the distance from the branch line to the sprinkler.

Flux: A chemical substance that aids the flow of solder or brazing metal by removing and preventing the formation of oxides on the pieces to be joined.

FM Approvals®: A testing laboratory and Listing agency that certifies new products for service in particular locations and for specific applications.

Galvanic: Refers to direct-current electricity, especially when produced chemically.

Galvanized: Dipped in zinc during the manufacturing process.

Gasket: A flat sheet or ring of rubber or other material used to seal a joint between metal surfaces to prevent gas or liquid from entering or escaping.

Haunching: The process of placing and consolidating material in areas that may remain void or loose when the rest of the backfill is added.

Hazard: A possible source of danger or potential injury.

Inside diameter (ID): The distance between the inner walls of a pipe. Used as the standard measure for tubing used in heating and plumbing applications.

Iron pipe size (IPS): The nominal pipe size of iron pipe.

K-factor: A mathematical constant that represents a sprinkler's discharge coefficient, which is the amount of water distributed through the sprinkler orifice at a given pressure. The larger the orifice, the larger the K-factor.

Labeled: Equipment or materials to which has been attached a label, symbol, or other identifying mark of an organization that is acceptable to the Authority Having Jurisdiction (AHJ).

Light hazard: An occupancy or portions of other occupancies where the quantity and/or combustibility of contents is low and fires with relatively low rates of heat release are expected.

Liquidus: The state of solder when it has melted by heating and is flowing readily.

Listed: Included in a list of products or services published by a recognized Listing agency and acceptable to the AHJ for use in sprinkler systems for a specified purpose.

Makeup: A term used to describe the amount of pipe that screws into a threaded fitting.

Malleable iron: A metallic material that is produced by slowly reheating and cooling cast iron. Malleable iron fittings have a classification of either 150 or 300.

Manufactured air gap: An air gap installed on a reduced-pressure zone (RPZ) backflow preventer assembly to prevent back pressure and back siphonage.

National Fire Protection Association (NFPA): An organization that publishes codes and standards with the goal of preventing loss of life and property.

National Pipe Thread (NPT): The US standard for pipe threads. NPT has a $^{1}/_{16}$" taper per inch from back to front.

NCCER Standardized Curricula: Standardized construction education materials produced by NCCER.

NFPA 13, Standard for the Installation of Sprinkler Systems: The organization (NFPA) that provides the basic reference to the requirements (NFPA 13) for installing sprinkler systems.

Nipple: A piece of pipe that is threaded on both ends and is less than 12" in length. Any pipe over 12" is referred to as *cut pipe*.

Occupational Safety and Health Administration (OSHA): The Occupational Safety and Health Administration (OSHA) is an agency of the US Department of Labor established by the Occupational Safety and Health Act of 1970. OSHA's mission is to ensure safe and healthful working conditions for working men and women by setting and enforcing standards as well as providing safety training, outreach, education, and assistance.

Office of Apprenticeship (OA): The US Department of Labor office that sets the minimum standards for training programs across the country.

On-the-job training (OJT): Job-related training acquired while working.

Ordinary hazard: Group 1 (OH1): An area in which the combustibility of the contents is low, the quantity of the combustibles is moderate, and stockpiles of contents with low combustibility do not exceed 8'. Group 2 (OH2): Occupancies or portions of other occupancies where the quantity and combustibility of contents are moderate to high, stockpiles of contents with moderate to high combustibility do not exceed 12', and stockpiles of contents with high rates of heat release do not exceed 8'.

Outside diameter (OD): The distance between the outer walls of a pipe. Used as the standard measure for air conditioning and refrigeration (ACR) tubing.

Oxidation: The process by which the oxygen in the air combines with metal to produce tarnish and rust

Oxyfuel equipment: A type of brazing equipment that uses oxygen and fuel gas mixed in a nozzle and directed at the work surface.

Pipe schedule: A pipe labeling system based on pipe wall thickness. As the schedule increases, the wall thickness also increases.

Plain-end pipe: Pipe that has no groove or threads on the end and has all imperfections removed.

Potable: Water that is fit for human consumption.

Preaction sprinkler system: A sprinkler system employing sprinklers attached to a piping system containing air or nitrogen that may or may not be under pressure. A supplemental detection system is installed in the same area as the system sprinklers.

Proctor curve: A test performed in a laboratory that determines the moisture content, cohesion factor, and soil gradation of a particular soil.

Rack: Device used to support pipe in tiers.

Recognized standards: Standards that are generally accepted as relevant in civil law, even in areas where no standards have been formally adopted.

Reduced-pressure zone (RPZ) backflow preventer assembly: A backflow preventer assembly that uses two spring-loaded check valves plus a hydraulic, spring-loaded pressure differential relief valve to prevent back pressure and back siphonage.

Regulations: Rules and specifications for the sprinkler industry provided by various organizations.

Roll groove: A groove that is formed by pressing the metal around the pipe.

Shielding: A structure that is able to withstand the forces imposed on it by a cave-in and thereby protect employees within the structure. Shielding can be permanent structures or designed to be portable and moved along as work progresses.

Shoring: A structure such as a metal hydraulic, mechanical, or timber system that supports the sides of an excavation and is designed to prevent cave-ins.

Sloping: A method of protecting employees working in excavations from cave-ins by cutting the excavation walls to a specific angle at which the soil will rest without sliding.

Solder: An alloy that is melted and used to join metallic surfaces.

Soldering: A method of joining metals with a nonferrous filler metal using heat below 840°F and below the melting point of the base metals being joined.

Solidus: The solid or stable phase of solder.

Sprinkler: A heat-sensing device used in sprinkler systems to discharge water onto a fire.

Standard dimensional ratio (SDR): The pipe wall thickness in proportion to its outside diameter.

Standpipe: A horizontal or vertical supply main running through buildings and having hose valves on each floor.

Subsidence: A depression in the earth that is caused by unbalanced stresses in the soil surrounding an excavation.

Sway bracing: Bracing of sprinkler pipe to minimize pipe movement during seismic activity.

Sweat fittings: Thin-wall copper fittings suitable for making soldered connections.

Sweat joint: A soft solder joint, or a joint soldered below 840°F.

Tee: A fitting that has one side outlet 90 degrees to the run.

Teflon® tape: Tape made of Teflon® that is wrapped around the male threads of a pipe before the pipe is screwed into a fitting. Teflon® tape serves as both a lubricant and a sealant.

Test and drain valve: A type of valve that includes both a test function and a drain function. The test function is used to simulate sprinkler activation to verify the operation of the waterflow alarm. The drain function can be used to drain all or part of a sprinkler system.

Test cocks: Valves in a backflow preventer assembly that permit the testing of individual pressure zones.

Thermal expansion: The change in pipe length due to changes in temperature.

Thermoplastic: Refers to plastic piping that is soft and pliable when heated, but hard when cooled.

Transverse: The lateral shifting of pipe.

Trench box: A pre-manufactured steel box that provides a shielding system by allowing workers to safely lay pipe within the box while it is inside an excavation.

Trench: A narrow excavation (in relation to its length) made below the surface of the ground. In general, the depth is greater than the width, but the width of a trench (measured at the bottom) is not greater than 15' (4.6 m).

Underwriters Laboratories® (UL): A testing laboratory and Listing agency that certifies new products for service in particular locations and for specific applications.

Uprights: The vertical members of a trench shoring system placed in contact with the earth and usually positioned so that individual members do not contact each other.

Wales: Horizontal members of a shoring system placed parallel to the excavation face whose sides bear against the vertical members of the shoring system or the earth.

Water hammer: Shock wave in piping caused by a sudden change in water flow.

Wet pipe sprinkler system: The simplest, most common type of sprinkler system. Wet pipe sprinkler systems are always filled with water, compared to dry pipe sprinkler systems that contain no water until the system is activated.

Wick: To draw solder into a joint.

Wrought fittings: Heavy-duty copper fittings suitable for making brazed connections.

Notes

Notes

Notes

Notes

Notes

Notes

Notes

Notes

Notes

Notes

Notes

Notes

Notes

Notes

Notes